WAVELET TRANSFORMS AND THEIR RECENT APPLICATIONS IN BIOLOGY AND GEOSCIENCE

Edited by **Dumitru Baleanu**

Wavelet Transforms and Their Recent Applications in Biology and Geoscience
Edited by Dumitru Baleanu

Published by InTech

Janeza Trdine 9, 51000 Rijeka, Croatia

Publishing Process Manager Vana Persen

Technical Editor Teodora Smiljanic

Cover Designer InTech Design Team

Additional hard copies can be obtained from orders@intechopen.com

Wavelet Transforms and Their Recent Applications in Biology and Geoscience,
Edited by Dumitru Baleanu
p. cm.
ISBN 978-953-51-0212-0

For EU product safety concerns :
IN TECH d . o . o . , Prolaz Marije Krucifikse Kozulić 3 , 51000 Rijeka , Croatia ,
info@intechopen . com or visit our website at intechopen . com .

Meet the editor

Dr Dumitru Baleanu holds his MSc in 1989 from the Faculty of Physics at University of Bucharest and PhD in 1996 from the Institute of Atomic Physics, Magurele-Bucharest, Romania. He worked as a postdoctoral researcher for two years in the Bogoliubov Laboratory of Theoretical Physics, JINR, Dubna, Russia. He joined the Department of Mathematics and Computer Sciences of Faculty of Art and Sciences at Cankaya University on February 2000. In 2006 he became a Professor at the Institute of Space Sciences, Magurele-Bucharest, Romania. He has written more than 200 papers published in journals indexed in SCI. He is member of several editorial boards and his h-index is 16.

Contents

Preface

Wavelet analysis is one of the most effective solutions to overcome the drawbacks of Fourier analysis, and since the seminal work of Ingrid Daubechies was published in the late 1980s, it has been applied very successfully in various branches of science and engineering. The fractional wavelet techniques started to play an important role in data treatment corresponding to complex systems during the last decade.

The wavelet transform and it generalization started to be an efficient mathematical tool which has been widely used to investigate a signal in a time-frequency representation to allow analyzing the parts of interest individually.

As it is well known the main difficulty in processing the biological signals is their extreme variability and the requirement to analyse on a case by case basis. The wavelet transform is recognized as an useful tool in the signal analysis corresponding to the biological data as well as to the complex seismic data.

This book aims to show the range of applications of wavelet analysis to data from the field of biology and geosciences.

Specifically, the book contains chapters showing how the wavelet transform was used for the analysis of EEG signals in patients with oral communications,how it was used to investigate the energy distribution of the EEG signal components as well as its use for the dimensional reduction in a gait recognition framework.

One of the chapters shows how we can convert EEG activity into cursor movement by a brain computer interface using an appropriate feature extraction scheme. The detection data of mammary carcinoma processing method based on the wavelet transformation was presented. The applications of the wavelet method in sport and hand palm texture biometric identification and verification were illustrated.

Improvement of shimmer parameter of oesophageal voices using wavelet transform and the application of this method for textile material feature extraction is another interesting topic covered by this book. The powerful utilization of the wavelet transform and its varieties in two important applications, namely biometrics and non-destructive testing is explained in detail.

In this book the reader can find new methods involving two spherical multiscale transforms, the wavelet and the curvelets, and the design for restoration of spherical data with noise following a Poisson distribution.

The book ends with two applications of 1D and 2D wavelet transforms used in treatments of the geophysical signals.

I strongly believe that the chapters of this book will be of great help for graduate and postgraduate students, as well as for researchers who will get a general view of the powerful wavelet method and its applications in treatment of biological and geoscience data .

Dumitru Baleanu
Cankaya University, Faculty of Art and Sciences,
Department of Mathematics and Computer Sciences, Ankara,
Turkey

Institute of Space Sciences, Magurele-Bucharest,
Romania

Part 1

Wavelet Transforms in Biology

1

Wavelet Transform for the Analysis of EEG Signals in Patients with Oral Communications Problems

Maria Viqueira, Begona García Zapirain,
Amaia Mendez Zorrilla and Ibon Ruiz
Deusto Institute of Technology, Deustotech-LIFE Unit, University of Deusto, Bilbao
Spain

1. Introduction

This study is a part of a developed project to help people who can´t talk or control their movements when communication.

There are different kinds of methods to improve the communication of these people. Some examples are:

1. Methods based in the ocular movement: for example the videooculography systems, where a little camera detects the eye position and place the mouse in the same position on a monitor.
2. Physical communicators based on boards: the user points, on a board, what he wants to communicate. This method involves that the user´s disability allows him to have some control over his movements.

In this case, the intention is to use a method which can be used with people whose disability do not let them utter a word (or they present a big difficult for that), neither take the whole control of their movements.

This kind of disability may be due to several cerebral palsy, one example of that could be athetosis. This disease is a kind of paralysis which also presents injuries in the extrapyramidal system, which is part of the nervous system. This illness causes that the one who suffers it cannot take the whole control of his movements, and the difficulties at the time to talk (in some cases it prevent them the speech totally).

For these end, we have chosen to work with the signals from an EEG, which is a brain activity record. When analyzing the signals, the intention is to detect what a person wants to say. Therefore it wouldn´t be necessary that the user had a hole control of the movements (or even eyes), because what is pretended is to interpret what the user wants to say, without having to express it externally. Nowadays EEG is a very used method applied to improve user interaction (see the literature [1-2]).

The brain shows electrical activities on different frequency bands and it can be useful analyze each frequency band separately.

Fourier transform is a very used method in frequency domain, but it presents the inconvenient that it does not show temporal information of the signal. It means that Fourier shows the signal frequency ranges, but it doesn´t inform when we are in a frequency or in another.

Wavelet transform, explained widely in [1], is a method to analyze the signal in time-frequency scale. This method is very useful in non stationary signals analysis, it means: those signals which don´t stay constant over time.

The use of the different frequency bands allows to separate the signal in different frequency components in order to eliminate noise signals and information which doesn´t provide anything significant.

In this case we want to analyze the EEG signal in different frequency bands to get the P300, which is a signal that appears as an answer to a light or sound stimulus.

The P300 wave is between the band frequencies corresponding to [0-8Hz]. The predominant component of the P300 is about 2.5Hz, existing also another theta component (6Hz approximately), which appears in a latter zone [4]. Analyzing this band, we can detect the presence or absence of the P300.

The method which has been used to help in the communication consists in illuminate different letter, so that, when the letter that the user is watching is illuminated (target letter), the brain sends the P300 wave as an answer to that stimulus.

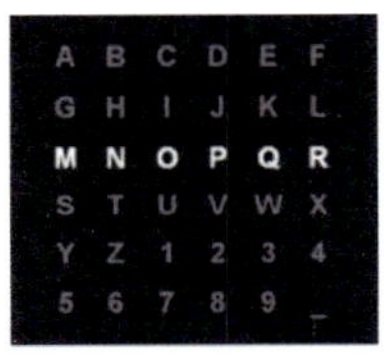

Fig. 1. Matrix shown on the screen of Competition BCI

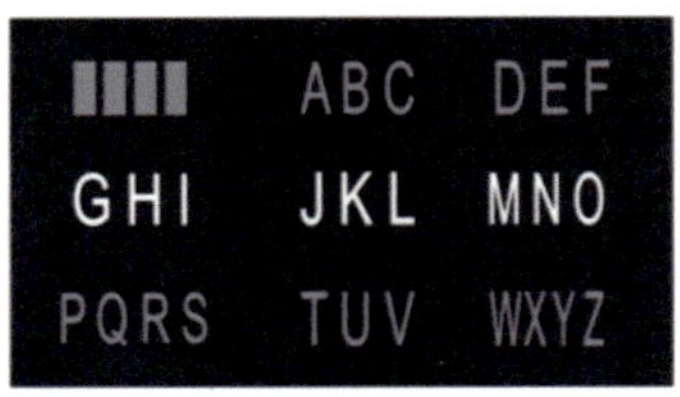

Fig. 2. Matrix shown on the screen of the final application of this project

If it illuminates the target row, it results the P300b, whereas if the illuminated row is not the one the user is watching, appears the P300a.

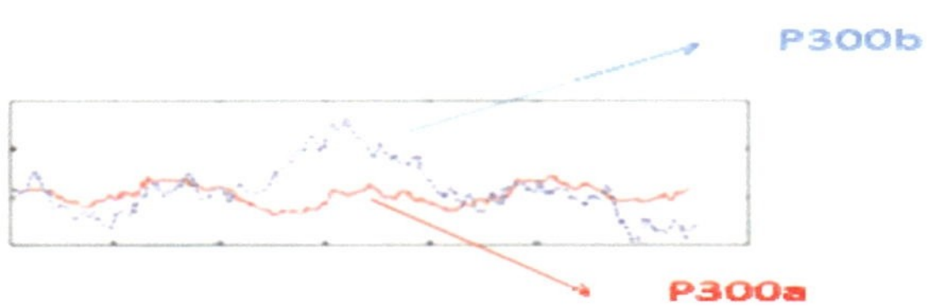

Fig. 3. P300a and P300b

Detecting the signal, which appears between 300 and 600 milliseconds after the target photostimulate, and knowing the sequence of letter illumination, we can predict the letter that the user wants to say.

We have decided working with the Discrete Wavelet Transform (DWT) because it doesn´t require so much memory as the Continuous Wavelet Transform (CWT), it is efficient if we use scales in powers of 2, it presents higher facilities to analyze numerically the transformed data and the calculations are faster.

This technique has been used in studies in order to get the P300, such as [21]and [22], and combining the DWT with another techniques [26] and [27].

The DWT performs basically two functions:

1. It filters the signal along a low-pass filter and another high-pass filter to obtain the approximate and detail signal coefficients in different ranges of frequency.
2. Sub-sampling of the output signals.

The successive decompositions are made from the signal obtained in the low pass filter.

When developing the algorithm, we have made a preliminary study to understand the behavior of the P300 and Wavelet Transform.

From this study, conclusions are reached concerning the shift of the signal according to the mother wavelet chosen and the frequency band where the P300 is best detected.

Subsequently, there has performed a second part to detect which wavelet is more appropriated to integrate at the final application and in which channels we have obtained higher success percentage, because the EEG to use has only 4 channels.

Fig. 4. EEG 4 channels

Fig. 5. Final application

2. State of the art

2.1 Technical

The Wavelet Transform results very useful in the analysis of non stationary signals [30-31], where the signal of the study presents parts with different frequencies over time.

Wavelet is not only useful in the study of signs, but it is also frequently used in digital image processing, although this study focuses on the first case.

In medicine applications, it is used in a case which is needed to detect any peculiarity in the registry of biomedical signals.

1. Noise removal: the Wavelet decomposition is produced without spaces and without overlap the data, so the decomposition process is totally reversible (the signal can be rebuilt). Thus, they result useful in the algorithms where it is wished to retrieve the original information with a minimal loss. Wavelet Transform is also used for removing the noise in images, such as the noise removal of planar images in nuclear medicine [5][32-34].
2. Feature extractions: as it has been said before, the DWT is used in non stationary analysis. It allows the detection of the "oddballs" that can be originated in EEG, EOG or ECG signals, among others[6][35].
3. Detection of tumors [23]: extracting texture features through the DWT, it can help in the diagnosis of certain tumors, as the breast [24].
4. Seizures:
   a. Epilepsy: it is possible to detect certain epilepsy cases using the DWT. There are studies where it has been used the Wavelet Transform to detect the seizures in newborn, like in [7] and [8]. Knowing in which moment the seizures are happening, it can help in the diagnosis of diseases [28].
5. Electrocardiograms: the use of the DWT allows to analyze the ECG signals and make a classification of them [9]. Similarly, it allows to detect if there exits an ECG abnormality [10].
6. Pregnancy: we can know if the fetus presents any problem during the pregnancy, working with the images obtained by ultrasounds [11].
7. P300 detection: Wavelet transform has been used in previous studies to detect the P300 in EEG signals, like in [12], [13] and [14].
8. Detecting the frame of mind: there are differences between thinking a movement and keeping the mind relaxed. The use of the wavelets indicates the behavior of the brain in the different bands, so it can determinate if an individual is thinking in something or not, coming to know which movement is been imagined.
9. Alzheimer: as study of this disease, there are methods which start from wavelet coefficient to test the difference between a healthy people EEG and EEG from Alzheimer patients [25].

2.2 Social

The analysis methods in the frequency domain are useful to detect possible anomalies, which is a great help in medicine applications. Wavelet is a method which represents signals in time and frequency scales, but there are methods which work in time-frequency distribution and they are also used in medical applications.

A new method which has acquired relevance in the signal analysis is the Matching Pursuit, which consists in select, in a signal dictionary, the signal which matches better with the one of the study.

Usually, the methods used in time-frequency distributions use classify algorithms to separate different kinds of signals [15].

3. Methods

3.1 Wavelet Transform

The Wavelet Transform is defined by:

$$W_f(s,\tau) = \int f(t)\psi^*_{(s,\tau)}(t)dt \qquad (1)$$

The wavelets are generated from a wavelet function $\psi(t)$, called "mother wavelet", which is defined as:

$$\psi_{s,\tau}(t) = \frac{1}{\sqrt{s}}\psi\left(\frac{t-\tau}{s}\right) \qquad (2)$$

being τ the translation factor and "s", the dilation factor.

The signal f(t) is sampled by mother wavelet versions, which are dilated and moved.

The DWT is defined by the following way:

$$W(j,k)=\sum_j \sum_k f(x)2^{\frac{-j}{2}}\psi\left(2^{-j}x-k\right) \qquad (3)$$

3.2 BCI 2003 database

We have used the database corresponding to the BCI competition. The data have been obtained with a sample frequency of 240Hz. They produced 12 illuminations (one for each row and column) and, for each letter to predict, they do 15 repetitions (each one is illuminated 15 times). Each illumination lasts 100 milliseconds and, between each illumination, there is a space of 75 milliseconds.

This database has been used in more occasions to make studies with EEG signals and detect the P300, like in [16] or [29].

There have been performed three sessions, with different trials, where the first two are destined as a part of the training, where they difference the illumination of a target or non target row or column. The third session corresponds to the words to predict.

The algorithms developed have been used initially with this database for test its efficacy. This database provides the result of 64 channels of study and it has applied the algorithm to all of them in order to know the behavior of the P300 in each one, and testing in which ones the P300 is best appreciated.

The Wavelet Transform performs a set of consecutive filters, leading to decomposition of the signal in different levels and frequency bands. In this case, the level decomposition results:

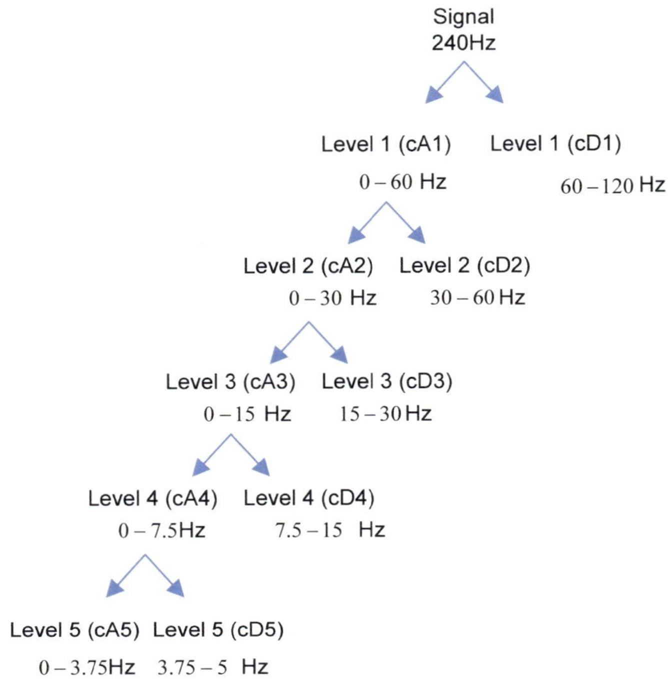

Fig. 6. Decomposition level

The result of the last low-pass filter corresponds to the approximate coefficients, where we should detect the P300 main component.

Initially, we have separated the target and the non target samples corresponding to the different trials of the season 1.

To check the algorithm efficacy, we have probed with the trials corresponding to the third session, which is the session of prediction.

3.3 Matlab Toolbox and mother wavelet

The algorithm has been developed entirely in Matlab and, for Wavelet Transform operations, we have used the Wavelet Toolbox from Matlab.

Moreover, the wavelets used have been the ones which this toolbox gives. The choice of one mother wavelet or another conditions the form of the low-pass and high-pass filters, so the Wavelet Transform will be different depending on the mother wavelet chosen.

The chosen mother wavelets have been the following ones: bior3.9, bior3.7, bior3.5, bior3.3, rbio1.5, rbio1.3 and DB9, as in [21], because of the similarity with the P300, [17] and [18].

It has been studied the behavior of the signal by the following way:

	Approximate coefficients	Detail coefficients
Frequency band level 4	0-7,5 Hz	7,5-15 Hz
Frequency band level 5	0-3,75 Hz	3,75-7,5 Hz

Table 1. Frequency ranges

4. System design

Below it is detailed the design of the algorithm that we pretend integrate into the final application.

In the section called "High level designs" it is shown a general scheme of the data acquisition and its treatment, and one scheme with de different modules and jobs which compose the study.

In the section called "Low level design" it will be concreted the reflected jobs in the previous section.

4.1 High level design

The main application makes the following operations:

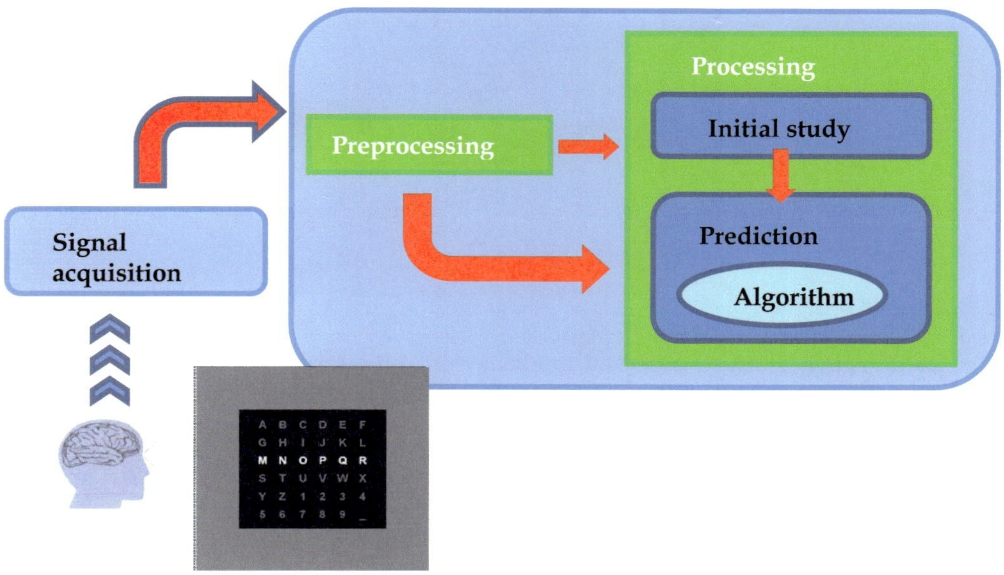

Fig. 7. Application scheme

To develop the algorithm, we make a first study and, then, we develop the method based in the conclusions obtained in the first part of the study.

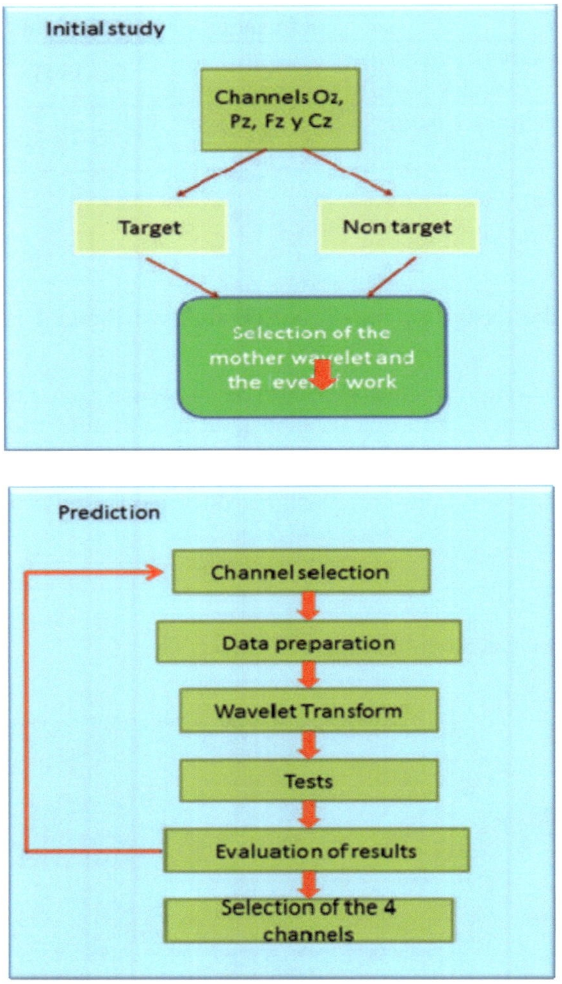

Fig. 8. General scheme

In the initial study it can be observed the behavior of the target and non target signals to check the shift that each wavelet produces to the signal. Subsequently, we work between the corresponding time interval to localize the P300 depending on the used mother wavelet.

4.2 Low level design

Initial study

The initial study provides information about the behavior of the P300 at the frequency domain. Studying the difference between target and non-target data, it is possible to obtain some initial conclusions about this behavior.

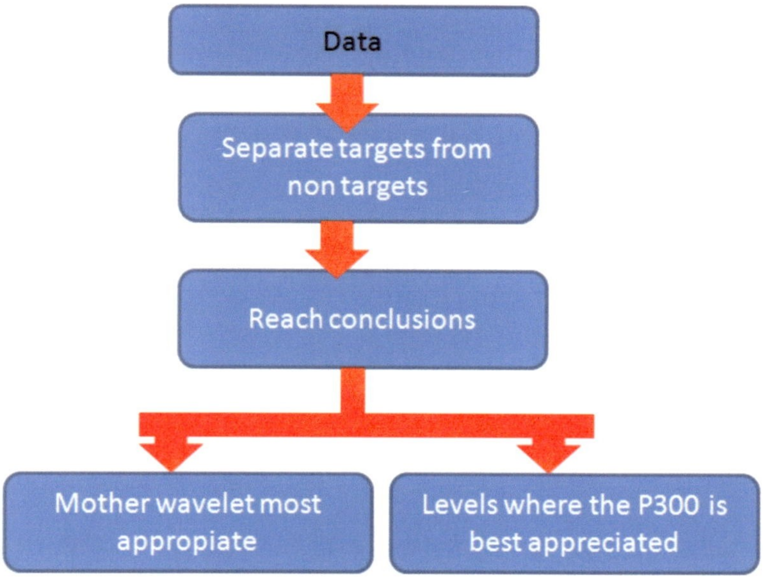

Fig. 9. Objectives of the initial study

It involves the following steps:

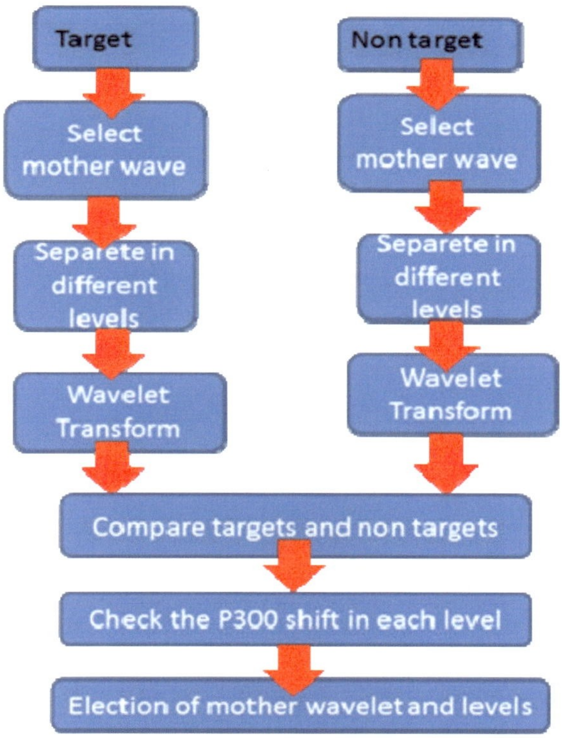

Fig. 10. Initial study

On the one side, separate the targets and non target data corresponding to Albany BCI competition. With these data, it performs the Wavelet Transform at the levels where the P300 should be appreciated to check the time intervals where the peak is appreciated.

As it has been said before, the interval where the peak is best appreciated is between 0 and 7, 5 Hz.

After doing this initial study it can be observed that there exists a big difference between the signals corresponding to the targets and the ones which are not target. This difference is well appreciated because the separation of the data has been performed by averaging the results of several sessions: the more signals are averaged; the greater will be difference between targets and non targets.

The prediction stage won´t show that difference so clearly, but with the initial study it can be known where will be best appreciated.

As it has been said before, the interval where the peak is best appreciated is between 0 and 7, 5 Hz.

After doing this initial study it can be observed that there exists a big difference between the signals corresponding to the targets and the ones which are not target. This difference is well appreciated because the separation of the data has been performed by averaging the results of several sessions: the more signals are averaged; the greater will be difference between targets and non targets.

The prediction stage won´t show that difference so clearly, but with the initial study it can be known where will be best appreciated.

Preparing data

This stage is done in the initial study and on the prediction of the words.

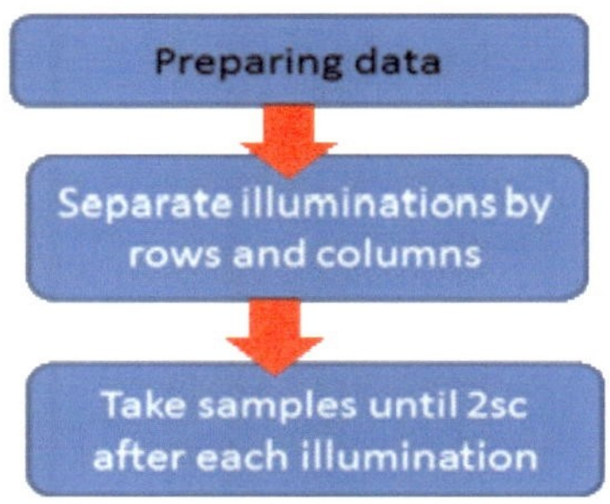

Fig. 11. Preparing data

Data are separated according to the row or column illuminated, obtaining in that way, 12 different data sets. It takes samples from the illumination until 2 seconds after in order to obtain the number of data which ensures the correct transformation at the last level (in this case, level 5). Using the Matlab command *wmaxlev*, it can check that the number of samples is correct.

The reason of separate columns and rows is because the cross of the target row and the target column will be the letter that the user is looking.

Wavelet Transform

The Wavelet Transform has been also used in the initial study and at the time of predicting the words. After the transformation, the data will be at the frequency domain.

The part which transforms the data makes the following steps:

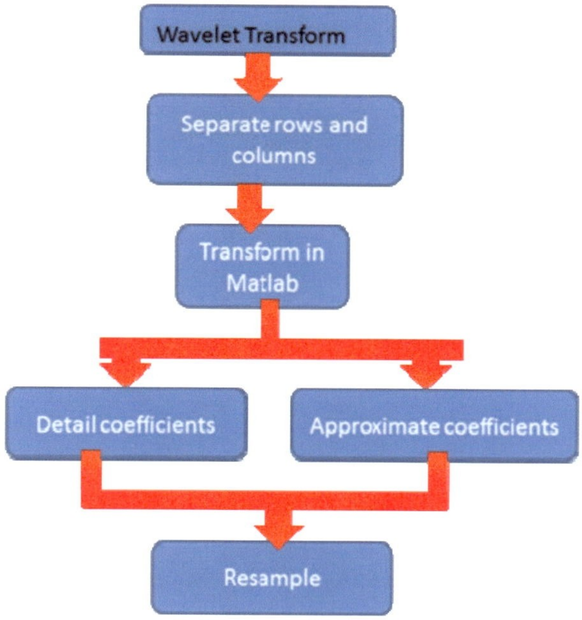

Fig. 12. Wavelet Transform

The transform of each row and column is made in Matlab (12 transformations in total for each letter, corresponding at 6 columns and 6 rows) to give rise to the different frequency bands. The detail coefficients are the high-pass filter result, whereas approximate ones are obtained from the low-pass filter. In this case:

Level 4→[3.75-7.5 Hz] (Aprox 4) and [7.5-15Hz] (Det 4)
Level 5→[0-3.75 Hz] (Aprox 5) and [3.75-7.5] (Det 5)

Finally, it is made a resample stage to add more samples.

Transform in Matlab

The first thing that is needed is to ensure that the Wavelet Transform will be correct with the number of samples which have been selected. To assure this, we use the command wmaxlev() in Matlab. With this command it is possible to know the highest level of work with guarantees by giving the number of samples and the mother wavelet.

In this case, taking two seconds after the illumination is enough to ensure the correct transform at level 5.

The Wavelet Transform has been performed for each one of the 64 channels. In the final application, there will be only four channels, so the transform will be much faster.

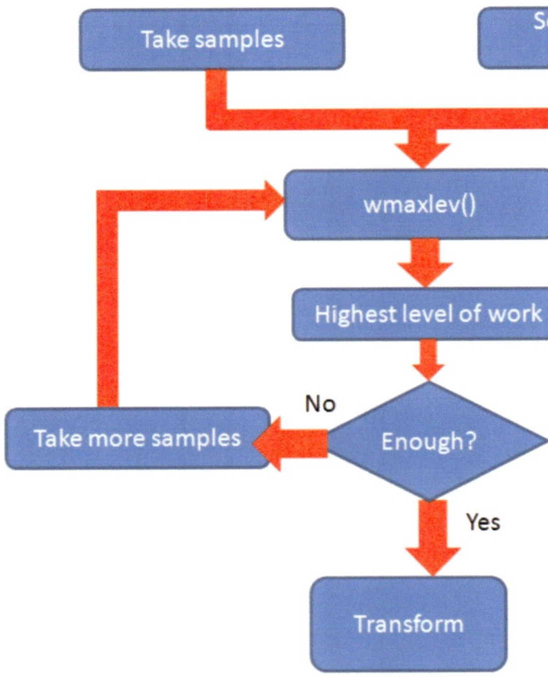

Fig. 13. Transform in Matlab

The following scheme shows the instructions used in Matlab:

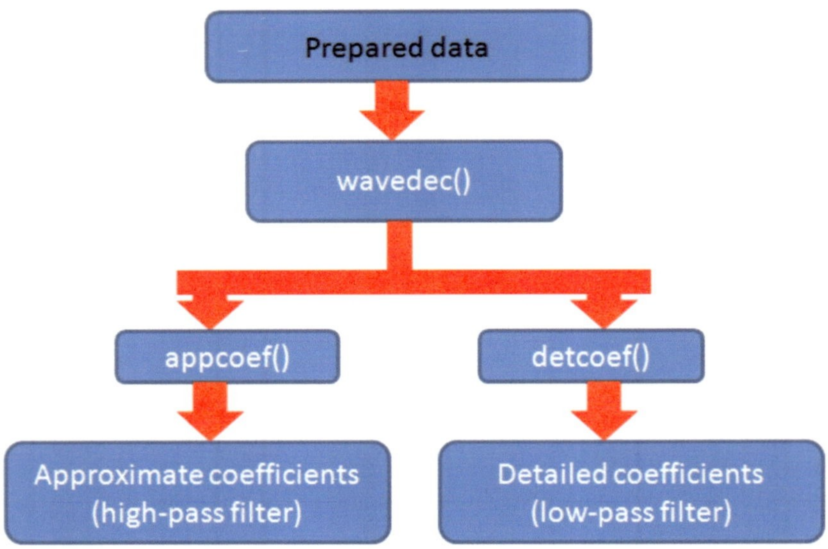

Fig. 14. Instructions in Matlab

To obtain the Wavelet Transform it is needed to give the signal to transform, the mother wavelet that will be used as filter, and the level of work.

In this case, as we need ranges of frequencies at level 4 and level 5, we have to do the transform twice (one for each level).

Test

To detect the letter where the P300 has been appreciated, it performs the following operations:

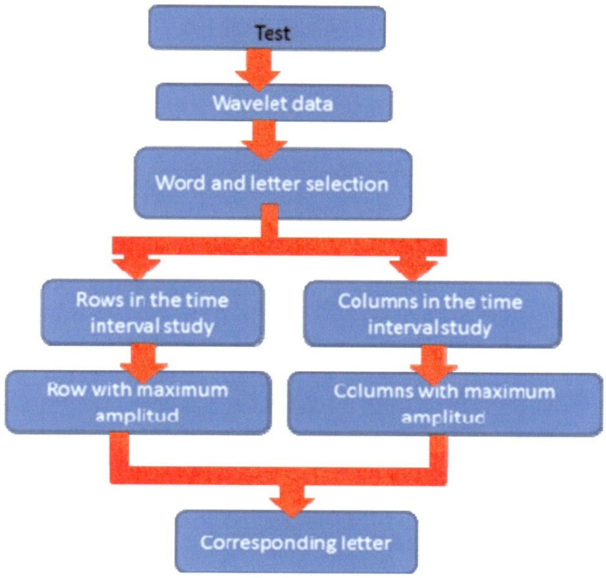

Fig. 15. Test

Once the transform is made, it takes the samples corresponding to the interval obtained in the study stage. On one side, it compares the 6 columns and, by the other side, the 6 rows; thus, it takes as target row and column the ones which present a higher peak than the rest.

Finally, it localizes the letter corresponding to these row and column.

After obtaining the letter, there is also another program which compares the theoretical letter with the real obtained and shows the average of success for each channel.

5. Results

Initial study

For this first part, we have used the following channels: FZ, CZ, PZ and OZ. We wanted to know the shift that means, for the P300, the election of one mother wavelet or another. The same way, we wanted to know in which levels is best appreciate the difference between a target and non target.

Mother wavelets: bior3.9, bior3.7, bior3.5, bior3.3, rbio1.5, rbio1.3 and DB9.

After the corresponding transformation, we observe a higher peak in the following time interval:

CZ channel:

Wavelet	Aprox 4	Det 4	Aprox 5	Det 5
bior3.9	[0.5 - 0.6]	[0.5 - 0.6]	[0.6 - 0.7]	-------
bior3.7	[0.4 - 0.6]	------------	[0.5 - 0.7]	-------
bior3.5	[0.4 - 0.6]	[1 - 1.2]	[0.4 - 0.7]	-------
bior3.3	[0.3 - 0.5]	[0.5 - 0.7]	[0.4 - 0.6]	-------
rbio1.5	[0.4 - 0.55]	------------	[0.4 - 0.6]	-------
rbio1.3	[0.3 - 0.5]	[0.4 - 0.5]	[0.3 - 0.5]	-------
db9	[0.75 - 0.9]	------------	[0.9 -1.2]	-------

Table 2. Time intervals for CZ channel

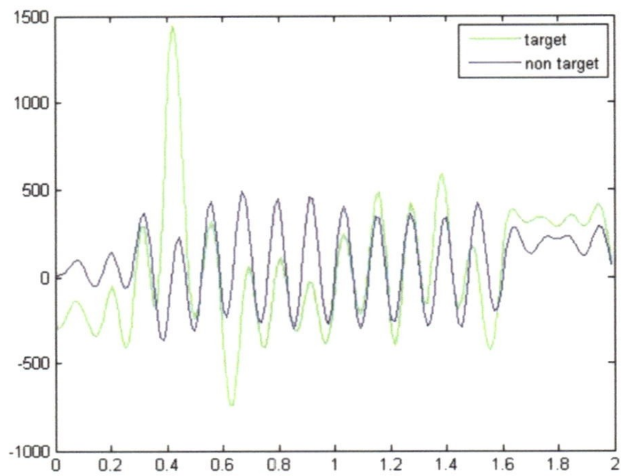

Fig. 16. bior3.7, level 4

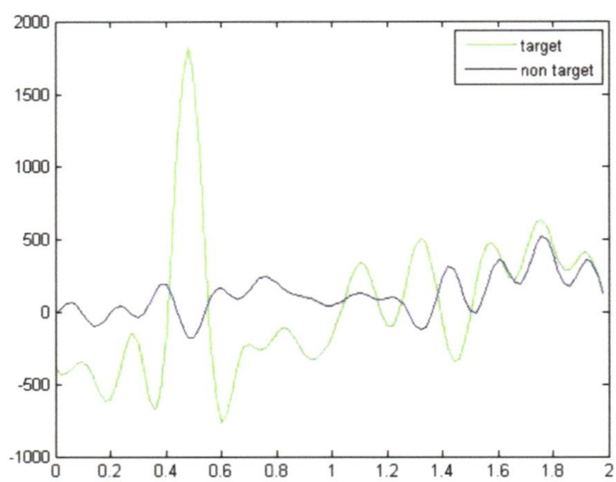

Fig. 17. bior3.5, level 5

PZ channel:

Wavelet	Aprox 4	Det 4	Aprox 5	Det 5
bior3.9	------------	------------	[0.6 - 0.75]	-------
bior3.7	------------	[0.4 - 0.6]	[0.5 - 0.7]	-------
bior3.5	------------	------------	[0.45 - 0.65]	-------
bior3.3	------------	------------	[0.35 - 0.6]	-------
rbio1.5	[0.4 - 0.55]	------------	[0.4 - 0.6]	-------
rbio1.3	[0.34 - 0.5]	------------	[0.3 - 0.55]	-------
db9	------------	------------	[0.9 - 1.1]	-------

Table 3. Time intervals for PZ channel

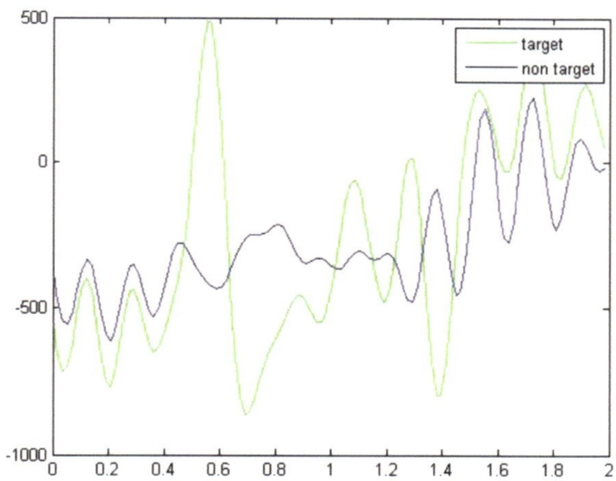

Fig. 18. bior3.7 level 5

FZ channel:

Wavelet	Aprox 4	Det 4	Aprox 5	Det 5
bior3.9	[0.5 - 0.6]	[0.5 - 0.6]	[0.6 - 0.7]	-------
bior3.7	------------	[0.4 - 0.6]	[0.45 - 0.65]	-------
bior3.5	[0.4 - 0.6]	[1 - 1.2]	[0.45 - 0.65]	-------
bior3.3	[0.4 - 0.6]	------------	[0.4 - 0.6]	-------
rbio1.5	[0.35 - 0.5]	[0.6 - 0.7]	[0.4 - 0.6]	-------
rbio1.3	[0.3 - 0.45]	------------	[0.3 - 0.5]	-------
db9	[0.75 -0.9]	------------	[0.9 -1.5]	-------

Table 4. Time intervals for FZ channel

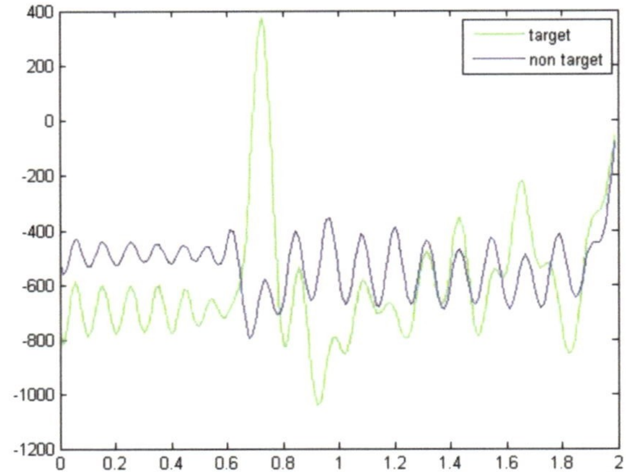

Fig. 19. DB9, level 4

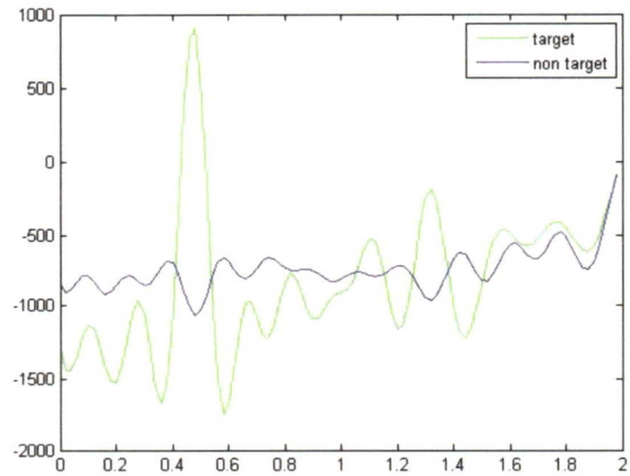

Fig. 20. bior3.5, level 5

The cells with green background show the frequency band where it is appreciated a higher peak. In case of FZ channel, the yellow background makes reference to time interval that, without been where the peak is the highest, show a considerable difference.

From OZ cannel we have not obtained significant differences, so the results are not shown.

Second stage

The Project where this part is integrated, consist in detect the P300 using an EEG with only 4 channels. For that, it is only shown the results of the 4 channels where the percentage of success has been higher.

It also shows the relative absolute error, the equations is the following:

AE = TV - RV
TV→Theoretical value (100)
RV→Real value

The hit rate is applied by letter, no by the whole word.

rbio1.5

Word	FC1	FCZ	CZ	FZ
FOOD	ECCC-	ECCC-	ECCB-	ECFA-
MOOT	MQOT-	MQOT-	MOOW-	MQOW-
HAM	BAM--	BAM--	BAM--	BAM--
PIE	PIE--	PIE--	PIE--	PIE--
CAKE	CAKE-	CAKE-	CFKE-	CAKE-
TUNA	TUNA-	TUNA-	SUNA-	TUBA-
ZYGOT	Z5GOX	45GOX	Z5GOX	4YGOX
4567	4567-	4567-	4567-	4567-
Success	74.2 %	71.97 %	67.75 %	67.74 %

Table 5. Predicted letters for rbio1.5

Average of the 4 channels: 70.415%

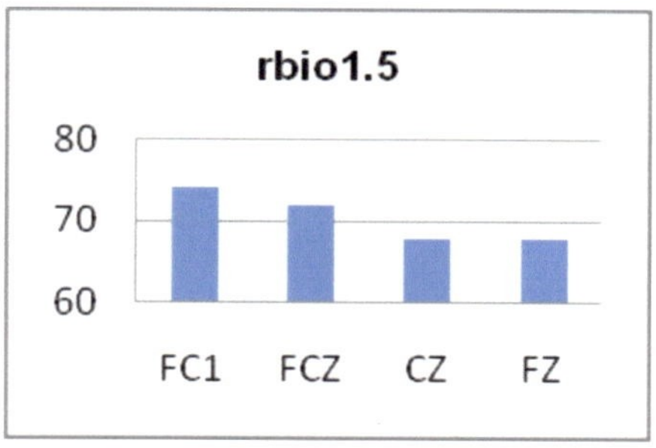

Fig. 21. Channel comparison for rbio1.5

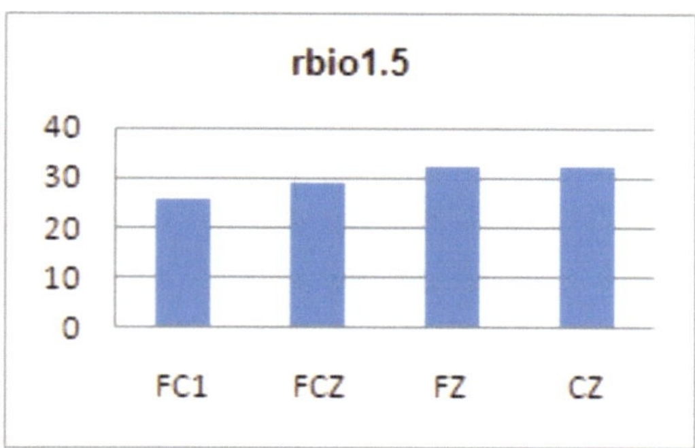

Fig. 22. Errors rbio1.5

None of the four channels have an error less than the 20%, so, although the success rate could be accepted around 65.

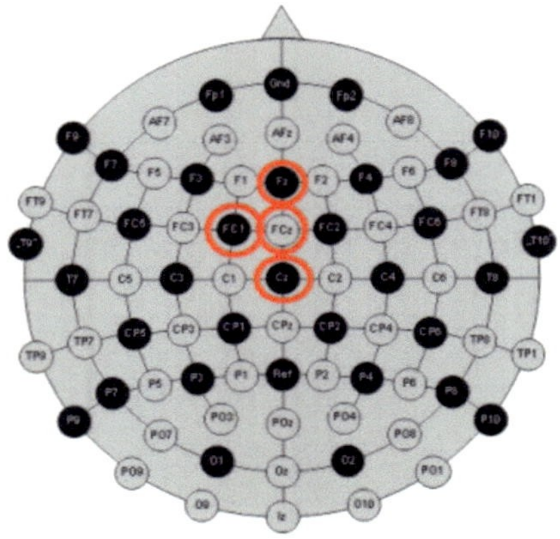

Fig. 23. Electrodes localization for rbio1.5

The obtained peaks have the following forms:

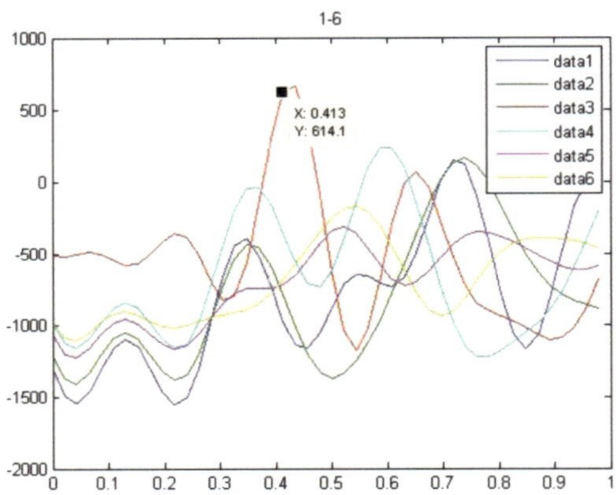

Fig. 24. Letter C, columns, FC1

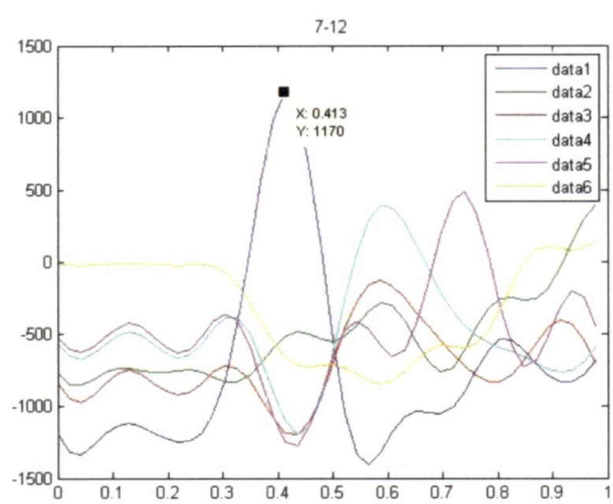

Fig. 25. Letter C, rows, FC1

rbio1.3

Word	FC1	FCZ	FZ	F2
FOOD	FCCC-	ECCC-	ECCC-	ECCC-
MOOT	GQOT-	MQOT-	GQOT-	GQOT-
HAM	BAM--	BAM--	BAM--	BAM--
PIE	PIE--	PIE--	PIE--	PIE--
CAKE	CAKE-	CAKE-	CAKE-	CAKE-
TUNA	TOBA-	TCBA-	TCBA-	TCBA-
ZYGOT	4YGOX	4YGOX	4YGCX	4YGIX
4567	4567-	4567-	4567-	4567-
Success	67.74 %	67.74 %	61.3 %	61.3 %

Table 6. Predicted letters for rbio1.3

Average of the 4 channels: 64.52%

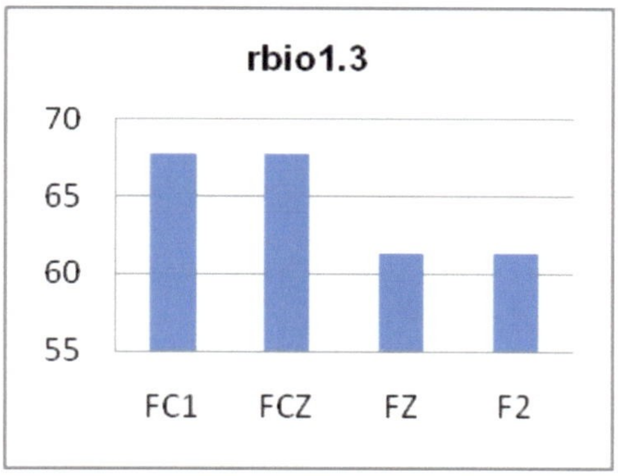

Fig. 26. Channel comparison for rbio1.3

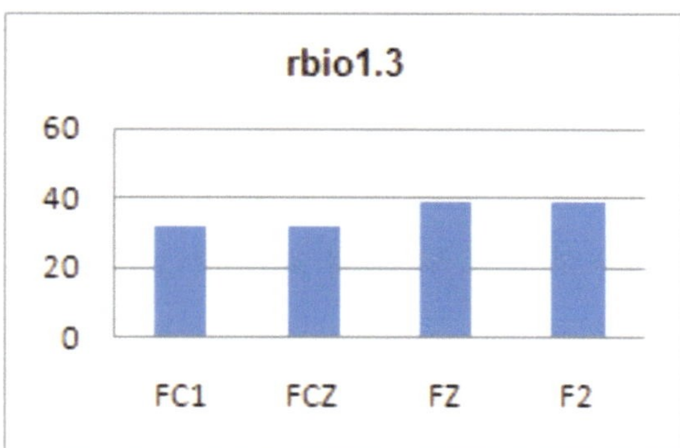

Fig. 27. Errors rbio1.3

The success rate is worse than the previous wavelet. In this case, the rbio family doesn´t arrive at the 80% of success.

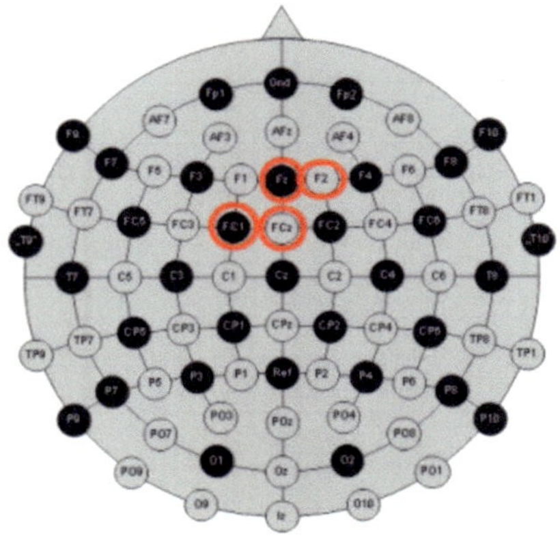

Fig. 28. Electrodes localization for rbio1.3

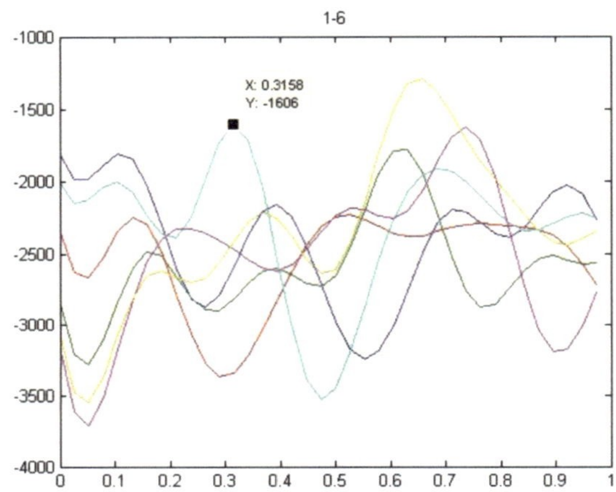

Fig. 29. Letter P, columns, FC1

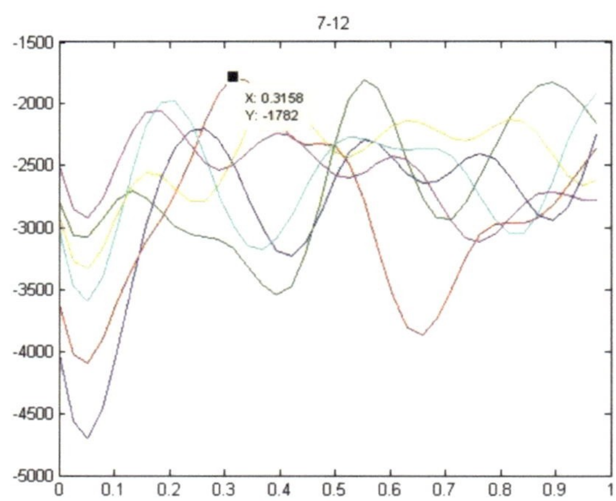

Fig. 30. Letter P, rows, FC1

bior3.5

Word	FCZ	FC1	F1	FZ
FOOD	FCCC-	FCCC-	FCCC-	FCCC-
MOOT	MOOT-	MOOT-	MQOT-	MQOT-
HAM	BAM--	BAM--	BAM--	BAM--
PIE	PIE--	PIE--	PIE--	PIE--
CAKE	CAKE-	CAKE-	CAKE-	CAKE-
TUNA	TCNA-	TON A-	TCNA-	TCNA-
ZYGOT	ZSGON	ZSAON	ZYGCX	ZYGCX
4567	4567-	4Y67-	4557-	XY67-
Success	77.42 %	70.97 %	70.97 %	70.97 %

Table 7. Predicted letters for bior3.5

Average of the 4 channels: 72.58%

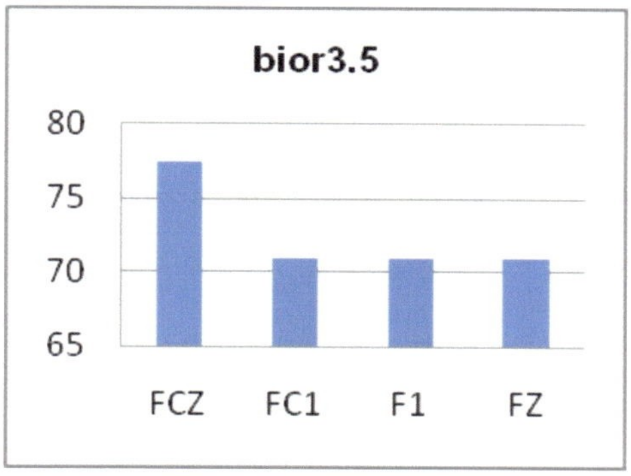

Fig. 31. Channel comparison for bior3.5

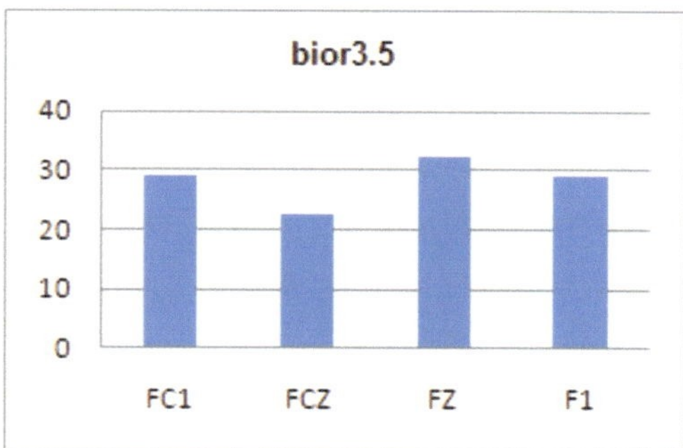

Fig. 32. Errors bior3.5

The error is still more than the 80%, but the success is higher than rbio family.

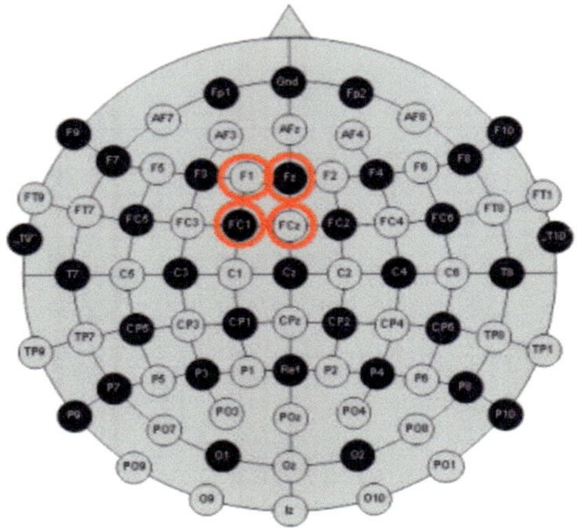

Fig. 33. Electrodes localization for bior3.5

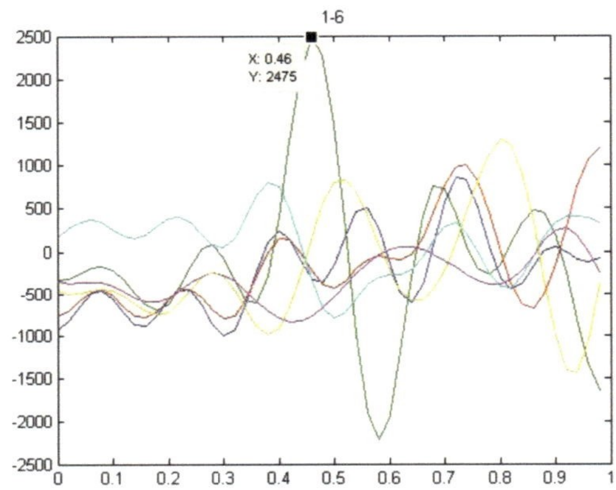

Fig. 34. Letter N, columns, FCZ

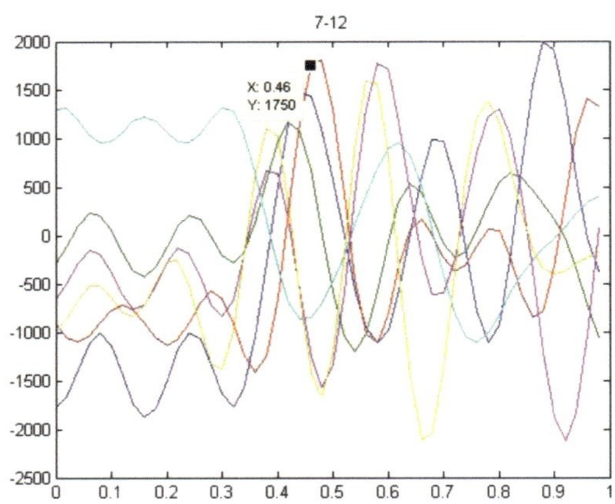

Fig. 35. Letter N, rows, FCZ

bior3.7

Word	FCZ	AFZ	FZ	F2
FOOD	FCCC-	ECCC-	FCCC-	FCCC-
MOOT	MQOT-	MQOT-	MQOT-	MQOT-
HAM	HAM--	HAM--	HAM--	HAM--
PIE	PIE--	PIE--	PIE--	PIE--
CAKE	CAKE-	CAKE-	CAKE-	CAKE-
TUNA	TUNA-	TUNA-	TUNA-	TUNA-
ZYGOT	ZSGOT	ZYGOX	ZYGOX	ZYGOX
4567	4Y67-	4567-	4Y67-	4Y67-
Success	80.65 %	80.65 %	80.65 %	80.65 %

Table 8. Predicted letters for bior3.7

Average of the 4 channels: 80.65%

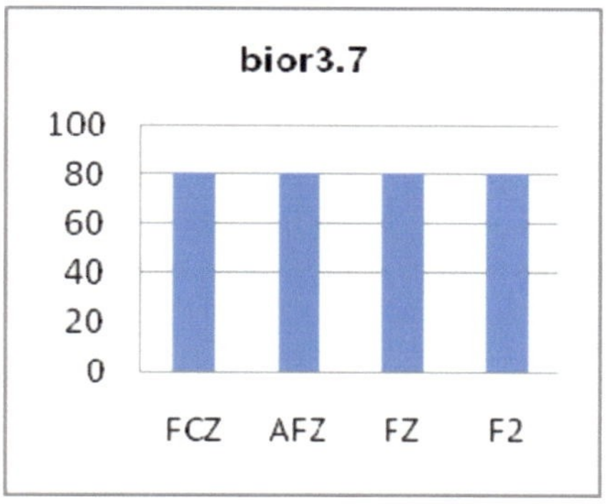

Fig. 36. Channel comparison for bior3.7

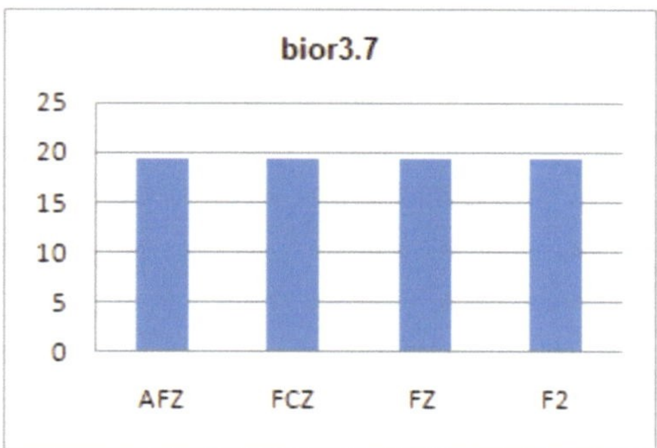

Fig. 37. Errors bior3.7

The four channels have an error less than the 80%, so the result is acceptable.

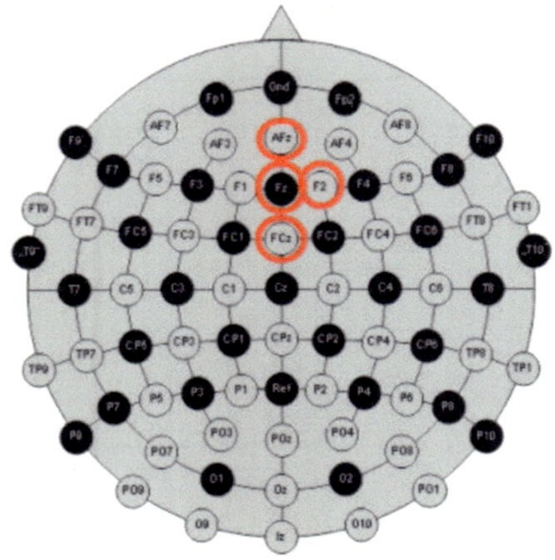

Fig. 38. Electrodes localization for bior3.7

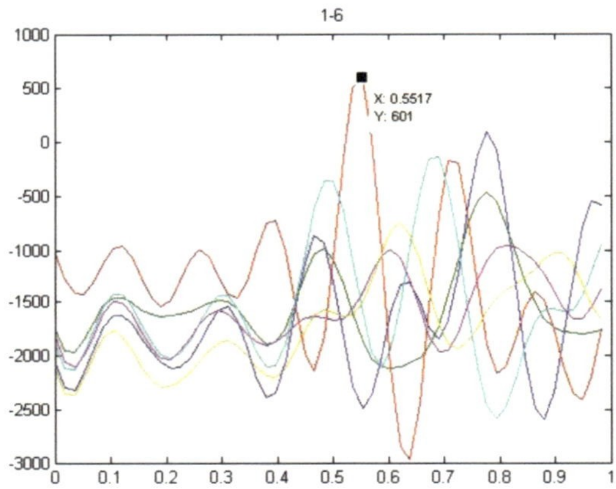

Fig. 39. Letter C, columns, FC2

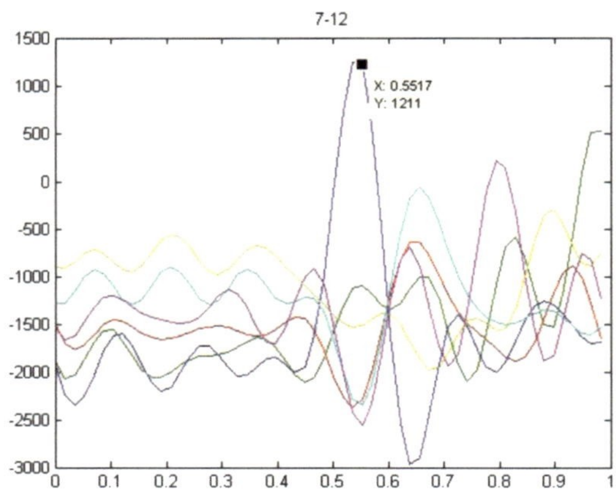

Fig. 40. Letter C, rows, FC2

bior3.9

Word	FC1	FCZ	FZ	F2
FOOD	FCCC-	ECCC-	ECCC-	ECCC-
MOOT	MQOT-	MQOT-	MQOT-	MQOT-
HAM	HAM--	HAM--	HAM--	HAM--
PIE	PIE--	PIE--	PIE--	PIE--
CAKE	CAKE-	CAKE-	CAKE-	CAKE-
TUNA	TUNA-	TUNA-	TUNA-	TUNA-
ZYGOT	ZYGOT	ZYGOX	ZYGOX	ZYGOX
4567	4567-	4567-	4557-	4Y67-
Success	87.1 %	80.65 %	77.42 %	77.42 %

Table 9. Predicted letters for bior3.9

Average of the 4 channels: 80.65%

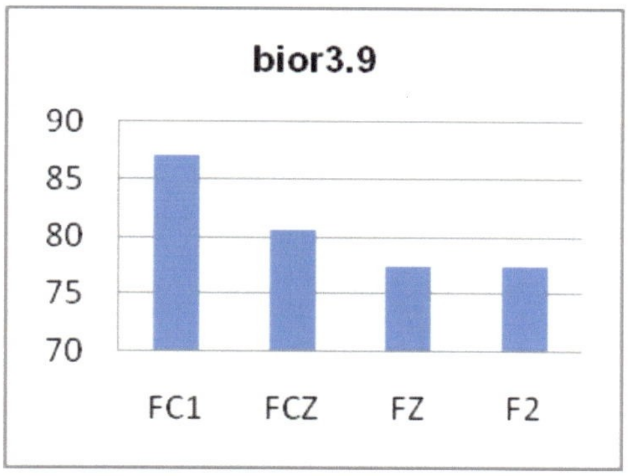

Fig. 41. Channel comparison for bior3.9

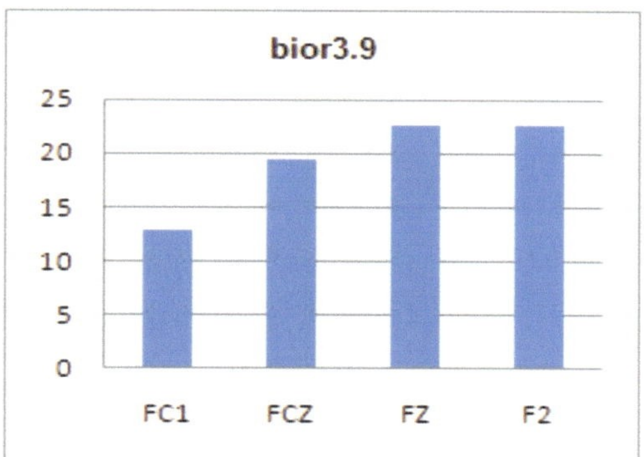

Fig. 42. Errors bior3.9

In this case, only two channels are over the 80% of success, but it can be observed that the channel FC1 has an error less than the 15%.

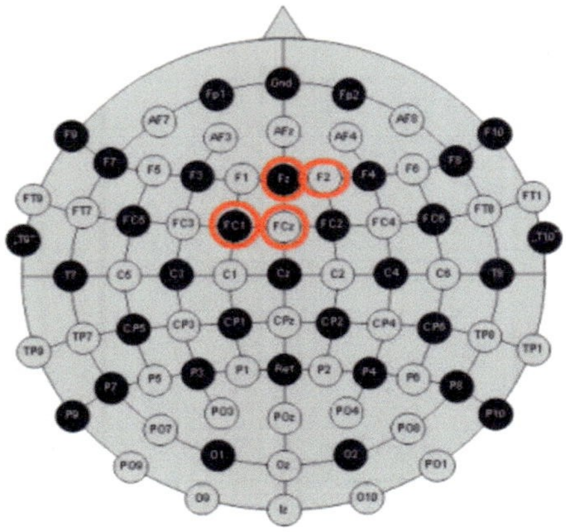

Fig. 43. Electrodes localization for bior3.9

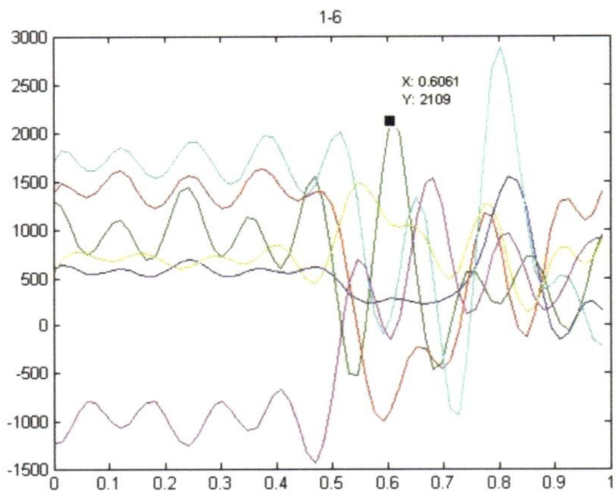

Fig. 44. Letter Z, columns, FC1

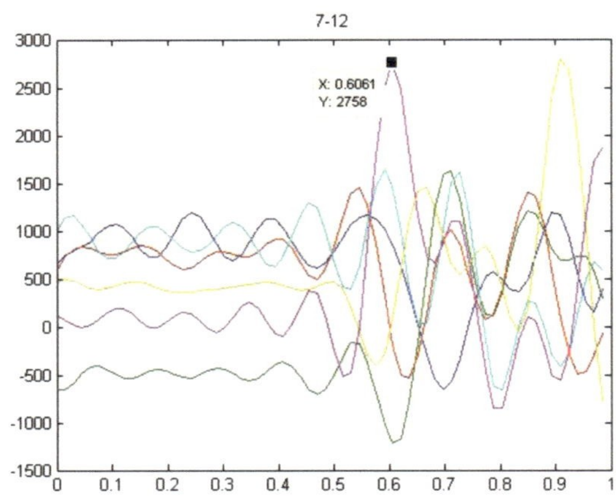

Fig. 45. Letter Z, rows, FC1

bior3.3

Word	FCZ	FC2	FZ	F2
FOOD	ECCC-	FCCC-	ECCC-	FCCC-
MOOT	MOOT-	GOOT-	GOOT-	GQOT-
HAM	AAM--	AAM--	AAM--	AAM--
PIE	PIE--	PIE--	PIE--	PIE--
CAKE	CAKE-	CFKE-	CAKE-	CAKE-
TUNA	TOBA-	TOBA-	TCBA-	TCBA-
ZYGOT	ZSAON	ZSAOR	ZSAON	Z5AOR
4567	4Y67-	4Y67-	XY57-	XY57-
Success	64.52 %	61.29 %	54.84 %	54.84 %

Table 10. Predicted letters for bior3.3

Average of the 4 channels: 58.87%

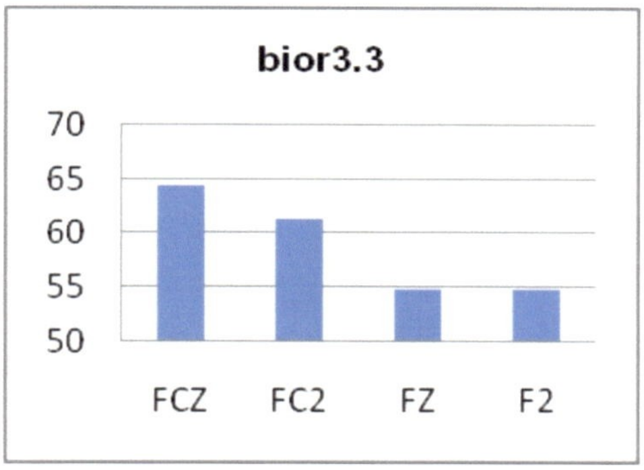

Fig. 46. Channel comparison for bior3.3

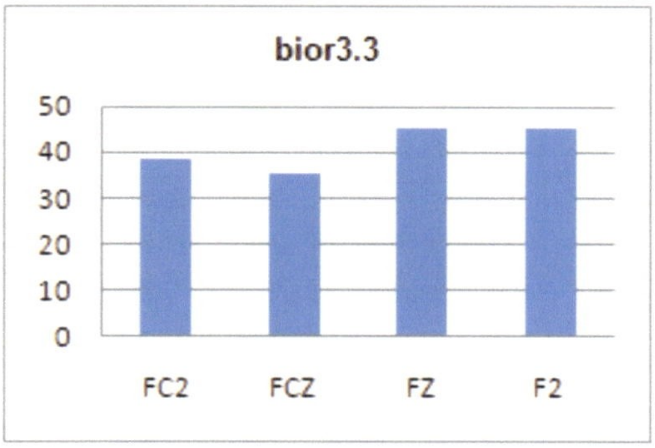

Fig. 47. Errors bior33

Within the family of the bior, the bior3.3 is the one which presents higher errors, exceeding all of them the 35%.

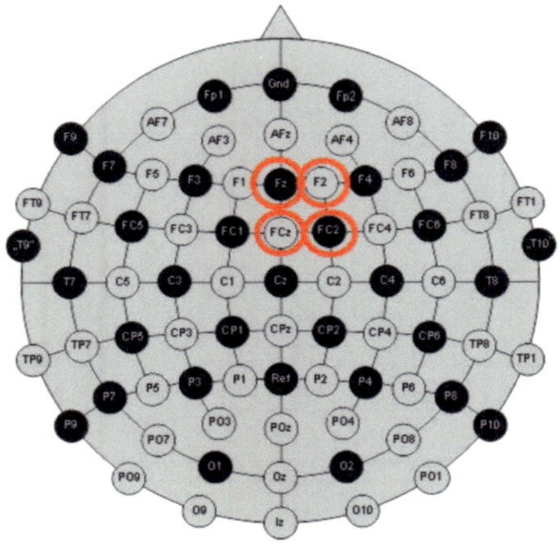

Fig. 48. Electrodes localization for bior3.3

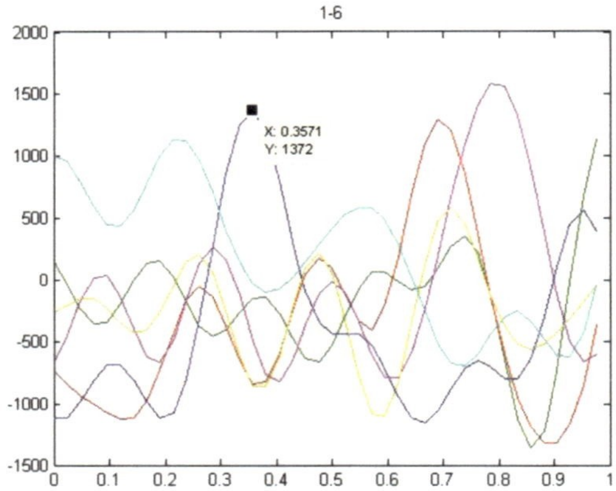

Fig. 49. Letter M, columns, FCZ

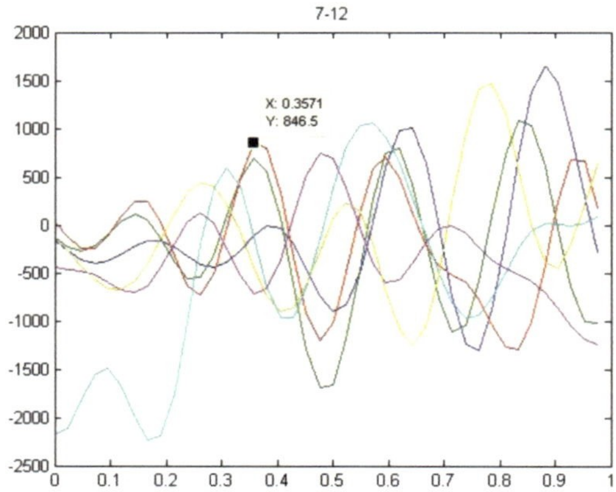

Fig. 50. Letter M, rows, FCZ

DB9

Word	FC1	FCZ	C1	CZ
FOOD	ECFC-	ECFD-	ECCD-	ECCD-
MOOT	MQOW-	MQOW-	MQOW-	MQOW-
HAM	6AM--	HAM--	HAM--	HAM--
PIE	PIB--	PIE--	PIE--	PIE--
CAKE	CAKE-	CAKE-	CAKE-	CAKE-
TUNA	XUNA-	XUNA-	XUNA-	XUNA-
ZYGOT	Z5GOX	45GOX	Z5GIX	Z5GOX
4567	4587-	4567-	4567-	4567-
Success	61.3 %	70.97 %	70.97 %	74.19 %

Table 11. Predicted letters for DB9

Average of the 4 channels: 69.36%

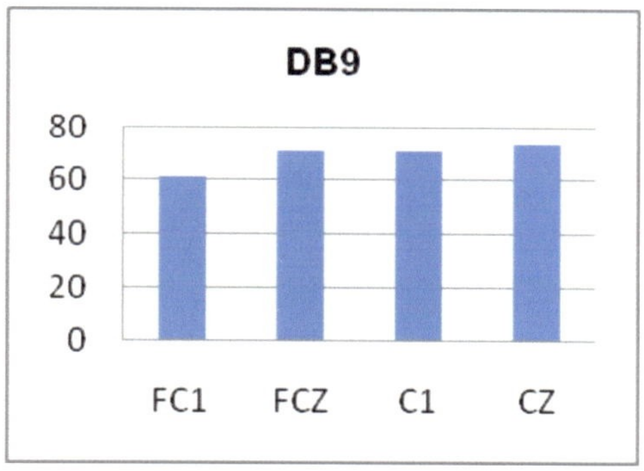

Fig. 51. Channel comparison for DB9

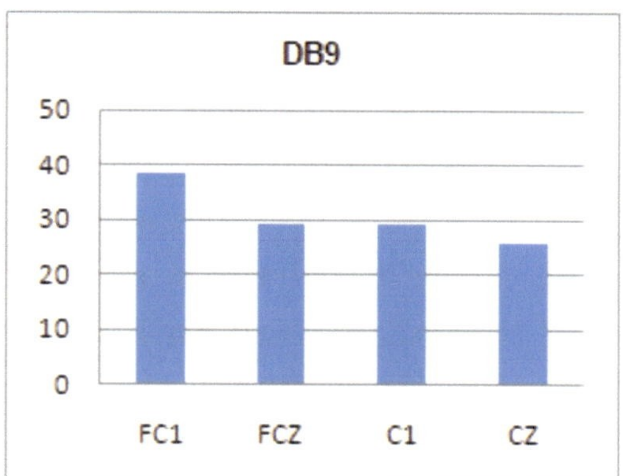

Fig. 52. Errors DB9

The success rate of the DB9 is higher than the rbio family, but it presents higher errors than the bior family (except for the bior3.3).

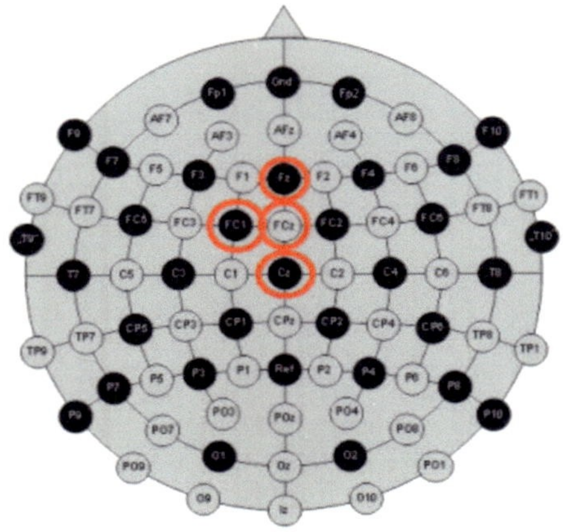

Fig. 53. Electrodes localization for DB9

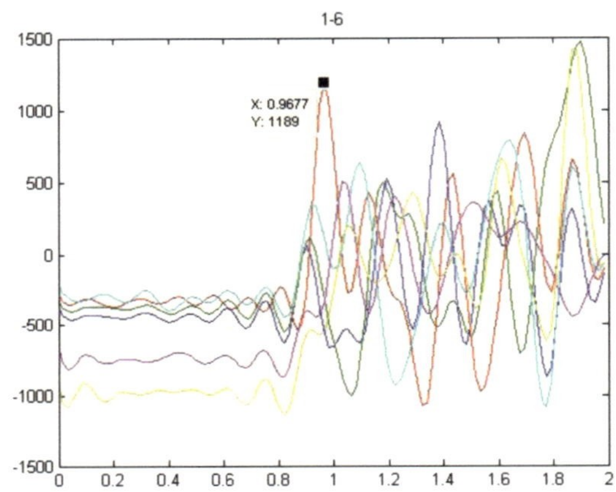

Fig. 54. Letter C, columns, CZ

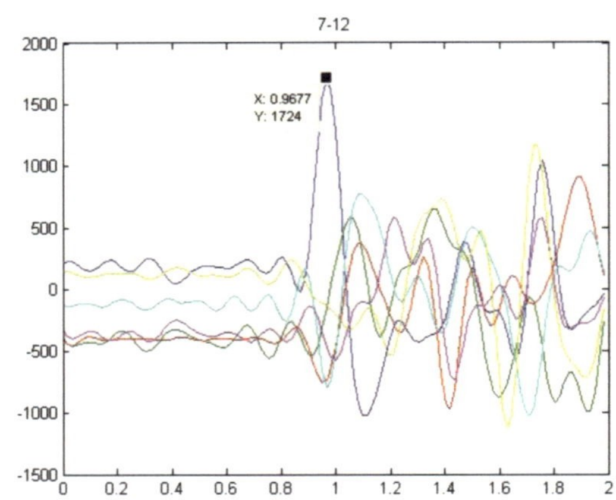

Fig. 55. Letter C, rows, CZ

Following it can be observed the typical deviation of each wavelet. The deviation has been obtained using the corresponding average of each one:

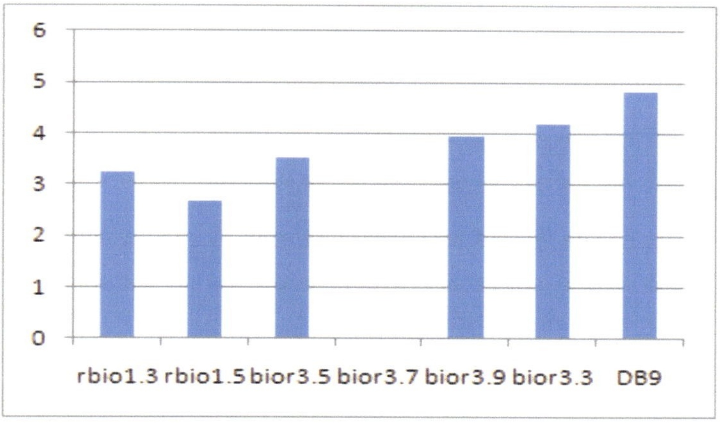

Fig. 56. Typical deviation from the average of each one

The value of the bior3.7 stands out from the rest because its four channels have the same value (80.65%), so the deviation is zero.

The DB9 presents de highest typical deviation because it has a big difference between FC1 and CZ.

We also present another graph with the typical deviation, but in this case, the deviation has been calculated respect to the average of the seven chosen wavelets (71%):

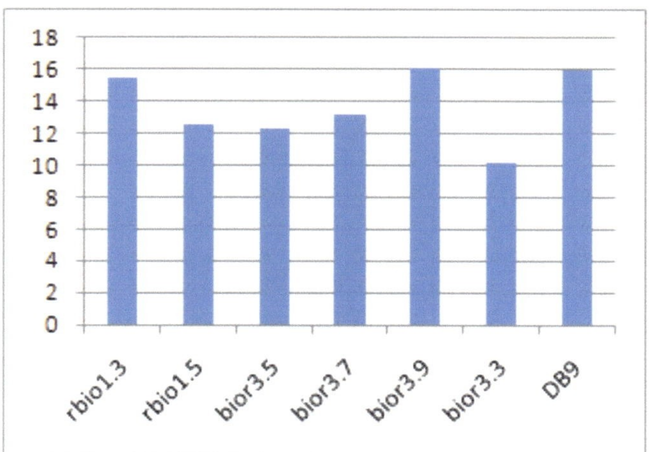

Fig. 57. Typical deviation from the average of the seven wavelets

This figure shows the average of the best four channels of each mother wavelet:

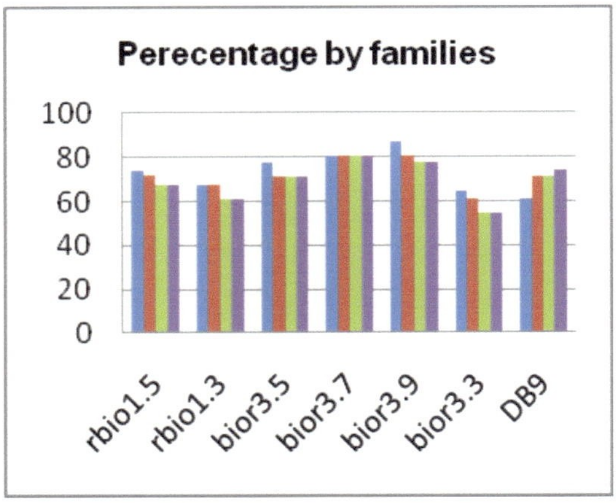

Fig. 58. Percentage of success by families

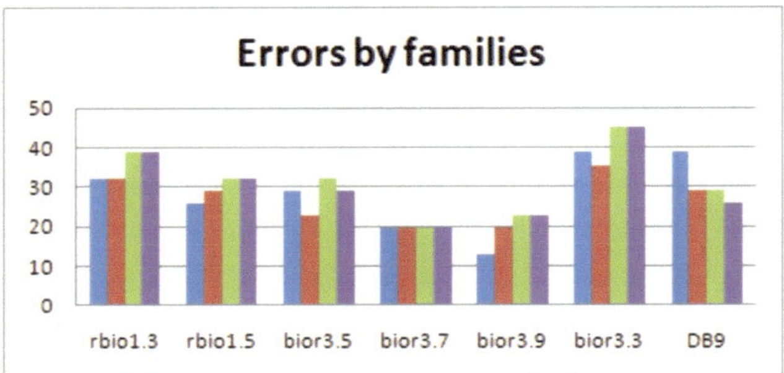

Fig. 59. Percentage of errors by families

As we can see, the best percentage corresponds to the channel FC1 with the wavelet bior3.9, but the wavelet with best average is the bior3.7, so this is the chosen one.

It only has been shown the best four channels for the bior3.7, but it also presents results that overcome the 65%.

FC1	77,41935484
FC2	74,19354839
C1	67,74193548
CZ	70,96774194
AF3	67,74193548
F1	74,19354839
F4	70,96774194

Table 12. Channels for the bior3.7

The electrodes are localized in the frontal lobe, which is the region that, among other functions, is responsible for the attention functions; and in the parietal lobe.

The order of success of the seven used wavelets is the following (from the best to the worst):

Order	Wavelet	Average(%)
1	bior3.7	80.65
2	bior3.9	80.65
3	bior3.5	72.58
4	rbio1.5	70.415
5	DB9	69.36
6	rbio1.3	64.52
7	bior3.3	58.87

Table 13. Order of success

The worst results corresponds to rbio1.3 and bior3.3, which are the wavelets with the lower order of the filter. The lower the filter is, the more frequencies are allowed by it.

The following table shows the different selected channels and the time where the peak is detected:

Wavelet	Channels	Time (sc)
bior3.7	FCZ-AFZ-FZ-F2	≈ 0.55
bior3.9	FC1-FCZ-FZ-F2	≈ 0.6
bior3.5	FCZ-FC1-F1-FZ	≈ 0.46
rbio1.5	FC1-FCZ-FZ-CZ	≈ 0.41
DB9	FC1-FCZ-C1-CZ	≈ 0.97
rbio1.3	FC1-FCZ-FZ-F2	≈ 0.32
bior3.3	FCZ-FC2-FZ-F2	≈ 0.37

Table 14. Channels and times

The difference of the time that the peak appears is due to the order of the filter: the higher the order is, the later the peak will appear.

This table contains the advantages and disadvantages comparing with each other:

Wavelet	Advantages	Disadvantages
bior3.7	High Success	Peak appears late
bior3.9	High Success	Peak appears late
bior3.5	Succes acceptable for BCI Peak appears before 0.5 seconds	Success less than the 80%
rbio1.5	Succes acceptable for BCI Peak appears before 0.5 seconds	Success less than the 80%
DB9	Succes acceptable for BCI	Peak appears late
rbio1.3	Peak appears early	Low success
bior3.3	Peak appears early	Low success

Table 15. Advantages and disadvantages

The ideal situation would be one wavelet with a high success and with the peak appearing as soon as it is possible.

This is difficult to obtain because to get precision, we need one mother wavelet with a high order, but this situation makes that the peak is delayed.

Although the chosen wavelet (bior3.7) shows the peak after 500 milliseconds, it does not mean that this delay represents any problem at the final application.

The first thing is because the P300 appears between 0.3 and 0.6 milliseconds since the stimulus is detected.

The second part is that, in order to ensure a good level decomposition , we have used samples until 2 seconds since the letter is illuminated. If the decomposition wasn´t conditioned by the number of samples, the sample frequency could be lower.

The bior3.5, rbio1.5 and DB9 don´t arrive at the established 80%, but their average is superior than 65, so the results are acceptable, although they are not enough in this application.

Following, it is presented the error of those mother wavelets that don´t arrive at the objective, being the theoretical value the 80%:

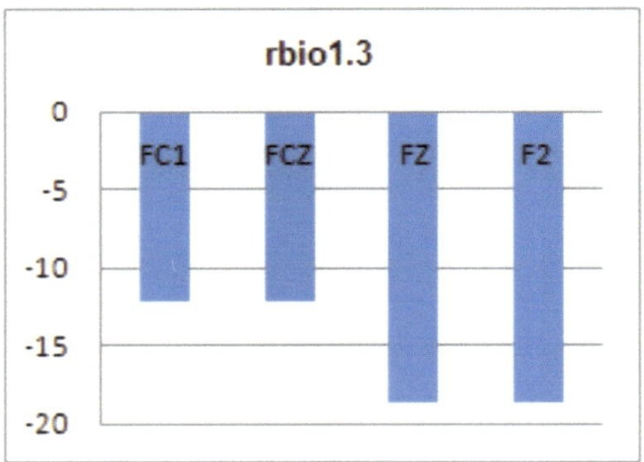

Fig. 60. Error rbio1.3 respect 80%

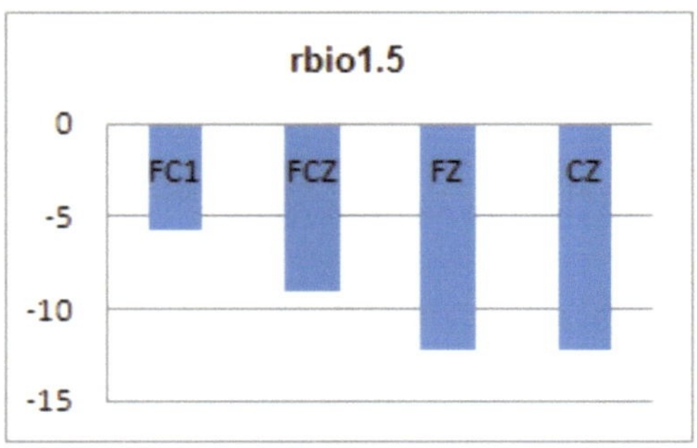

Fig. 61. Error rbio1.5 respect 80%

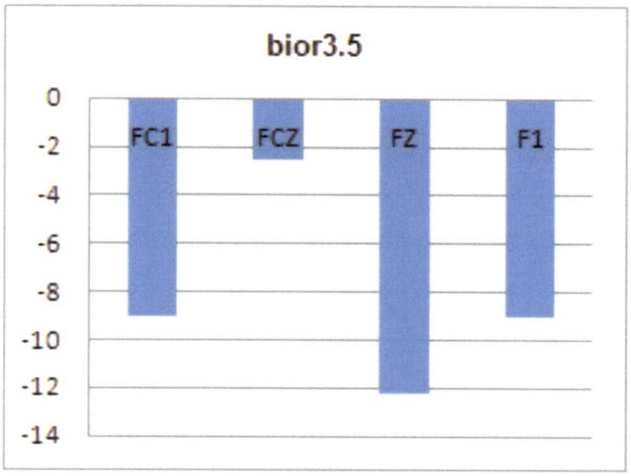

Fig. 62. Error bior3.5 respect 80%

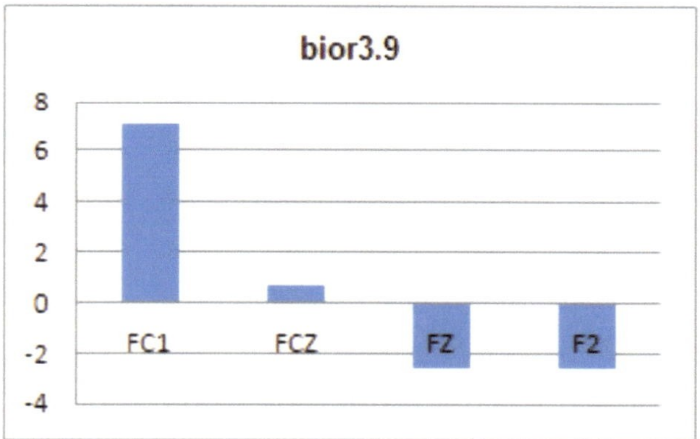

Fig. 63. Error bior3.9 respect 80%

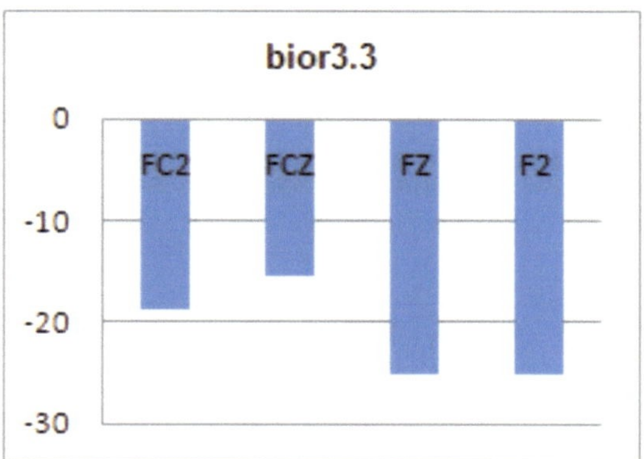

Fig. 64. Error bior3.3 respect 80%

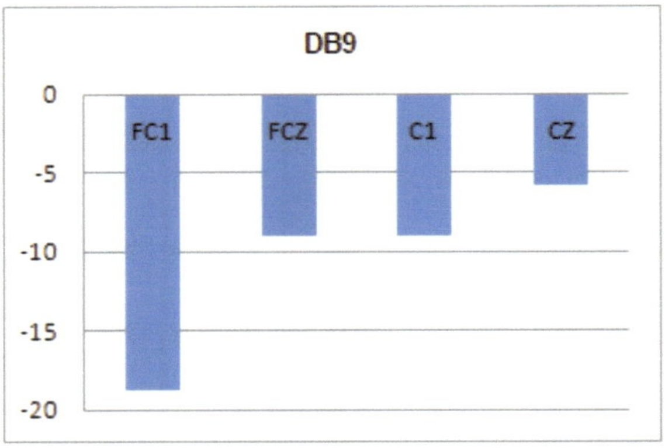

Fig. 65. Error DB9 respect 80%

The error is negative to indicate that the success doesn´t arrive at the established objective. It can be observed at bior3.9, where FC1 and FCZ overcome the 80% and the error (which is not an error in this case) is positive.

Bior3.3 is the one which is the farthest, with two channels with a distance higher than the 20%.

On the contrary, the bior3.9 practically arrives at the 80%, because the error of FZ and F2 is less than a 3%.

The following figures show the Wavelet Transform for the word HAT, with the wavelet bior3.7.

Letter H: column 2 and row 2:

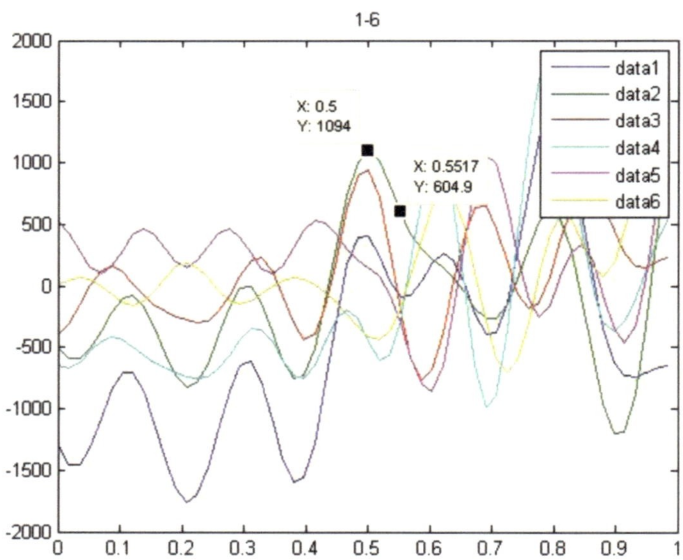

Fig. 66. Letter H, columns

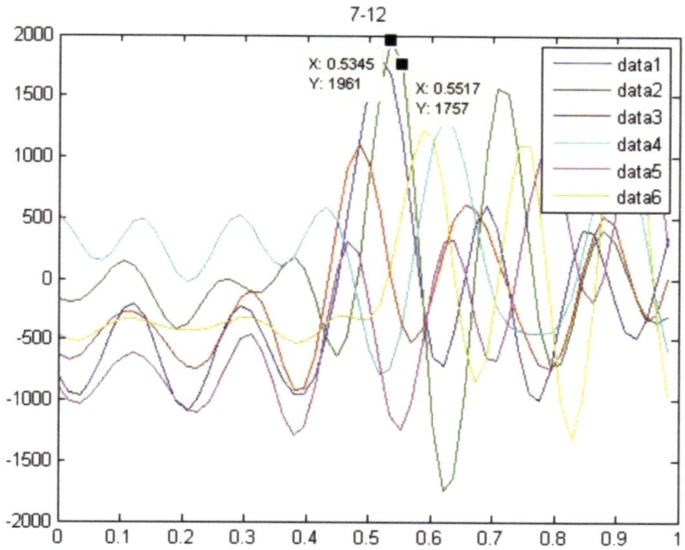

Fig. 67. Letter H, rows

Letter A: column 1 and row 1:

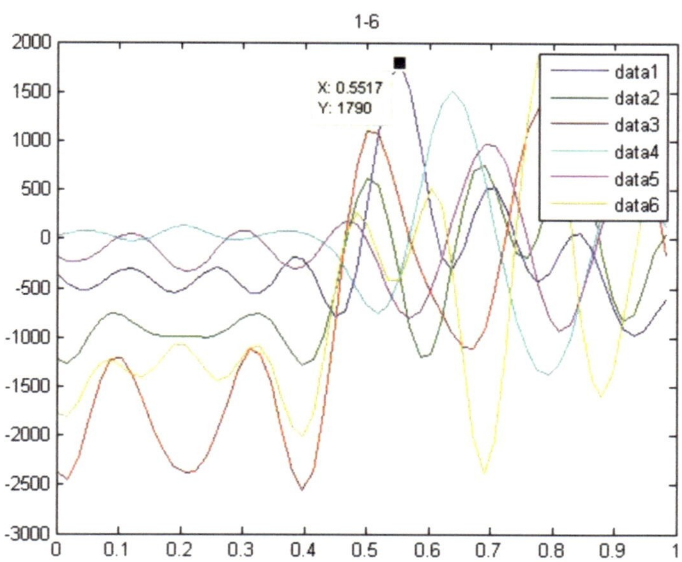

Fig. 68. Letter A, columns

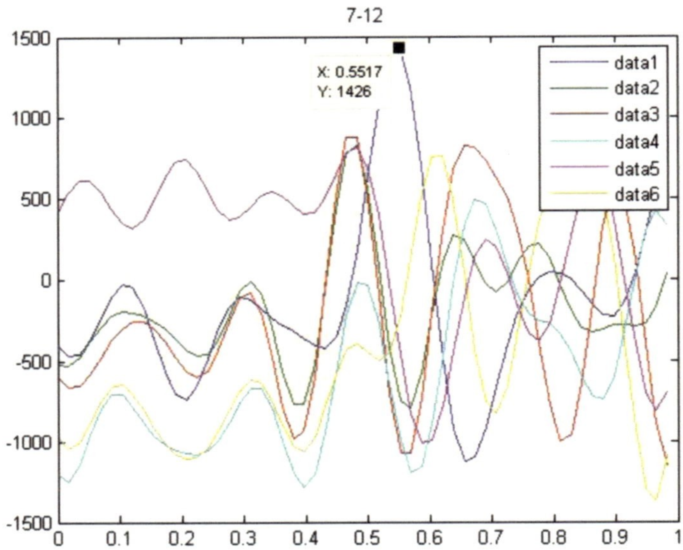

Fig. 69. Letter A, rows

Letter T: column 1, row 2

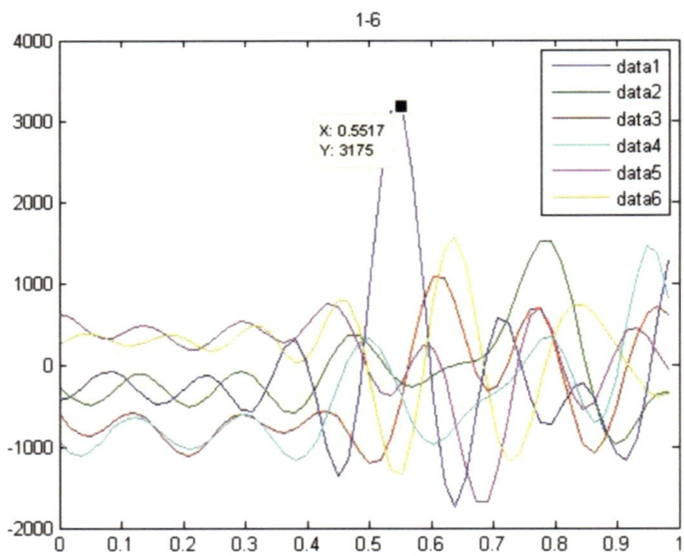

Fig. 70. Letter T, columns

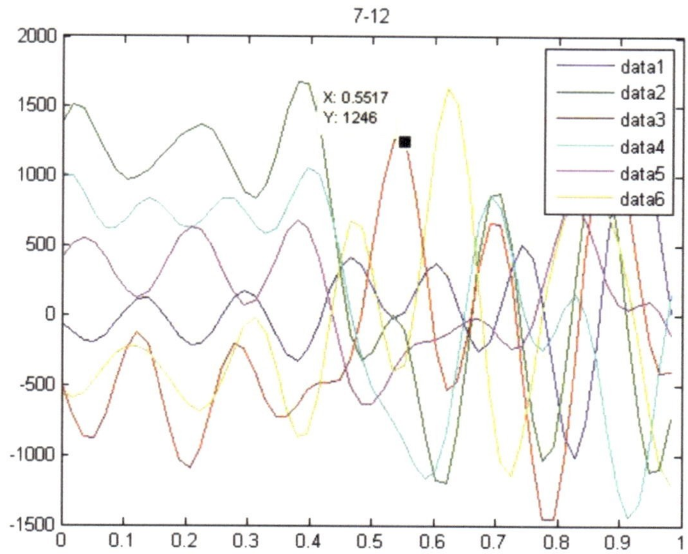

Fig. 71. Letter T, rows

6. Discussion

The performed algorithm is intended to be integrated in an application where the temporal answer must be the minimum possible. Although it´s an initial study of the P300 after the Wavelet Transform, the method focus to the posterior real application, so, even existing methods and calculations that would improve the hit rate, they are not applied due to the time that it suppose.

It is considerate as valid those results with a hit rate higher than an 80%.

It has made only comparisons between the different rows and columns to know which ones differ more than the rest in its behavior. It considers as target row and target column the two ones which show more difference respect to the rest.

It has been made entirely with the wavelets given by Matlab, being the bior3.7 the wavelet with best results.

Using other statistical methods after the transformation, and also with a quadratic B-Sprow, it can obtain higher results [19], but it supposes employ more time in calculations.

For the final application, it wouldn´t need a previous stage of training to difference between targets and non targets, because there has not been used any classifier.

In this study it has been used the 64 channels to know in which ones of them exists higher difference between the P300b and the P300a, in order to chose the best four channels that will be used in the application. In many studies the occipital region shows higher hit rates [18], although it can also be appreciated some differences in the frontal part [20]. Nevertheless, in [4] they point that the P300 can appear in the parietal and temporal zone.

7. Future research

There can differentiate three lines of actuation:

1. Improve the success of the algorithm in order to be more reliable without increasing the time due to the calculations that it supposes. This could be possible, among other things, by making our own wavelet, so that the noise that the used wavelets add to the signal could be minimized.
2. Use the algorithm in other applications where is necessary analyze signals in frequency bands, such as epilepsy problems.
3. Use other methods to work with EEG signals, such as Matching Pursuit.

8. Acknowledgment

Finally I would like to thank to those people who have collaborated in the elaboration of the study and the final application: Begoña García Zapirain, director of the Final Project where the study is integrated; Eneko Lopetegui, main creator of the application and is the one who proposed me using wavelets; John O'Toole, who gave me good advices at the time of working with wavelets; and at last, to the Arrieta sisters, who showed interest for the project since the first time.

Special mention should also go to the Education, University and Research Department from the Basque Government for their support of the project.

9. References

[1] Ibáñez J, Serrano J.I, del Castillo M.D, Barrios L, Gallego J.A, and Rocon E.(2011). *An EEG-Based Design for the Online Detection of Movement Intention.* LNCS 6691, p. 370 ff.

[2] Lopez-Gordo M.A, Ron-Angevin R, and PelayoValle F. (2011). Auditory Brain-Computer Interfaces for Complete Locked-In Patients. LNCS 6691, p. 378 ff.

[3] Mallat S. (2009). *A Wavelet Tour of signal processing.* Third edition: the Sparse Way. Burlington.

[4] Cardinali D. P. (2007). *Neurociencia aplicada: sus fundamentos* .Edit: Médica panamericana.

[5] Perez A, La Mura G, Piotrkowski R, Serrano E. (2002). *Procesamiento no lineal con Wavelet para la eliminación del ruido en imágenes planares de medicina nuclear.* Elsevier.

[6] Prochazka Al, and Kukal J. (2008). *Wavelet Transform Use for Feature Extraction and EEG Signal Segments Classification.* in 3rd International Symposium on Communications, Control and Signal Processing, 2008.

[7] Kitayama M, Otsubo H, Parvez S, Lodha A, Ying E, Parvez B, iIshii R, Mizuno-Matsumoto Y, Zoroofi R. A, Snead O.C. (2003). *Wavelet analysis for neonatal electroencephalographic seizures.* In Elseiver.

[8] Zarjam P. (2003). *EEG Data acquisition and automatic seizure detection using wavelet transforms in the newborn EEG.* Thesis

[9] Chazal P, Celler B. G. and Reilly R. B. (2000). *Using Wavelet Coefficients for the Classification of the Electrocardiogram.* In Proceedings of World Congress on Medical Physics and Biomedical Engineering, Chicago, July 2000.

[10] Mahmoodabadi S.Z, Alirezaie J, Babyn P. (2007). *Bio-signal Characteristics Detection Utilizing Frequency Ordered Wavelet Packets*. In Proceedings of ISSPIT 2007.

[11] Noble E. (2010). *Wavelet analysis of Malaria in pregnancy*. In proceedings of MIUA 2010.

[12] Demiralp T, Ademoglu A, Schürrmann M, Basar-Eroglu C, Basar E. (1999). *Detection of P300 Waves in Single Trials by the Wavelet Transform (WT)*. Brain Lang. 1999 Jan;66(1):108-28.

[13] Demiralp T, Ademoglub A, Istefanopulosb Y, Basar-Eroglu C, Basar E. (2001). *Wavelet analysis of oddball P300*. In International Journal of Psychophysiology 39 (2001).

[14] Ramírez-Cortes J. M, Alarcon-Aquino V, Rosas-Cholula G, Gomez-Gil P, Escamilla-Ambrosio J. (2010). *P-300 Rhythm Detection Using ANFIS Algorithm and Wavelet Feature Extraction in EEG Signals*. In Proceedings of the World Congress on Engineering and Computer Science 2010.

[15] Sorensen T.L, Olsen U.L, Conradsen I, Henriksen J, Kjaer T.W, Thomsen C.E, Sorensen H.B.D. (2010). *Automatic epileptic seizure onset detection using Matching Pursuit: A case study*. In Engineering in Medicine and Biology Society (EMBC), 2010 Annual International Conference of the IEEE.

[16] Hsua WY, Lin CC, Ju MS, Suna YN. (2007). *Wavelet-based fractal features with active segment selection: Application to single-trial EEG data*. In Journal of Neuroscience Methods 163 (2007) 145–160.

[17] Costagliola S, Dal Seno B, Matteucci M. (2009). *Recognition and Classification of P300s in EEG Signals by Means of Feature Extraction Using Wavelet Decomposition*. In Proceedings of International Joint Conference on Neural Networks, Atlanta, Georgia, USA, June 14-19, 2009.

[18] Gareis I. E, Gentiletti G. G, Acevedo R, Rufiner L. (2009). *Extracción de características en interfaces cerebro computadoras mediante transformada wavelet discreta: Resultados preliminares*. Memorias del XVII Congreso Argentino de Bioingenieria (SABI 2009), Number 167, page 58 – 62.

[19] Z. Seyyedsalehi, A.M. Nasrabadi, V. Abootalebi "Committee Machines and Quadratic B-sprow Wavelet for the P300 Speller Paradigm "

[20] Rakotomamonjy A. (2008). *Ensemble of SVMs for BCI P300 Speller*. Biomedical Engineering Transactions. Volume: 55 Issue: 3 On page(s): 1147 - 1154.

[21] Markazi S.A, Stergioulas S, Ramchurn L.S, Bunce A. (2006). *Wavelet Filtering of the P300 Component in Event-Related Potentials*. Engineering in Medicine and Biology Society, 2006. EMBS '06. 28th Annual International Conference of the IEEE

[22] Rosas-Cholula G, Ramírez-Cortes J.M, Alarcón-Aquino V, Martinez-Carballido J, Gómez-Gil P. (2010). *On Signal P-300 Detection for BCI Applications Based on Wavelet Analysis and ICA Preprocessing*. 2010 IEEE Electronics, Robotics and Automotive Mechanics Conference.

[23] Barbosa D.J, Ramos J, Lima C. (2008). *Detection of small bowel tumors in capsule endoscopy frames using texture analysis based on the discrete wavelet transform*. Engineering in Medicine and Biology Society, 2008. EMBS 2008. 30th Annual International Conference of the IEEE.

[24] Zhen L, Chan A.K. (2001). *An artificial intelligent algorithm for tumor detection in screening mammogram*. Transactions on Medical Imaging, IEEE. Volume: 20 Issue:7 On page(s): 559 - 567.

[25] Polikar R, Keinert F, Greer M.H. (2001). *Wavelet analysis of event related potentials for early diagnosis of Alzheimer's disease.* In: A. Wavelets in Signal and Image Analysis, From Theory to Practice, Kluwer Academic Publishers, Boston, 2001.

[26] Ishita H, Sakai M, Watanabe J, Chen W, Wei D. (2007). *Development of P300 Detection Algorithm for Brain Computer Interface in Single Trial.* In Proceedings of 7th IEEE International Conference on Computer and Information Technology, 2007.

[27] Salvaris M, Sepulveda F. (2009). *Wavelets and ensemble of FLDs for P300 classification.* In Proceedings of 4th International IEEE/EMBS Conference on Neural Engineering 2009.

[28] Adeli H, Ghosh-Dastidar S, Dadmehr S. N. (2007). *A Wavelet-Chaos Methodology for Analysis of EEGs and EEG Subbands to Detect Seizure and Epilepsy.* In IEEE Transactions on Biomedical Engineering. Volume: 54 Issue: 2 On page(s): 205 - 211.

[29] Ting W, Guo-Zehng Y, Bang-Hua Y, Hong S. (2008). *EEG feature extraction based on wavelet packet decomposition for brain computer interface.* Measurement Volume 41, Issue 6, July 2008, Pages 618-625.

[30] AlMuhit A, Islam S. and Othman M. (2004). *VLSI Implementation of Discrete Wavelet Transform (DWT) for Image Compression.* 2nd International Conference on Autonomous Robots and Agents December 13-15, 2004 Palmerston North, New Zealand.

[31] Secker, A;Taubman, D. (2001). *Motion-compensated highly scalable video compression using an adaptive 3D wavelet transform based on lifting.* Proceedings of ICIP 2001. Greece.

[32] Easwaramoorthy, D; Uthayakumar, R. (2010). *Analysis of biomedical EEG signals using Wavelet Transforms and Multifractal Analysis.* Proceedings of IEEE International Conference on Communication Control and Computing Technologies (ICCCCT), 2010. Ramanathapuram.

[33] Ayhan, T, Seker, S. (2011). *Detemination of decision moments in EEG signals using DWT.* Proceedings of International Symposium on Innovations in Intelligent Systems and Applications (INISTA). Istanbul.

[34] Shaker Maan M. (2005). *EEG Waves Classifier using Wavelet Transform and Fourier Transform.* International Journal of Biological and Life Sciences 1:2 2005.

[35] Cvetkovic, D, Übeyli E.D, Cosic I. (2008). *Wavelet transform feature extraction from human PPG, ECG, and EEG signal responses to ELF PEMF exposures: A pilot study.* Journal Digital Signal Processing, Volume 18 Issue 5, September, 2008.

2

Energy Distribution of EEG Signal Components by Wavelet Transform

Ibrahim Omerhodzic[1], Samir Avdakovic[2],
Amir Nuhanovic[3], Kemal Dizdarevic[1] and Kresimir Rotim[4]
[1]*Clinical Center University of Sarajevo, Department of Neurosurgery, Sarajevo*
[2]*EPC Elektroprivreda of Bosnia and Herzegovina, Sarajevo*
[3]*Faculty of Electrical Engineering, University of Tuzla, Tuzla*
[4]*University Hospital "Sisters of Charity", Department of Neurosurgery, Zagreb*
[1,2,3]*Bosnia and Herzegovina*
[4]*Croatia*

1. Introduction

Wavelet theory is a natural extension of the Fourier transformation and its modified Short-Term Fourier transformation (STFT). Over the years, wavelets developed independently in mathematics, quantum physics, electrical engineering, as well as in other areas of science. The result is their significant application in all branches of science. Due to its advantages over other techniques of signal processing, Wavelet Transform (WT) in recent years has attracted considerable attention in signal processing in medicine. The advantage of WT over the Fourier transformation is reflected in the time-frequency analysis (Daubechies, 1992; Mallat, 1998; Mertins, 1999; Vetterli & Kovacevic, 1995; Wang & Xu 2009).

On the other hand, epilepsy is the second most prevalent neurological disorder in humans after stroke. It is characterized by recurring seizures in which abnormal electrical activity in the brain causes altered perception or behavior. As one of the world's most common neurological diseases, it has affected more than 40 million people worldwide. Epilepsy's hallmark symptom, seizure, is manifestations of epilepsy and can have a broad spectrum of debilitating medical and social consequences (Aylward, 2008; Lefter et al., 2010; McHugh & Delanty, 2008; Ngugi, 2011; Tong & Thacor, 2009). Although antiepileptic drugs have helped treat millions of patients, roughly a third of all patients are unresponsive to pharmacological intervention. An area of great interest is the development of devices that incorporate algorithms capable of detecting an early onset of seizures or even predicting the hours before seizures occur. This lead time will allow for new types of interventional treatment. Intention is, in the near future, that a patient's seizure may be detected and aborted before physical manifestations begin (Latka et al., 2005; Saiz Díaz et al., 2007).

Electroencephalogram (EEG) established itself in the past as an important means of identifying and analyzing epileptic seizure activity in humans. It serves as a valuable tool for clinicians and researchers to study the brain activity in a non-invasive manner. Careful analyses of the electroencephalograph (EEG) records can provide valuable insight into and improved understanding of the mechanisms causing epileptic disorders. Detection of epileptiform discharges in the EEG is an important component in the diagnosis of epilepsy. In most cases, identification of the epileptic EEG signal is done manually by skilled professionals, who are small in number (Adeli et al., 2003; Patnaik & Manyamb, 2008; Wang & Xu 2009). The diagnosis of an abnormal activity of the brain functionality is a vital issue. The clinical interests (in EEG) are, for example, the sleep pattern analysis, cognitive task registration, seizure and epilepsy detection, and other states of the brain, both normal and pathophysiological (Asaduzzaman et al., 2010; Ernst et al., 2007; Leise & Harrington, 2011; Subasi et al., 2005)

EEG signals involve a great deal of information about the function of the brain. But classification and evaluation of those signals are limited. Since there is no definitive criteria established by experts, visual analysis of EEG signals in time domain may be insufficient. The routine clinical diagnosis needs the analysis of EEG signals. Therefore, some automation and computer techniques are used for this aim. Recent applications of the WT and Neural Network (NN) to engineering-medical problems can be found in several studies that refer primarily to signal processing and classification in different medical areas. Several authors used WT in different ways to analyze EEG signals and combined WT and NN in the process of classification (Adeli et al., 2007; Guo et al., 2010; Leung et al., 2009; Mirowski et al., 2009; Subasi et al., 2005; Zandi et al., 2008).

As EEG signals are non-stationary, the conventional method of frequency analysis is not highly successful in diagnostic classification (Subasi & Erçelebi, 2005). A few papers recently published have reported on the effectiveness of WT applied to the EEG signal for representing various aspects of non-stationary signals such as trends, discontinuities, and repeated patterns where other signal processing approaches fail or are not as effective (Adeli et al., 2003; Asaduzzaman et al., 2010; Guo et al., 2009; Lessa, 2011), but there are still some problems with classical EEG analysis and classification (Arab et al., 2010; Bauer et al., 2008; Oehler et al., 2009). It is important to emphasize the algorithm for classification of EEG signals based on WT and Patterns Recognize Techniques. Discrete Wavelet Transform (DWT) with the Multi-Resolution Analysis (MRA) is applied to decompose EEG signal at the resolution levels of the EEG signal components (δ, θ, α, β and γ), and Parseval's theorem is employed to extract energy distribution percentage features of the EEG signal at different resolution levels. The neural network classifies those extracted features to identify the EEG type according to the energy distribution percentage.

2. Energy distribution of the EEG signal components

Some results of our previous research, shown in this chapter, were published recently (Omerhodzic et al., 2010), and the datasets were originally selected from the Epilepsy Center in Bonn, Germany, collected by Ralph Andrzejak (Andrzejak et al., 2001). The datasets we particularly used and denoted consisting of three groups of EEG signals, were basically extracted from both normal subjects and epileptic patients. The first group was recorded from healthy subject (A set), the second group was recorded prior to a seizure (steady state) from part of the brain of the patient with epilepsy syndrome (C set) and the third group (E set) was recorded from the patient with the epilepsy syndrome during the seizure. Each set contains 100 single channel EEG segments of 23.6-sec duration at a sampling rate of Fs = 173.61 Hz. Set A consisted of segments taken from surface EEG recordings that were obtained from five healthy volunteers using a standardized electrode placement. Set E only contained seizure activity.

As is well known, the EEG signal contains a several spectral components. The magnitude of a human brain surface EEG signal is in the range of 10 to 100 μV. The frequency range of the EEG has a fuzzy lower and upper limit, but the most important frequencies from the physiological viewpoint lie in the range of 0.1 to 30 Hz. The standard EEG clinical bands are the delta (0.1 to 3.5 Hz), theta (4 to 7.5 Hz), alpha (8 to 13 Hz), and beta (14 to 30 Hz) bands. EEG signals with frequencies greater than 30 Hz are called gamma waves (Schiff et al., 1994; Tong & Thacor, 2009; Vetterli & Kovacevic, 1995).

Generally, a wavelet is a function $\psi \in L^2(R)$ with a zero average

$$\int_{-\infty}^{+\infty} \psi(t)dt = 0 . \tag{1}$$

The Continuous Wavelet Transformation (CWT) of a EEG signal $x(t)$ is defined as:

$$CWT_{\psi}x(a,b) = \frac{1}{\sqrt{|a|}} \int_{-\infty}^{+\infty} x(t)\psi^*\left(\frac{t-b}{a}\right)dt \tag{2}$$

where $\psi(t)$ is called the 'mother wavelet', the asterisk denotes complex conjugate, while a and b $(a,b \in R)$ are scaling parameters, respectively (He & Starzyk, 2006; Mei et al., 2006;. Omerhodzic et al., 2010). The scale parameter a determines the oscillatory frequency and the length of the wavelet, and the translation parameter b determines its shifting position.

In practise, the application of WT in engineering areas usually requires the discrete WT (DWT). The DWT is defined by using discrete values of the scaling parameter a and the translation parameter b. Adjustment: $a = a_0^m$ and $b = nb_0a_0^m$, we obtain the following

$$\psi_{m,n}(t) = a_0^{-m/2}\psi\left(a_0^{-m}t - nb_0\right),$$

where $m,n \in Z$, and m is indicating frequency localization and n is indicating time localization. Generally, we can choose $a_0 = 2$ and $b_0 = 1$. This choice will define a dyadic-orthonormal WT and provide the basis for multi-resolution analysis (MRA).

In MRA, any EEG signal $x(t)$ can be completely decomposed in terms of approximations, provided by scaling functions $\phi_m(t)$ (also called father wavelet) and the details, provided by the wavelets $\psi_m(t)$. The scaling function is closely related with the low-pass filters (LPF), and the wavelet function is closely related with the high-pass filters (HPF). The decomposition of the signal starts by passing a signal through these filters. The approximations are the low-frequency components of the time series or signal and the details are the high-frequency components of the signal. The signal passes through a HPF and a LPF. Then, the outputs from filters are decimated by 2 to obtain the detail coefficients and the approximation coefficients at level 1 (A1 and D1). The approximation coefficients are then sent to the second stage to repeat the procedure. Finally, the signal is decomposed at the expected level (Avdakovic et al., 2009, 2010; Mallat, 1998; He & Starzyk, 2006; Mei et al., 2006).

The frequency band $[F_m/2 : F_m]$ of each detail scale of the DWT is directly related to the sampling rate of the original signal, which is given by $F_m = F_s/2^{l+1}$, where F_s is the sampling frequency, and l is the level of decomposition. In this study, the sampling time is 0.00576 sec or sampling frequency is 173.6 Hz of the EEG signals. The highest frequency that the signal could contain, from Nyquist' theorem, would be $F_s/2$ i.e. 86.8 Hz. Frequency bands corresponding to five decomposition levels for wavelet Db4 used in this study, with sampling frequency of 173.6 Hz of EEG signals were listed in Table 1.

Decomposed signals	Frequency bands (Hz)	Decomposition level
D1	43.4-86.8	1 (noises)
D2	21.7-43.4	2 (gama)
D3	10.8-21.7	3 (beta)
D4	5.40-10.8	4 (alpha)
D5	2.70-5.40	5 (theta)
A5	0.00-2.70	5 (delta)

Table 1. Frequency bands corresponding to different decomposition levels.

The Db4 transform has four wavelet and scaling function coefficients. The scaling function coefficients are:

$$h_0 = \frac{1+\sqrt{3}}{4\sqrt{2}},\ h_1 = \frac{3+\sqrt{3}}{4\sqrt{2}},\ h_2 = \frac{3-\sqrt{3}}{4\sqrt{2}},\ h_3 = \frac{1-\sqrt{3}}{4\sqrt{2}}.$$

The wavelet function coefficient values are:

$$g_0 = h_3,\quad g_1 = -h_2,\quad g_2 = h_1,\quad g_3 = -h_0$$

or:

$$g_0 = \frac{1-\sqrt{3}}{4\sqrt{2}},\ g_1 = \frac{\sqrt{3}-3}{4\sqrt{2}},\ g_2 = \frac{3+\sqrt{3}}{4\sqrt{2}},\ g_3 = \frac{-1-\sqrt{3}}{4\sqrt{2}}.$$

Below, based on Parseval's theorem, the energy of EEG signal can be partitioned at different resolution levels. Mathematically this can be presented as:

$$ED_i = \sum_{j=1}^{N} \left|D_{ij}\right|^2,\quad i = 1,......,l \tag{3}$$

$$EA_l = \sum_{j=1}^{N} \left|A_{lj}\right|^2 \tag{4}$$

where $i = 1,......,l$ is the wavelet decomposition level from level 1 to level l. N is the number of the coefficients of detail or approximate at each decomposition level. ED_i is the energy of the detail at decomposition level i and EA_l is the energy of the approximate at decomposition level l (Avdakovic et al., 2011; Jaffard et al., 2001; Mertins, 1999; Omerhodzic et al., 2008, 2011). Figure 1 shows the three signals from the analyzed database.

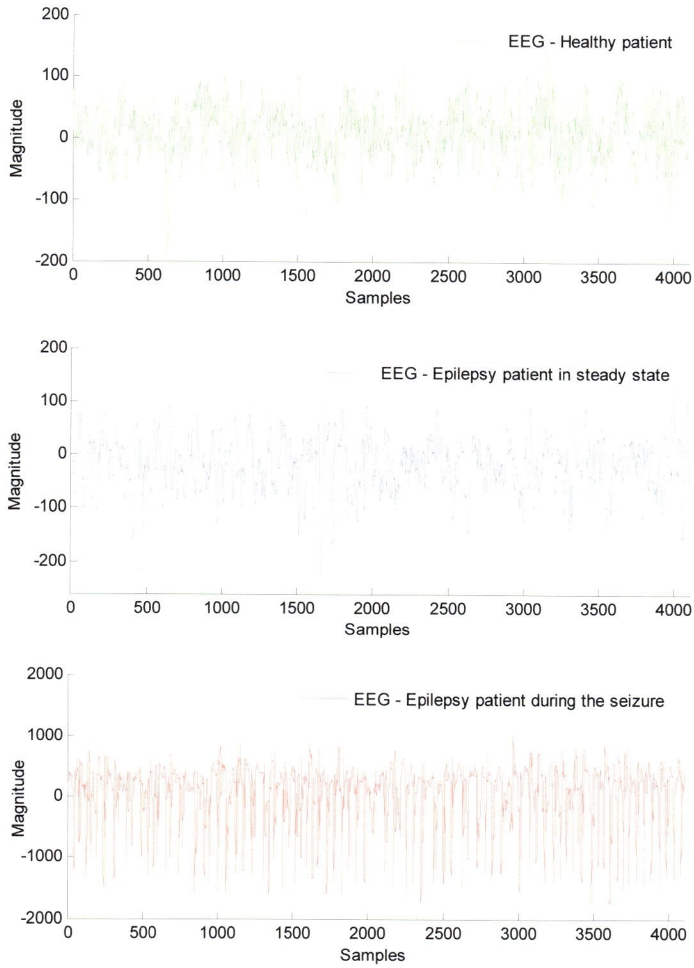

Fig. 1. The first EEG signal presents the EEG of a healthy patient, the second EEG signal presents the EEG of an epilepsy patient in steady state and the third EEG signal presents the EEG of an epilepsy patient during the seizure.

It is obvious (Fig. 1) that the magnitudes of the EEG signal of a patient with epilepsy and during the seizure are much larger than those of the other two EEG signals. Also, components of the EEG signal (δ, θ, α, β and γ) of a patient with epilepsy and during the seizure have much larger magnitudes than the other two EEG signals. The magnitude of the EEG signals of the healthy patient and the EEG signals of the epilepsy patient in steady state have approximately the same values. The activity (magnitude) of the components of these two signals will be determined by using DWT. The activity of the components (δ, θ, α, β and γ) of the EEG signals of the healthy patient and EEG signals of the epilepsy patient in steady state (signals in Fig. 1), after DWT and MRA and the use of Db4 wavelet functions, are shown in Fig. 2. After MRA of signals from Fig. 1, physical characteristic components of EEG signals are identified. It is obvious that the magnitude of the EEG signals of the epilepsy patient in steady state in frequency ranges [21.7-43.4] Hz and [10.8-21.7] Hz (γ and β waves) is much lower than the magnitude of the EEG signals of the healthy patient. The magnitudes of the signals in the frequency range [5.40-10.8] Hz (α waves) have roughly the same characteristics, while the magnitude of the EEG signal of the epilepsy patient in steady state in frequency ranges [2.70-5.40] Hz (θ wave) is much higher than the magnitude of the EEG signal of the healthy patient. The magnitude of the EEG signal of the epilepsy patient in steady state in the frequency range and [0.00-2.70] Hz (δ waves) has a higher magnitude than the EEG signal of the healthy patient. Energy distribution diagrams of EEG signals for different analysis cases are shown in Fig. 2.

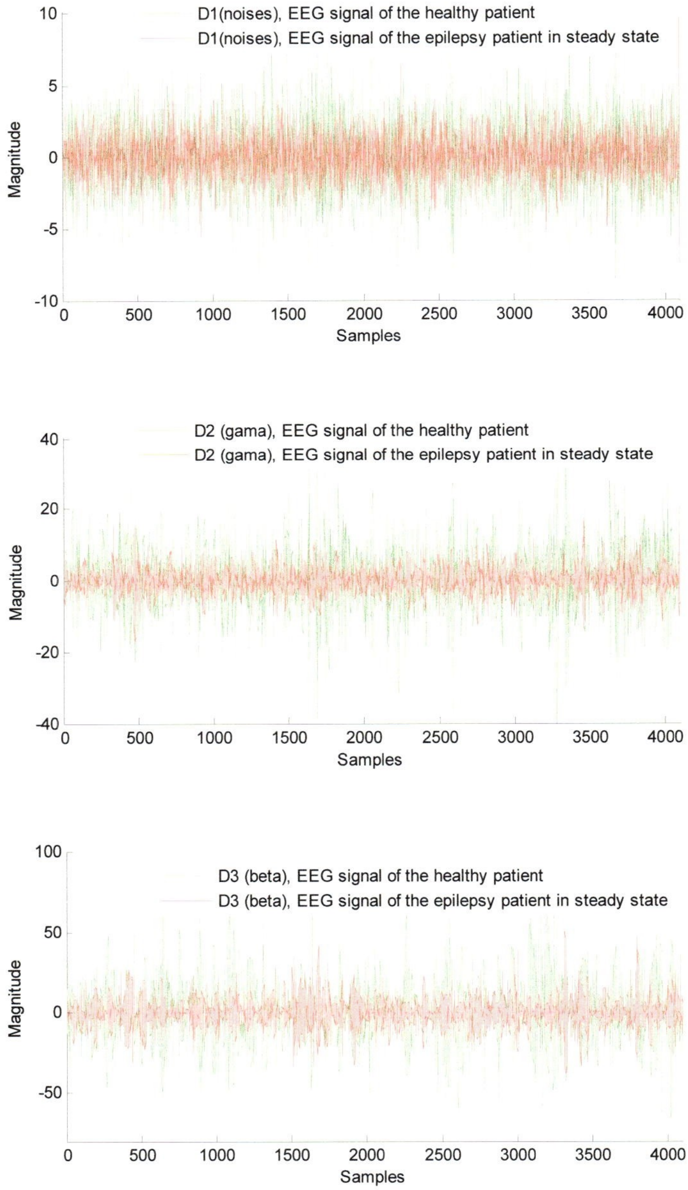

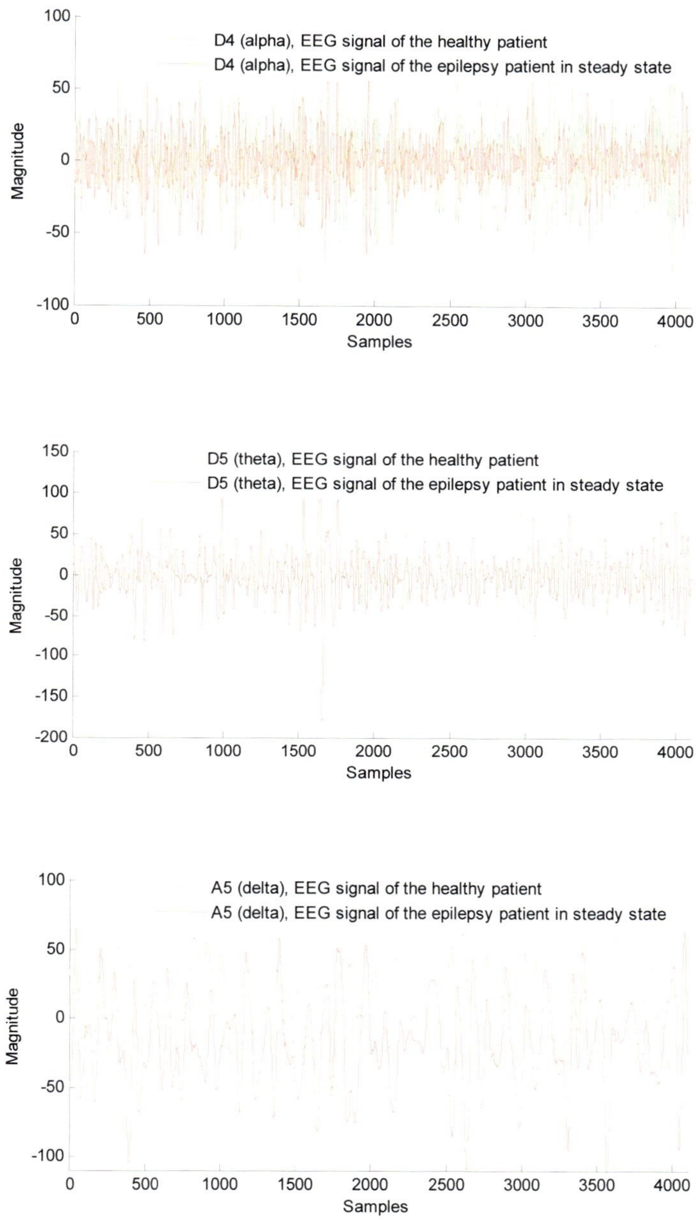

Fig. 2. Identification of components of EEG signals and their activity.

We could recognize different distribution of energy of the analyzed signals, which was generally quite similar for each group of EEG signals. The results showed that different groups of the analyzed signals (sets A, C and E) are obviously different in energy distribution of signals in the frequency bands of decomposition of EEG signals (Fig. 3).

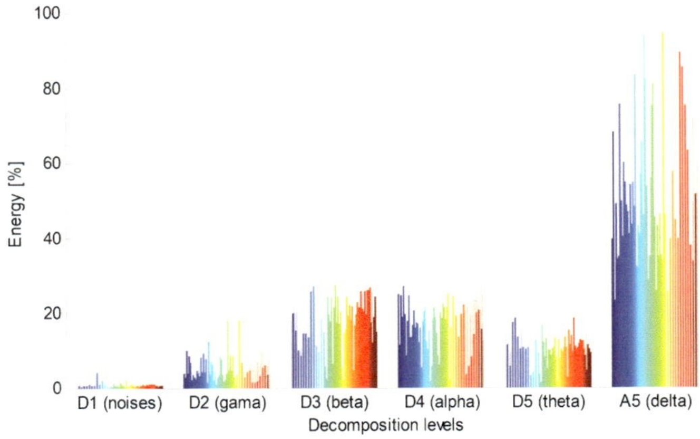

a)

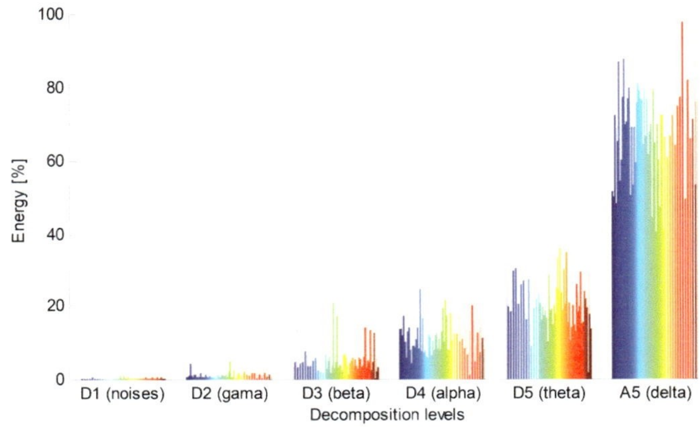

b)

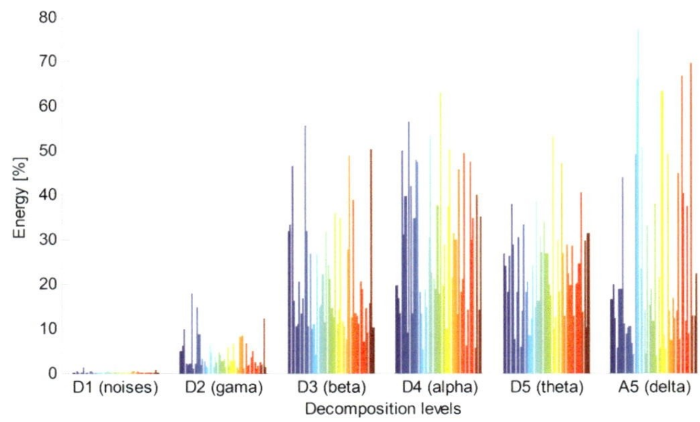

c)

Fig. 3. Energy distribution diagram (%) of a) A set - 100 EEG signals of a healthy patient, b) C set - 100 EEG signals of an epilepsy patient in steady state and c) E set - 100 EEG signals of an epilepsy patient during a seizure (Omerhodzic et al., 2010)

It was noted in the EEG signal of healthy subjects that energy activity in the frequency components of D3 and D4 (beta and alpha) waves was quite similar, and percentage of total energy value of the signal was around 20%. Energy activity in the frequency range D5 component (theta wave) was slightly lower in intensity and percentage of its value in total energy signal value was around 10%, while value percentage of D2 (gamma wave) was approximately 5%. Noise was negligibly small (D1) while the value of the frequency components of A5 was about 45%, although for some samples it had a much higher value. Unlike the distribution of EEG signals of healthy subjects, energy distribution of the signal of patients with epilepsy syndrome was obviously different. In comparison with EEG signals of healthy subjects, D2, D3 and D4 components of EEG signals have a significantly lower percentage of total energy distribution than D5 and A5 signals. Energy distribution of EEG signals where epileptic seizure was registered was significantly different from the first two cases. Energy activity of D3, D4 and D5 components was dominant, while A5 component was somewhat lower.

3. Indicators of epilepsy based on WT

EEG is the recording of electrical activity along the scalp of head, produced by the firing of neurons within the brain. It refers to the recording of the brain's spontaneous electrical activity over a short period of time, usually 20–40 minutes, as recorded from multiple electrodes placed on the scalp (Adeli et al., 2003; Niedermeyer & da Silva, 2004). In neurology, the main diagnostic application of EEG is in the case of epilepsy, as epileptic activity can create clear abnormalities on a standard EEG study (Abou-Khalil & Musilus, 2006). Well-known causes of epilepsy may include: genetic disorders, traumatic brain injury, metabolic disturbances, alcohol or drug abuse, brain tumor, stroke, infection, and cortical malformations (dysplasia). Therefore, EEG activity always reflects the summation of the synchronous activity of thousands or millions of neurons that have similar spatial orientation. Because voltage fields fall off with the square of the distance, activity from deep sources is more difficult to detect than currents near the skull (Klein & Thorne, 2007). Scalp EEG activity shows oscillations at a variety of frequencies. Several of those oscillations have characteristic frequency ranges, spatial distributions, and are associated with the different states of brain functioning. These oscillations represent synchronized activity over a network of neurons. Daubechies wavelets are the most popular wavelets representing foundations of wavelet signal processing, and are used in numerous applications (Daubechies, 1992). Daubechies 4 (Db4) is selected because its smoothing feature was suitable for detecting changes of the EEG signals.

In the context of a better understanding of the consequences of epilepsy, but also drawing some conclusions, which may indicate the development of the disease prior to the event, in the form of attack (seizure), a detailed analysis of A and C sets of EEG signals was carried out. Below, in the same way, using MRA and Db4 wavelet function, two sets of EEG signals (A set - 100 EEG signals of the healthy patient and C set - 100 EEG signals of the epilepsy patient in steady state) were partitioned at five resolution levels. Thereafter, the energy values of the components of the EEG signals were determined using Parseval's theorem (Eq. 3 and Eq. 4). Fig. 4 shows energy values of individual components of EEG signal for set A and set C respectively.

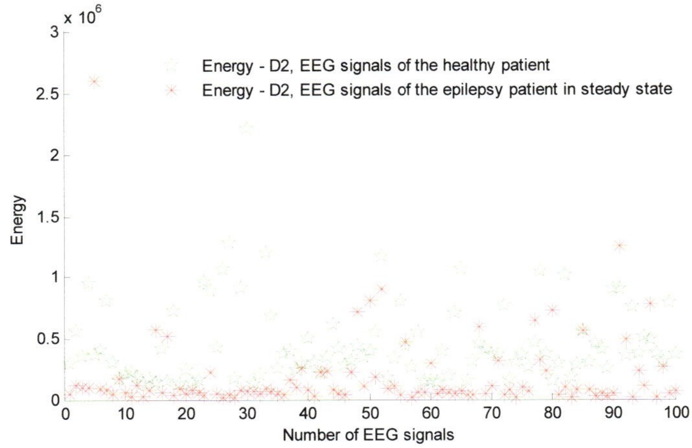
x 10^6
Energy - D2, EEG signals of the healthy patient
Energy - D2, EEG signals of the epilepsy patient in steady state
Energy
Number of EEG signals
0
0.5
1
1.5
2
2.5
3
0
10
20
30
40
50
60
70
80
90
100

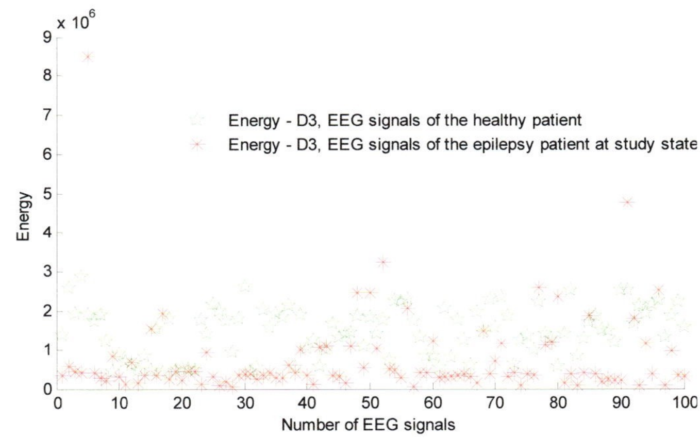
x 10^6
Energy - D3, EEG signals of the healthy patient
Energy - D3, EEG signals of the epilepsy patient at study state
Energy
Number of EEG signals
0
1
2
3
4
5
6
7
8
9
0
10
20
30
40
50
60
70
80
90
100

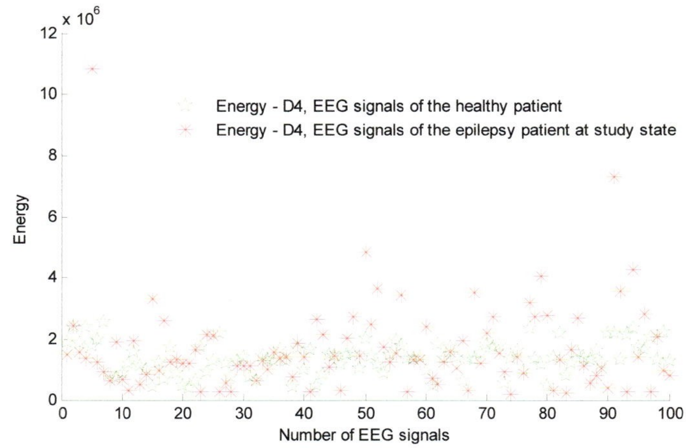
x 10^6
Energy - D4, EEG signals of the healthy patient
Energy - D4, EEG signals of the epilepsy patient at study state
Energy
Number of EEG signals
0
2
4
6
8
10
12
0
10
20
30
40
50
60
70
80
90
100

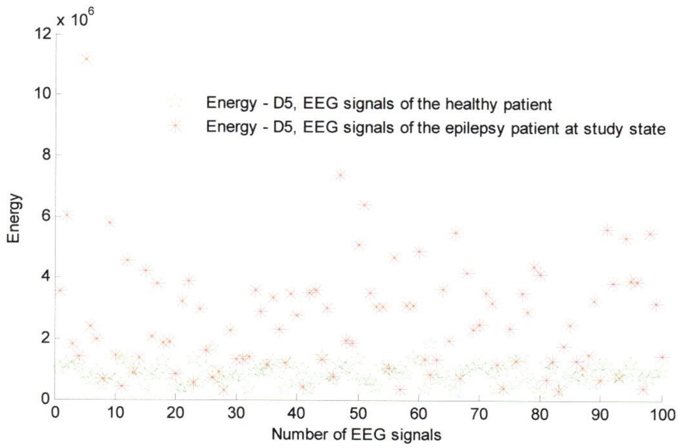

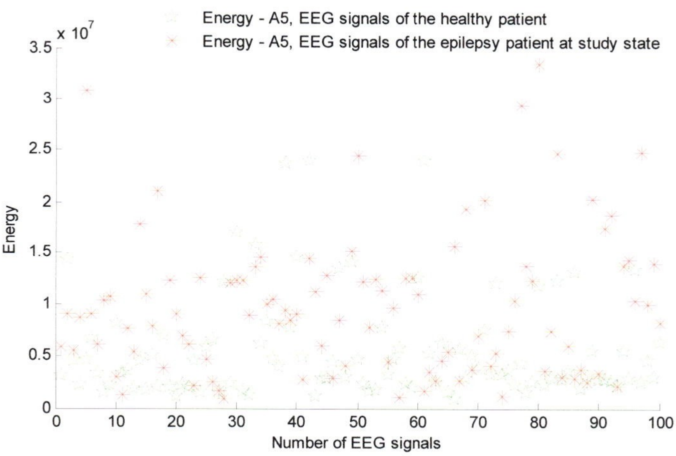

Fig. 4. Energy distribution diagram of EEG signals: comparison of A and C sets (A set - 100 EEG signals of the healthy patient and C set - 100 EEG signals of the epilepsy patient in steady state).

It may be noted that in the patients diagnosed with epilepsy, D2 component activity (γ waves) is quite low and on average it is by 58.26% lower than the EEG signals of the healthy patient. D3 activity component (β waves) was on average lower by 48.22%, while the activity of D4 component (α waves) was quite similar to the EEG signals of the healthy patient. The activity of D5 component (θ waves) was about 200% higher than the EEG signals of the healthy patient, and the activity of A5 components (δ waves) was on average higher approximately 77.32%. On average, the analyzed signals, the energy value of set C (epilepsy patients) was 82% higher than in set A (healthy subjects). However, it is possible to observe the different activities of individual components of the EEG signals for healthy and epilepsy patients, which indicates different physical processes. Weakening of or a decrease in magnitudes, over time, of some components of EEG signals (β and γ waves), or strengthening of or an increase in the magnitude of θ wave over time can be reliable indicators of the development of epilepsy. This finding indicates that a timely analysis of energy values of the components of EEG signal, if made in the same patient at regular intervals, could lead to timely detection of the developing of disease before it manifests itself clinically. For a better insight into the results of an analysis of the minimum and maximum values of the components of EEG signals are used to establish thresholds (limits) of energy value where EEG signal is normal. Energy thresholds at different frequency bands (decomposition levels) based on the calculation of energy value of the EEG signals from A set and comparison with the results based on the calculation of energy value of the EEG signals from C set is presented in Table 2. Each of the 100 analyzed EEG signals from C set had 'punching' at set thresholds of the decomposition level. Signals marked with N005 and N097 had values outside the limits at each level of decomposition. Ten (10) signals had values beyond the boundaries of just one level of decomposition, while 90 signals had energy values beyond the boundaries at two or more levels of decomposition.

Decomposition level	A set		C set		
	Threshold $(\mu V)^2$		No. of EEG signals		
	min	max	< min	> max	total
D2 [21.7-43.4] Hz	124.524	2.211.951	72	1	73
D3 [10.8-21.7] Hz	406.110	2.871.387	60	3	63
D4 [5.40-10.8] Hz	395.460	2.549.130	13	19	32
D5 [2.70-5.40] Hz	240.620	1.872.889	0	58	58
A5 [0.00-2.70] Hz	846.406	23.970.453	0	6	6

Table 2. Energy thresholds at different frequency bands (decomposition levels) based on the calculation of the energy value of the EEG signals from A set and comparison with results based on the calculation of the energy value of EEG signals from C set.

4. EEG signal classifier based on percentage of energy distribution

The percentage of energy distribution can be used for classification of EEG signals. One of the common tools used for classification are Artificial Neural Networks (ANN). Details on the mathematical background of ANN can be found in many books and papers (Subasi & Erçelebi, 2005; Dreiseitl & Ohno-Machado, 2002; Basheer & Hajmeer, 2000; Chaudhuri & Bhattacharya, 2000). In the classifier based on percentage of energy distribution of EEG signals (Omerhodzic et al., 2010) the Feed-Forward Neural Network (FFNN) is used to classify different EEG signals. FFNN model was provided in Matlab. The algorithm structure is based on two stages: feature extraction stage (FES) and classification stage (CS). The input of CS is a pre-processed signal. In this case, EEG signal in the time domain is transformed into the wavelet domain before applying as input to CS. Based on the feature extraction, 6-dimensional feature sets (D1, D2, D3, D4, D5 and A5) for training and testing data were constructed. The dimensions here describe different features resulting from the WT, that is to say, the total size of training data or testing data set is 6×300. Considering the classification performance of this method, this input vector is applied as the input to the WNN structure. The training parameters and the structure of the WNN used in this study are shown in Table 3.

Architecture	
The Number of Layers	3
The Number of Neuron on the Layers	Input: 6, Hidden: 5, Output: 1
The Initial Weights and Biases	Random
Activation Functions	Tangent Sigmoid
Training Parameters	
Learning Rule	Levenberg-Marquardt Back-Propagation
Mean-Squared Error	1E-01

Table 3. NN Architecture and Training Parameters

They were selected to obtain the best performance, after several different experiments, such as number of hidden layers, size of hidden layers, value of the moment constant and learning rate, and type of activation functions. Data for each experiment were selected randomly. Table 4 presents classification results of WNN algorithm where 250 data sets were used to train the NN model, and 50 data sets were used for the testing process. The system can correctly classify 47 of the 50 different EEG signals in the testing set, as shown in Table 4. The classified accuracy rate of EEG signals of the proposed approach was 94.0%. A hundred percent correct classification rates were obtained for normal EEG signals.

Class	Healthy	Epilepsy Syndrome	Seizure	Accuracy [%]
Healthy	16	0	0	100.0
Epilepsy Syndrome	2	17	0	88.2
Seizure	0	1	14	92.9
Overall Success Rate				94.0

Table 4. EEG classification results of WNN algorithm

This approach presents relativly simple WNN classifier with high accuracy of EEG signal classification and could be compared with findings of other authors (Adeli et al., 2007; Ghosh-Dastidar et al., 2007). The DWT-based method proposed in this chapter was applied to three sets of EEG signals for identification of components of EEG signals and their activity. Frequency band of signal decomposition corresponded to the frequency range of individual components of EEG signal (gamma, beta, alpha, theta and delta). The results showed that different groups of the analyzed signals (sets A, C and E) are obviously different in energy distribution of signals in the frequency bands of decomposition of EEG signals.

5. Conclusion

WT, due to its advantages over the other techniques of analyzing and processing of signals, found its application in medicine. EEG signals provide important information for several types of neurological diseases. The presented methods for the analysis of EEG signal did not give at full capacity the necessary information that would help secure confirmation or exclusion of certain diseases (e.g. epilepsy). We believe that analysis of EEG signals using

WT can be a suitable method for precise and reliable identification of bioelectric state of the cerebral cortex, both in healthy patients and epilepsy patients in steady state. Finally, it can be quite a reliable indicator of epilepsy. The example of analysis of EEG signals presented here, using the discrete WT, allows identification of components of EEG signals and determines their energy value. Monitoring and analysis of the patient over a longer period of time can give us more information concerning the development of epilepsy. WT in combination with ANN allows implementation of quite a simple classifier based on energy distribution of the EEG signal components. Identification of activities of individual components of EEG signals, as well as the physicality of the processes that occur at the source of these waves, should be subject of the future research.

6. References

Abou-Khalil, B. & Musilus, K.E. (2006). *Atlas of EEG and Seizure Semiology*. Philadelphia: Butterworth-Heinemann/Elsevier

Adeli, H.; Ghosh-Dastidar, S. & Dadmehr, N. (2007). A wavelet-chaos methodology for analysis of EEGs and EEG subbands to detect seizure and epilepsy. *IEEE Trans Biomed Eng*, Vol.54, No.2, pp. 205-211

Adeli, H.; Zhou, Z. & Dadmehr, N. (2003). Analysis of EEG records in an epileptic patient using wavelet transform. *J Neurosci Methods*, Vol.123, pp. 69-87

Andrzejak, R.G.; Lehnertz, K.; Rieke, C.; Mormann, F.; David, P. & Elger, C.E. (2001). Indications of nonlinear deterministic and finite dimensional structures in time series of brain electrical activity: Dependence on recording region and brain state. *Phys Rev E*, Vol. 64 (6 Pt 1):061907, 10.10.2011, Available from: http://epileptologie-bonn.de/cms/front_content.php?idcat=193&lang=3&changelang=3

Arab, M.R.; Suratgar, A.A. & Ashtiani, A.R. (2010). Electroencephalogram signals processing for topographic brain mapping and epilepsies classification. *Comput Biol Med*, Vol.40, pp. 733-739

Asaduzzaman, K.; Reaz, M.B.; Mohd-Yasin, F.; Sim, K.S. & Hussain, M.S. (2010). A study on discrete wavelet-based -noise removal from EEG signals. *Adv Exp Med Biol*, Vol.680, pp. 593-599

Avdakovic, S.; Music, M.; Nuhanovic, A. & Kusljugic, M. (2009). An identification of active power imbalance using wavelet transform, *Proceedings of The Ninth IASTED European Conference on Power and Energy Systems*, Palma de Mallorca, Spain, September 7-9, paper ID 681-019

Avdakovic, S.; Nuhanovic, A. & Kusljugic, M. (2011). An estimation rate of change of frequency using wavelet transform. *International Review of Automatic Control (Theory and Applications)*, Vol. 4, No. 2, pp. 267-272

Avdakovic, S.; Nuhanovic, A.; Kusljugic, M. & Music, M. (2010). Wavelet transform applications in power system dynamics. *Electric Power Systems Research*, Elsevier, doi: 10.1016/j.epsr.2010.11.031

Aylward, R.L. (2008). Epilepsy: a review of reports, guidelines, recommendations and models for the provision of care for patients with epilepsy. *Clin Med*, Vol. 4, pp. 433-438

Basheer, I.A. & Hajmeer A. (2000). Artificial neural networks: fundamentals, computing, design, and application, *J Microbiol Methods*, Vol. 43, pp. 3-31

Bauer, G.; Bauer, R.; Dobesberger, J.; Unterberger, I.; Ortler, M.; Ndayisaba, J.P. & Trinka, E. (2008). Broad sharp waves-an underrecognized EEG pattern in patients with epileptic seizures. *J Clin Neurophysiol*, Vol. 25, pp. 250-254

Chaudhuri, B.B. & Bhattacharya, U. (2000). Efficient training and improved performance of multilayer perceptron in pattern classification, *Neurocomputing*, Vol. 34, pp. 11-27

Daubechies, I. (1992). *Ten Lectures on Wavelets.* Philadelphia: Society for Industrial and Applied Mathematics

Dreiseitl, S. & Ohno-Machado, L. (2002). Logistic regression and artificial neural network classification models: a methodology review, *J Biomed Inform*, Vol. 35, pp. 352-359

Ernst, F.; Schlaefer, A. & Schweikard, A. (2007). Prediction of respiratory motion with wavelet-based multiscale autoregression. *Med Image Comput Comput Assist Interv*, Vol.10 (Pt 2), pp. 668-675

Ghosh-Dastidar, S; Adeli, H. & Dadmehr, N. (2007). Mixed-Band Wavelet-Chaos-Neural Network Methodology for Epilepsy and Epileptic Seizure Detection. *IEEE Trans Biomed Eng*, Vol. 54, No. 9, pp. 1545-1551

Guo, L.; Rivero, D.; Dorado, J.; Rabuñal, J.R. & Pazos, A. (2010). Automatic epileptic seizure detection in EEGs based on line length feature and artificial neural networks. *J Neurosci Methods*, Vol.191, pp. 101-109

Guo, L.; Rivero, D.; Seoane, J.A. & Pazos A. (2009). Classification of EEG signals using relative wavelet energy and artificial neural networks. *GEC'09*, Shanghai, China

He, H. & Starzyk, J.A. (2006). A self-organizing learning array system for power quality classification based on wavelet transform. *IEEE Transaction On Power Delivery*, Vol. 21, No. 1, pp. 286-295, ISSN 0885-8977

Jaffard S., Meyer Y. & Ryan R.D. (2001). *Wavelets - Tools for Science and Technology*, Philadeplhia: SIAM

Klein, S. & Thorne, B.M. (2007). *Biological psychology*, New York: Worth

Latka, M.; Kozik, A.; Jernajczyk, J.; West, B.J. & Jernajczyk, W. (2005). Wavelet mapping of sleep spindles in young patients with epilepsy. *J Physiol Pharmacol*, Vol. 56 (Suppl 4), pp. 15-20

Lefter, S.; O'Toole, O. & Sweeney B. (2010). Epilepsy and driving: new European Union guidelines. *Ir Med J*, Vol. 103, No. 3, pp. 86-88

Leise, T.L. & Harrington, M.E. (2011). Wavelet-based time series analysis of circadian rhythms. *J Biol Rhythms*, Vol.26 No.5, pp. 454-463

Lessa, P.S.; Sato, J.R.; Cardoso, E.F.; Neto, C.G.; Valadares, A.P. & Amaro, E. Jr. (2011). Wavelet correlation between subjects: a time-scale data driven analysis for brain mapping using fMRI. *J Neurosci Methods*, Vol. 194, No. 2, pp. 350-357

Leung, H.; Schindler, K.; Chan, A.Y.; Lau A.Y.; Leung, K.L.; Ng, N.G. & Wong, V.S. (2009). Wavelet-denoising of electroencephalogram and the absolute slope method: a new tool to improve electroencephalographic localization and lateralization. *Clin Neurophysiol*, Vol.120, No.7, pp. 1273-1281

Mallat, S. (1998). *A Wavelet Tour of Signal Processing*, San Diego: Academic Press, Inc., ISBN 0-12-466606-X

McHugh, J.C. & Delanty, N. (2008). Epidemiology and classification of epilepsy: gender comparisons. *Int Rev Neurobiol*, Vol. 83, pp. 11-26

Mei, K.; Rovnyak, S. M. & Ong, C-M. (2006). Dynamic Event Detection Using Wavelet Analysis, *Proceedings of IEEE PES General Meeting*, pp. 1-7, ISBN 1-4244-0493-2, Montreal, Canada, June 18-22, 2006

Mertins, A. (1999). *Signal analysis: Wavelets, Filter Banks, Time-Frequency, Transforms and Applications*, New York: John Wiley&Sons Ltd, ISBN 0471986267

Mirowski, P.; Madhavan, D.; Lecun, Y. & Kuzniecky R. (2009). Classification of patterns of EEG synchronization for seizure prediction. *Clin Neurophysiol*, Vol. 120, No. 11, pp. 1927-1940

Ngugi, A.K.; Kariuki, S.M.; Bottomley, C.; Kleinschmidt, I.; Sander, J.W. & Newton, C.R. (2011). Incidence of epilepsy: A systematic review and meta-analysis. *Neurology*, Vol. 77, No. 10, pp. 1005-1012

Niedermeyer, E. & da Silva, F.L. (2004). *Electroencephalography: Basic Principles, Clinical Applications, and Related Fields*. New York: Lippincot Williams & Wilkins

Oehler, M.; Schilling, M. & Esperer, H.D. (2009). Capacitive ECG system with direct access to standard leads and body surface potential mapping. *Biomed Tech*, Vol.54, pp. 329-335

Omerhodzic I.; Avdakovic, S.; Dizdarevic, K. & Nuhanovic, A. (2011). Wavelet transform based analysis of EEG signals and indicators of epilepsy. *Acta Clin Croat*, Vol. 50 (Suppl 1), p. 42

Omerhodzic, I.; Avdakovic, S.; Nuhanovic, A. & Dizdarevic K. (2010). Energy distribution of EEG signals: EEG signal Wavelet-Neural Network classifier. *World Academy of Science, Engineering and Technology*. Vol.61, pp.1190-1195

Omerhodzic, I.; Causevic, E.; Dizdarevic, K.; Avdakovic, S.; Music, M.; Kusljugic, M.; Hajdarpasic, E. & Kadic, N. (2008). First neurosurgical expirience with the wavelet based EEG in diagnostic of concussion. *Abstarct book*, p. 118, 11th Congress of Neurosurgeons of Serbia with International participation. Nais, Serbia, Oct. 2008

Patnaik, L.M. & Manyamb, O.K. (2008). Epileptic EEG detection using neural networks and post-classification" *Comput Methods Programs Biomed*, Vol. 91, pp. 100-109

Saiz Díaz, R.A.; Martínez Bermejo, A. & Gómez Alonso J. (2007). Diagnosis of epilepsy over the course of the disease. *Neurologist*, Vol. 13, No. 6 (Suppl 1), pp. S11-19

Schiff, S.J.; Aldroubi, A.; Unser, M. & Sato, S. (1994). Fast wavelet transformation of EEG. *Electroencephalogr Clin Neurophysiol*, Vol. 91, No 6, pp. 442-455

Subasi, A. & Erçelebi, E. (2005). Classification of EEG signals using neural network and logistic regression. *Comput Methods Programs Biomed*, Vol.78, pp. 87-99

Subasi, A.; Alkan, A.; Koklukaya, E. & Kiymik, M.K. (2005). Wavelet neural network classification of EEG signals by using AR model with MLE preprocessing. *Neural Netw*, Vol.18, No.7, pp. 985-997

Tong, S. & Thacor, N.V. (2009). *Engineering in Medicine & Biology- Quantitative EEG Analysis Methods and Clinical Applications*, Boston/London: Artech House

Vetterli, M. & Kovacevic, J. (1995). *Wavelets and subband coding*, New York: Prentice-Hall, Inc., ISBN 0-13-097080-8

Wang, J. & Xu G. (2009). Some Highlights on Epileptic EEG Processing. *Recent Pat Biomed Eng*, Vol. 2, pp. 48-57

Zandi A.S.; Dumont G.A.; Javidan M.A; Tafreshi, R.; MacLeod, B.A.; Ries, C.R. & Puil, E.A. (2008). A novel wavelet-based index to detect epileptic seizures using scalp EEG signals. *Conf Proc IEEE Eng Med Biol Soc*, Vol. 2008, pp. 919-922

3

Using Wavelet Transforms for Dimensionality Reduction in a Gait Recognition Framework

Milene Arantes and Adilson Gonzaga
School of Engineering of São Carlos, University of São Paulo
Brazil

1. Introduction

This work proposes a novel computer vision approach that processes video sequences of people walking and then recognizes those people by their gait. Human motion carries different information that can be analyzed in various ways.

Gait can be defined by motor behavior consisting of integrated movements of the human body. It is a cyclical pattern of corporal movements that are indefinitely repeated every cycle. Gait analysis is very important in the medical field both for detecting and treating locomotion disorders. Historically, gait analysis was restricted to medical contexts, but now it has been extended to other applications, such as biometry. Research has proven that human beings have special and distinct ways of walking (Winter, 1991; Sarkar et al., 2005; Havasi et al., 2007; Boulgouris, 2007). Given this premise, a human being's gait can be understood as an important biometric characteristic. Arantes and Gonzaga (Arantes & Gonzaga, 2010,2011) have proposed a new framework for gait recognition called Global Body Motion (GBM). This framework was developed as a fusion of four models of human movement. Each model was based on specific image segmentation of the human silhouette and extracted global information about tri-dimensional, bi-dimensional, boundary and skeleton motion. That work applied the Haar Wavelet Transform (WT) for image dimensionality reduction with reduced loss of movement information. However, they did not analyze which wavelet family could perform better, maintaining the discriminant information of the human body movement in spite of the image scale reduction. Wavelet Transforms can be seen as mechanisms to decompose or break signals into their constituent parts. Thus, you can analyze data in different frequency domains with the resolution of each component adjusted to their scale. In this chapter, we analyze several wavelet families, choosing the best one, where "best" is defined as the Wavelet Transform that maintains the movement information after scale reduction for each model (Arantes & Gonzaga, 2010,211).

2. Objectives

There are differences in the way each person walks, and these differences can be significant in terms of identifying an individual. In a video sequence with only one person walking, the movement of this person, even in images with a complex background, generates valuable data among the highly correlated frames. In this work, we assumed that a frame-by-frame video sequence of a person walking forms one class, where each frame is an element of this class. Thus, our objective is to establish a methodology that can recognize a person from the way he/she walks. Movement of the human body can be interpreted in various ways using standard techniques of image processing. Our system obtains global information about the body's movement as a whole, from four models of segmented video images of the human being, before merging the results into a single model that we call the GBM (Global Body Motion). This model should improve the rates of biometric recognition.

In this work, we propose to determine the best family of wavelets that maintains the characteristics of human body movement in scale for each previously published model (SGW, SBW, SEW and SSW) (Arantes & Gonzaga, 2010,2011). Because each family of wavelets has distinct characteristics, applications of low-pass and high-pass filters will generate different discriminant features. For gait recognition improvements, we developed a fusion of human movement models using the framework proposed by Arantes and Gonzaga (Arantes & Gonzaga, 2010,2011), and the fusion model results will be compared with the previously published models to determine the best-suited model.

The analyzed wavelets families in this work were as follows:

- Haar: Is the simplest wavelet family. This wavelet has a linear phase, is discontinuous and equal to Daubechies 1. The wavelet also has only two filter coefficients, and thus, a long-range transition is guaranteed. The Haar wavelet function is represented by a square wave where soft signs are not well reconstructed. This wavelet is the only symmetric and orthogonal wavelet (Burrus et al., 1998).
- Daubechies: Has a non-linear phase. The response to impulse is maximally flat. This wavelet is quite compact in time, but within the frequency domain, it has a high degree of spectrum superposition between scales (Burrus et al., 1998). These wavelets were the first to make discrete analysis practical. Ingrid Daubechies constructed these models with a maximum orthogonal relationship in the frequency response and half of the sampling rate, imposing a restriction on the amount of decay in a certain range, thereby obtaining a better resolution in the time domain; $2n$ filter coefficients are produced given the wavelet order, n (Burrus et al., 1998).
- Symlets: Have a non-linear phase. The response to impulse is more symmetrical. This model was proposed by Daubechies as a modification to the family "dbN" because of its similar properties and the fact that it tends to be symmetric (Burrus et al., 1998).
- Bi-Orthogonal: Has a linear phase. This family uses two wavelets: one for decomposition and another for reconstruction. The Bi-Orthogonal wavelet family has compact support and is symmetric (Burrus et al., 1998).

3. Methodology

The proposed framework is shown in Figure 1. The extracted features create independent models (SGW, SBW, SEW and SSW) of the global movement of the human body, and these are compared separately using distinct wavelet families.

To eliminate the background and consequent segmentation of the movement, we have used the algorithm based on the Gaussian Mixture Model (GMM), originally proposed by Stauffer and Grimson (Staufer & Grimson, 1999) and modified by KaewTrakulpong and Bowden (KaewTraKulPong & Bowden, 2001).

Two types of images were generated by segmentation:

The first image corresponds to the image in segmented movement in grayscale. This sequence is called Silhouette-Gray (SG).

The second image is obtained from the binary mask generated by the GMM. This sequence is called Silhouette-Binary (SB).

We have proposed four families with the goal of achieving scale reduction without significant loss of information. Each WT performance is tested, taking into account the previously proposed framework for gait recognition (Arantes & Gonzaga, 2010, 2011), which was studied only for the Haar family.

The images of people walking are decomposed into four sub-bands with different information in terms of both content and detail. For each level of decomposition, four new images are generated, each with half the special resolution and scale. Each decomposition level outputs one image from the low-pass filtering stage and three images from the high-pass filtering stage. The low-pass filter generates the approximation coefficient's image, and the high pass-filter outputs the vertical, horizontal and diagonal details. The approximation coefficients contain information about the human body shape and grey-level variations, and the detail coefficients furnish information about the silhouette contour.

Given that the original segmented image contains all the information about the global movement of the human body when walking and that this information does not change significantly with scale, the four families of wavelets are applied at two levels for each of the segmented sequences.

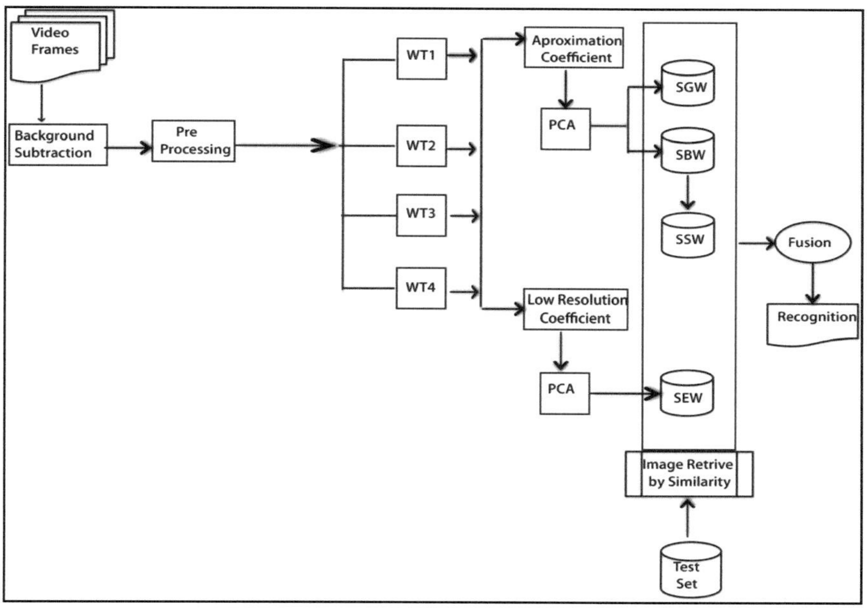

Fig. 1. Outline of the proposed framework for gait recognition.

3.1 Scale reduction

The SG sequence, after applying WT, generates the SGW sequence. The SB sequence, also after applying WT, generates the SBW sequence. The segmented sequences, which constitute each class of subjects, represent the output as images of 31 x 60 pixels with the subject walking in the center of each frame. The scale reduction is basic in terms of reducing the amount of data without decreasing the amount of global information contained in the movement, thereby optimizing the computational effort of the recognition. Thus, the representation of two models of human gait is obtained: SGW and SBW models. The SGW model is derived from the SG sequence, after application at two levels of WT and SBW is generated from the SB sequence, also after WT.

Scale reduction is performed for the four wavelet families: Haar, Symlets second-order, Daubechies second-order and Bi-Orthogonal 1.1 and 1.3. Thus, there are five databases for SGW and SBW, one for each order wavelet family.

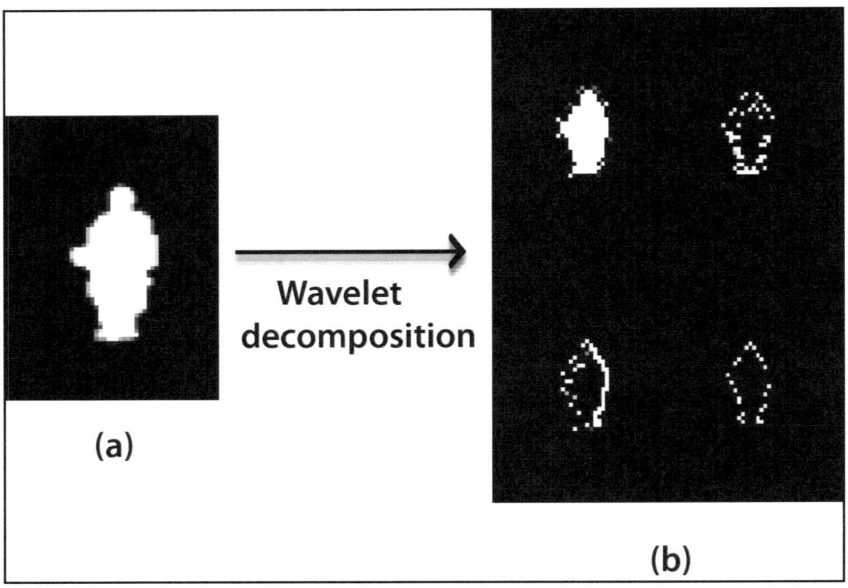

Fig. 2. Wavelet second-level decomposition using Bi-Orthogonal family.

The Figure 2(a) shows the original image size of 124 x 240 pixels. Figure 2(b) shows the second level of wavelet decomposition, broken down into four components: low frequency coefficient (approximation coefficient's image) and the coefficients with horizontal, vertical and diagonal details, respectively. At this stage, the size of the generated image is 31 x 60 pixels. This figure represents the WT Bi-Orthogonal. The same process is applied to the other wavelet families considered in this work.

3.2 Movement of the contour and the skeleton

Aiming to capture the global variations of the movement of the human body contained only in the contour of the silhouette, we used the horizontal, vertical and diagonal details generated by the WT. We applied the algorithm proposed by Lam (Lam et al., 1992) for the silhouette skeletonization. This procedure generates the skeleton sequence class of global movements called SSW (Arantes & Gonzaga, 2010, 2011). We generated five complete models for each wavelet family:

SGW Model - Silhouette-Gray-Wavelet: each class is represented by a grayscale silhouette sequence using the WT applied to moving objects segmented by GMM. The SGW model has information about the three-dimensional global movement of the human gait grayscale variations, but it can be quite sensitive to variations in light.

SBW Model - Silhouette-Binary-Wavelet: each class is represented by a sequence of binary silhouettes generated using the WT, applied to moving objects and segmented by GMM. The SBW Model provides information about the two-dimensional global movement of the silhouette of the human body while walking. The SBW Model reduces the sensitivity to the variation of light, but the clothes remain a variable that can negatively impact system performance.

SEW Model - Silhouette-Edge-Wavelet: each class is represented by a sequence of silhouettes of images of edges obtained from the horizontal, vertical and diagonal coefficients of WT. The SEW model carries information about the global behaviors of the contours while walking. The SEW model is even less sensitive to light variations than the SBW model. However, the contour is insufficient for satisfactory recognition.

SSW Model - Silhouette-Skeleton-Wavelet: each class is represented by a sequence of skeleton silhouettes obtained from the SBW method. The SSW model contains information about the global movement of joints of the human body and how they behave while walking.

3.3 Feature extraction – *EigenGait*

The *EigenGait* captures the temporal features (or temporal differences) of the human gait among the frames within each class and projects these features in a prototype vector.

Because each frame sequence represents a corresponding class of a person walking, the intra-class variance is small, and the inter-class variance is large. Therefore, the PCA (*Principal Component Analysis*) technique is used to extract relevant characteristics for recognition. The PCA technique is applied to all frames of all classes belonging to the four models (SGW, SBW, SEW and SSW) for the wavelet families with the best individual result for gait recognition.

The data dimensionality is also reduced in relation to the original variables, but maintains the relevant information. The main extracted characteristic is the feature vector that will be used for silhouette classification in its respective class.

3.4 Fusion

Different motion representation options carry distinct information about human body movement and silhouettes. Aside from being vulnerable in different situations (presence of shadows, change lighting, changes in dress, etc.), the proposed fusion model can add both static (for the SGW, SBW and SEW models) and dynamic characteristics of the movement (for the SSW model).

The proposed fusion approach assumes that the output of each model (SGW, SBW, SEW and SSW), trained individually with different wavelet families, yields a similarity score between each frame and the classes to be classified. This score similarity is obtained through the Nearest Neighbor (NN) classifier. Thus, we obtain the percentage of the correct answer for each model individually in each wavelet family. The model representation of the individual gait that yields a better performance will have a greater weight in the frame classification decision.

The following steps can describe the algorithm:

1. Calculate the similarity between each jth frame belonging to the test set and *EigenGait* of each class c, of model i, given by equation 1:

$$S^i_{j,c} = Min\left(\left|frame_j - EigenGait_c\right|\right) \tag{1}$$

where $S^i_{j,c}$ is the smallest Euclidean distance between each jth frame and the *EigenGait* of each class c for each model i, with i varying from 1 to 4 (SGW, SBW, SEW and SSW). The frame will be classified in the class such that the mean distance is minimized.

2. Calculate the average precision of correct answer for each model i, given by equation 2:

$$\rho_i = \frac{TP_i}{TG_i} \tag{2}$$

where TP_i is the number of correctly classified frames (true positives) of the model i, and TG_i is the total number of samples of the test set of model i.

3. Calculate the fusion score ($\phi_{j,c}$) between the jth frame and class c, given by equation 3:

$$\phi_{j,c} = \frac{\sum_{i=1}^{4} \rho_i S^i_{j,c}}{\sum_{i=1}^{4} \rho_i} \tag{3}$$

3.5 Materials

We have used the "Gait Database" of the National Laboratory of Pattern Recognition (NLPR) from the Institute of Automation at the Chinese Academy of Science (CASIA, 2010) in this work. These images were generated outside in an environment with natural light. The images include three views: side, oblique and front (0°, 45° and 90°).

Each class has three views and four sequences per view (two sequences walking from the left to the right and two sequences walking from the right to the left). These are numbered sequence 1, sequence 2, sequence 3 and sequence 4 with the following respective directions: right-left, left-right, right-left and left-right. Each variation of angles in each of the four sequences is illustrated in Figure 3.

The sequences of videos were assembled from the available images. Altogether, 20 classes were obtained with 240 video sequences and 8,400 frames.

Fig. 3. Samples of frames from the Gait Database with angle variations of 0°, 90° and 45° (CASIA, 2010).

3.6 Evaluation methods

To evaluate the Wavelet Transform performance for human gait recognition in this framework, independent tests were carried out for each type of sequence (SGW, SBW, SEW and SSW) of each wavelet family. The results for each wavelet family were analyzed, taking into account the False Acceptance Rate (FAR) and False Rejection Rate (FRR).

Each image of each frame was projected into a PCA sub-space and compared with the *EigenGait* prototype of each class. For each experiment, confusion matrices were generated, and the FAR and FRR rates were calculated for each class. We used the Feret protocol (Philips et al., 2000) with the leave-one-out cross-validation rule for evaluation of the results. After computing the similarity between the test sample and the training set, the nearest neighbor (NN) was applied to the classification.

To evaluate the performance of the GBM model with the Gait Database for each wavelet family and each type of sequence (SGW, SBW, SEW and SSW), four independent experiments were carried out. In each of these sequences, the direction of the movement of the subject was restricted to angles 0°, 45° and 90°. For each direction of movement (angle), we used four sequences of frames for twenty different subjects. For these experiments, the number of elements in each class is the sum of the frames of the individual sequences.

For the combination of four sequences, 144 elements of the class were used. For one test, we have used the combination of four directions of movement.

These are the combinations of sequences:

- Angle 0°, four sequences;
- Angle 45°, four sequences;
- Angle 90°, four sequences; and
- Angles 0°, 45° and 90°, two sequences from left-right and right-left of each angle.

The correct answer from each model is used as the weight in the weighted mean for the fusion process.

4. Results

Table 1 shows the percentages of matches with their respective FAR and FRR for the SGW model.

SGW Model				
Wavelet Family	**Angle**	**Match %**	**FAR%**	**FRR%**
Haar		70.5	5.9	29.5
Daubechies		69.8	5.3	30.2
Symlets		71.1	5.07	28.9
Bi-Orthogonal 1.1		70.4	5.2	29.6
Bi-Orthogonal 1.3	0°	75.8	4.25	24.2
Haar		55.0	7.9	45.0
Daubechies		62.8	6.5	37.2
Symlets		71.6	5.0	28.4
Bi-Orthogonal 1.1		73.4	4.7	26.6
Bi-Orthogonal 1.3	45°	81.2	3.3	18.8
Haar		60.2	6.9	39.8
Daubechies		65.8	6.0	34.2
Symlets		74.8	4.4	25.2
Bi-Orthogonal 1.1		80.5	3.4	19.5
Bi-Orthogonal 1.3	90°	83.7	2.9	16.5
Haar		47.9	9.1	52.1
Daubechies		51.2	8.8	50.7
Symlets		65.4	6.0	34.6
Bi-Orthogonal 1.1		67.3	5.7	32.7
Bi-Orthogonal 1.3	0°, 45° and 90°	69.4	5.3	30.6

Table 1. Percentages of matches with their respective FAR and FRR rates for the SGW model.

Analyzing the results obtained for the SGW model, angle 0°, the percentage of correct answers for each wavelet family is very similar. Though the SGW model carries more information about the human silhouette than the other models, the type of wavelet should not interfere in the classification model. In light of these results, the difference in the classification is determined by the camera angle in relation to the walker. For the 45° and 90° angles, the best hit rate is for the Bi-Orthogonal 1.3 WT.

For the test in which the three angles are used together to form a single base, the wavelet family that best extracts the coefficients of low frequency (approximation image) and the details of the high-pass filtering is the Bi-Orthogonal 1.3. This fact implies that the Bi-Orthogonal 1.3 WT captured more information on the global movement of the object. Another detail to consider is the low False Acceptance Rate (FAR) in the Bi-Orthogonal wavelet family. Table 2 shows the percentages of matches with their respective FAR and FRR considering the SBW model.

The SBW model carries global information of human motion contained in the binary images resulting from the coefficient approximation of the transformed wavelet. For this model, the original proposal [5] suggested the Haar WT would have the best rate of correct classifications. Our results for the Haar WT show 69.7% accuracy at an angle of 0°. However, other wavelet families performed better than the Haar WT.

SBW Model				
Wavelet Family	**Angle**	**Match %**	**FAR%**	**FRR%**
Haar	0°	69.7	5.3	30.3
Daubechies		72.1	4.9	27.9
Symlets		73.2	4.7	26.8
Bi-Orthogonal 1.1		68.8	5.5	31.2
Bi-Orthogonal 1.3		73.4	4.7	26.6
Haar	45°	48.4	9.0	51.6
Daubechies		61.3	6.8	38.7
Symlets		64.4	6.3	35.6
Bi-Orthogonal 1.1		61.8	6.7	38.2
Bi-Orthogonal 1.3		71.6	2.2	28.4
Haar	90°	53.8	8.1	46.2
Daubechies		54.7	7.9	45.3
Symlets		55.1	7.9	44.9
Bi-Orthogonal 1.1		50.3	8.7	49.7
Bi-Orthogonal 1.3		74.5	4.5	25.5
Haar	0°, 45° and 90°	45.3	9.6	54.7
Daubechies		54.3	9.1	52.1
Symlets		56.2	9.1	51.9
Bi-Orthogonal 1.1		57.4	9.7	55.3
Bi-Orthogonal 1.3		61.4	8.5	48.7

Table 2. Percentages of matches with their respective FAR and FRR rates for the SBW model.

When the amount of information present in the image is reduced, we can infer that the length of the wavelet filter has a very important role. The Bi-Orthogonal wavelet has the low-pass filter of length 6, which causes the detail level obtained for this family to be much larger than that of the Haar family. Because the images of the model are derived from SBW low-pass filtering, a higher level of detail is obtained, resulting in a better image. For an angle of 0°, Symlets and Bi-Orthogonal wavelets provided equivalent results. For the other angles, the best performance occurs with the Bi-Orthogonal family. FAR for this family are also lower than that of WT Haar and Daubechies. Table 3 shows the percentages of matches with their respective FAR and FRR for the SEW model.

The SEW model carries information about the global movement related to the human contours of the silhouettes. The SEW model is generated from coefficients with horizontal, vertical and diagonal details. The best performance for this model is with the Bi-Orthogonal wavelet at an angle of 90°. Table 4 shows the percentages of matches with their respective FAR and FRR considering the SSW model.

The SSW model carries the global information about movement of the body's joints. This model provides the least amount of information about the movement. Nevertheless, its rate of correct classifications using the families of Symlet and Bi-Orthogonal wavelets is good and far exceeds that of the Haar wavelet.

SEW Model				
Wavelet Family	**Angle**	**Match %**	**FAR%**	**FRR%**
Haar		52.1	8.4	47.9
Daubechies		72.8	4.8	27.2
Symlets		71.5	5.0	28.5
Bi-Orthogonal 1.1		53.6	8.1	46.4
Bi-Orthogonal 1.3	0º	74.8	4.4	25.2
Haar		54.6	8.0	45.4
Daubechies		68.5	8.0	31.5
Symlets		70.7	5.5	29.3
Bi-Orthogonal 1.1		68.4	5.1	31.6
Bi-Orthogonal 1.3	45º	70.8	5.6	29.2
Haar		58.3	5.1	41.7
Daubechies		65.8	7.3	34.2
Symlets		68.3	6.0	31.7
Bi-Orthogonal 1.1		71.3	5.6	28.7
Bi-Orthogonal 1.3	90º	80.3	5.1	19.7
Haar		45.3	3.4	54.7
Daubechies		56.2	9.6	43.8
Symlets		57.3	7.7	42.7
Bi-Orthogonal 1.1		60.1	7.5	39.9
Bi-Orthogonal 1.3	0º, 45º and 90º	62.7	7.0	37.3

Table 3. Percentages of matches with their respective FAR and FRR rates for the SEW model.

Analyzing the overall performance of the wavelet families, WT Bi-Orthogonal maintains good performance regardless of the type of movement or angle used. A Haar WT, for this study, is very susceptible to the motion model and the steering angle of the walker.

The average of the correct answers of each wavelet in each model is used to calculate the weighted mean within the fusion schema. The Feret protocol (Philips et al., 2000) is used to evaluate the results. The statistical performance of this method is reported as the CMS (*Cumulative Match Score*), which is defined as the cumulative probability of a correct classification of a test object within the top *k* hits.

The CMS curves in Figure 4 were obtained through the fusion of the SGW, SBW, SEW and SSW models. The models used in the fusion process were those that achieved the best results for the analyzed wavelet families.

The CMS curves in Figure 5 were obtained through the fusion of the SGW, SBW, SEW and SSW models, using the combination of the three views with two sequences each.

4.1 Comparative results

The results from the GBM model for this angle were compared with the results obtained from the previous work of Arantes and Gonzaga (Arantes & Gonzaga, 2010,2011). Four combined sequences were used, for angles 0º, 45º and 90º, as in the previous publication.

SSW Model				
Wavelet Family	**Angle**	**Match %**	**FAR%**	**FRR%**
Haar		49.6	8.8	50.4
Daubechies		51.3	8.5	48.7
Symlets		60.3	7.0	39.7
Bi-Orthogonal 1.1		56.8	7.6	43.2
Bi-Orthogonal 1.3	0°	63.2	6.5	36.8
Haar		42.7	10.0	57.3
Daubechies		50.3	8.7	49.7
Symlets		60.2	10.1	39.8
Bi-Orthogonal 1.1		59.4	8.7	40.6
Bi-Orthogonal 1.3	45°	61.8	7.0	38.2
Haar		52.5	7.1	47.5
Daubechies		56.3	6.7	43.7
Symlets		62.7	8.3	37.3
Bi-Orthogonal 1.1		63.7	7.7	36.3
Bi-Orthogonal 1.3	90°	78.4	6.6	21.6
Haar		42.7	6.3	57.3
Daubechies		45.8	3.8	54.2
Symlets		50.2	10.0	49.8
Bi-Orthogonal 1.1		56.8	9.5	43.2
Bi-Orthogonal 1.3	0°, 45° and 90°	61.7	8.7	38.3

Table 4. Percentages of matches with their respective FAR and FRR rates for the SSW model.

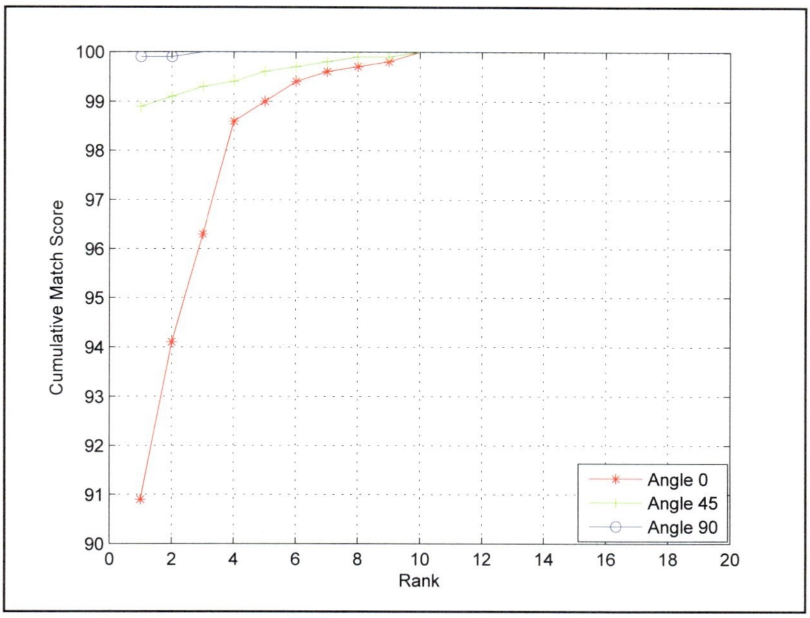

Fig. 4. CMS curves after fusion by combining the four sequences of walking.

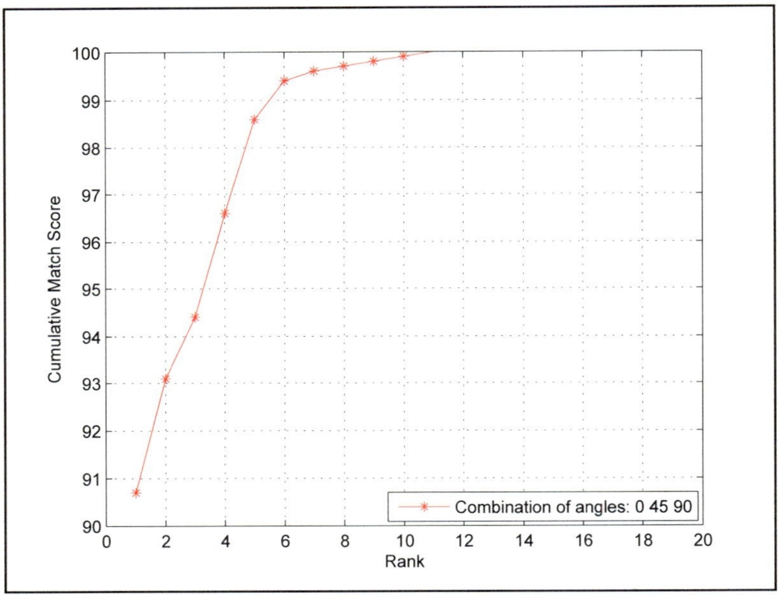

Fig. 5. CMS curves after fusion by combining two walking sequences and the combination of the three views.

In table 5, the best results for the correct answer are presented based on the CMS. The CMS ranks vary among 1, 5 and 10 at an angle of 0º. For this angle, the best results were obtained for the families of Symlets and Bi-Orthogonal wavelets.

Table 6 shows the best results for the correct answer, based on the CMS. The CMS ranks vary among 1, 5 and 10 at an angle of 45º. For this angle, the best results were obtained for the families of Symlets and Bi-Orthogonal wavelets.

Angle	Model	Method	Rank		
			1	5	10
0º	SGW	Symlets	91.8	99.0	100.0
	SBW	Symlets	90.7	99.3	100.0
	SEW	Bi-Orthogonal 1.3	85.5	94.7	99.7
	SSW	Bi-Orthogonal 1.3	80.8	92.8	97.5
	GBM Fusion		99.1	99.8	100
	SGW	GBM (Arantes & Gonzaga, 2011)	90.9	99.0	100
	SBW	GBM (Arantes & Gonzaga, 2011)	88.0	98.7	99.9
	SEW	GBM (Arantes & Gonzaga, 2011)	78.8	94.2	97.9
	SSW	GBM (Arantes & Gonzaga, 2011)	54.2	67.2	73.9
	GBM Fusion	GBM (Arantes & Gonzaga, 2011)	97.1	99.4	100

Table 5. Comparative table of the GBM model – angle 0º.

Angle	Model	Wavelet Family	Rank		
			1	5	10
45º	SGW	Symlets	97.8	99.0	100
	SBW	Bi-Orthogonal 1.3	95.3	97.7	99.1
	SEW	Bi-Orthogonal 1.3	90.6	96.5	98.7
	SSW	Bi-Orthogonal 1.3	88.3	92.8	97.9
	GBM Fusion		98.9	99.6	100
	SGW	GBM (Arantes & Gonzaga, 2011)	96.7	99.9	100
	SBW	GBM (Arantes & Gonzaga, 2011)	92.4	99.9	100
	SEW	GBM (Arantes & Gonzaga, 2011)	80.4	95.2	08.8
	SSW	GBM (Arantes & Gonzaga, 2011)	50.3	82.6	90.0
	GBM Fusion	GBM (Arantes & Gonzaga, 2011)	97.4	99.6	100

Table 6. Comparative table of the GBM model - angle 45º.

Table 7 shows the best results for the correct answer, based on the CMS. The CMS ranks vary among 1, 5 and 10 at an angle of 90º. For this angle, the best results were obtained for the families of Haar and Bi-Orthogonal wavelets.

Angle	Model	Wavelet Family	Rank		
			1	5	10
90º	SGW	Haar	99.9	99.9	100
	SBW	Haar	99.6	99.8	100
	SEW	Bi-Orthogonal 1.3	98.1	99.4	100
	SSW	Bi-Orthogonal 1.3	90.8	98.5	99.1
	GBM Fusion		99.9	100	100
	SGW	GBM (Arantes & Gonzaga, 2011)	99.9	100	100
	SBW	GBM (Arantes & Gonzaga, 2011)	99.4	99.9	100
	SEW	GBM (Arantes & Gonzaga, 2011)	95.6	98.5	100
	SSW	GBM (Arantes & Gonzaga, 2011)	72.8	82.2	91.7
	GBM Fusion	GBM (Arantes & Gonzaga, 2011)	99.9	100	100

Table 7. Comparative table of the GBM model - angle 45º.

In table 8, the best results for the correct answer are presented, based on the CMS. The CMS ranks vary among 1, 5 and 10 for the combination of angles at 0°, 45° and 90°. Two combined sequences were used for each angle: right-left and left-right. For this combination, the best results were obtained for the families of Symlets and Bi-Orthogonal wavelets.

Angle	Model	Wavelet Family	Rank		
			1	5	10
0° 45° 90°	SGW	Bi-Orthogonal 1.3	78.4	89.1	99.4
	SBW	Bi-Orthogonal 1.3	74.9	96.3	99.6
	SEW	Symlets	71.0	86.3	94.9
	SSW	Symlets	70.5	84.9	96.1
	GBM Fusion		90.7	98.6	99.9

Table 8. CMS best rates for angles 0°, 45° and 90°.

5. Conclusions

To evaluate the Wavelet Transform performance for human gait recognition in the proposed framework, independent tests were carried out for each type of sequence (SGW, SBW, SEW and SSW) for each wavelet family. The results for each wavelet family were analyzed, taking into account the FAR and FRR. Each image of each frame was projected into a PCA sub-space and compared with the *EigenGait* prototype of each class. For each experiment, the confusion matrices were generated and the FARs and FRRs were calculated for each class. The Feret protocol (Philips et al., 2000) with a leave-one-out cross-validation rule was used to evaluate the results. The fusion process, carried out with the best performance wavelet family, is compared with the original GBM (Arantes & Gonzaga, 2010,2011).

For the SGW model, at an angle of 0°, the average hit rate is similar for each of the wavelet families analyzed. The best rate of correct classifications is for WT Bi-Orthogonal 1.3, and the difference in the hit rate over Haar WT is 5.3%. The Daubechies WT, with second-order, obtained a lower rate of correct classifications. The SGW model carries the most information; however, it is also the model that is the most sensitive to interference from the external environment.

For the 45° and 90° angles, considering the SGW model, the Haar WT had the lowest rate of correct classifications in relation to other models. The Bi-Orthogonal 1.3 WT obtained 81.2% of corrected matches for the average angle of 45° and 83.7% for the angle of 90°. For these angles, the best choice is the Bi-Orthogonal family. This improvement in the hit rate can be attributed to many details that the family can capture with Bi-Orthogonal WT when compared with Haar WT.

The SBW model carries global information about human movement, present in binary images. For the three views in this model, there was an increase in the hit rate of approximately 23%, in the best case.

For the 45° angle, the Haar WT obtained a FRR higher than the rate of correct answers. When all the views are combined into a single base, the WT Bi-Orthogonal 1.3, also performs well.

The SEW model is obtained from the horizontal, vertical and diagonal coefficients generated from the WT implementation. This model carries fewer details compared with the SGW and SBW models. Thus, the greater the number of details that the WT can capture the better is the model performance. The match score is similar for Symlets and Daubechies families; this may be due to the fact that these families have the same length filter.

The SSW model provides the global information of the human movement contained in the skeleton of the body. The SSW model carries an even smaller number of details in relation to the other models, but they are less susceptible to changes in the external environment. The best results for the average hit rate are for the Bi-Orthogonal Wavelets 1.3 and Symlets.

The highest rates of correct classifications are chosen as the weights in the fusion process. Therefore, rates are chosen from the Bi-Orthogonal 1.3 family. This led to better performance in the system, which can be observed in the CMS curves. The amount of detail that each wavelet family captures is closely related to the system performance.

6. References

Arantes, M. & Gonzaga, A. (2011) Human Gait Recognition using Extraction and Fusion of Global Motion Features. *Multimedia Tools and Applications*, Springer, ISSN 1380-7501, vol.55, no.3, pp. 655-675. doi:10.1007/s11042-10-0587-y.

Arantes, M. & Gonzaga, A. (2010). Recognition of Human Silhouette Based on Global Features. *International Journal of Natural Computing Research (IJNCR)*, vol.1, no.4, pp.47-55. doi: 10.4018/jncr.2010100105

Boulgouris, N.V. & Zhiwei, X.C. (2007) Gait Recognition using Radon Transform and Linear Discriminant Analysis. *IEEE Transactions on Image Processing*, vol.16, n.3, pp.731-740.

Burrus, S.; Gopinath, R. & Guo, H. (1998) *Introduction to Wavelets and Wavelet Transform. A Primer*. Electrical and Engineering Department and Computer and Information Technology Institute, Rice University, Houston, Texas. Prentice Hall.

CASIA - Chinese Academy of Sciences (2010). Date of access: July, 210. Available from: http://www.cbsr.ia.ac.cn.

Havasi, L.; Zoltán, S. & Szirányi, T. (2007) Detection of Gait Characteristics for Scene Registration in Video Surveillance System. *IEEE Transactions on Imagem Processing*, vol 16, n. 2, pp.503-510.

KaewTraKulPong, P. & Bowden, R. (2001) An Improved Adaptative Background Mixture Model for Real-Time Tracking with shadow Detection. *Proc. 2nd European Workshop on Advanced Video Based Surveillance Systems, AVBS01. Sept 2001. Video Based Surveillance Systems: Computer Vision and Distributed Processing*, Kluwer Academic Publishers, pp. 135-144.

Lam, L.; Lee, S-W. & Suen, Y.S. (1992) Thinning Methodologies - A Comprehensive Survey. *IEEE Transactions on Pattern Analysis and Machine Intelligence*, September 1992, vol. 14, no. 9, pp. 869-884.

Philips, P.; Moon, H. ; Rizvi, S. & Rauss, P.(2000) The FERET evaluation methodology for face-recognition. *IEEE Transactions on Pattern Analysis and Machine Intelligence*, vol. 22, pp. 1090-1104.

Sarkar, S.; Philips, P.; Liu, Z.; Vega, I.; Grother, P. & Boweyer, K. (2005). The Human Id Gait Challenge Problem: Data Sets, Performance and Analysis. *IEEE Trans. On Pattern Analysis And Machine Inteligence,* vol. 27, n. 2, pp. 162-177.

Staufer, C. & Grimson, W. (1999) Adaptive Background Mixture Models for Real Time Tracking. *IEEE Computer Society Conf. on Computer Vision and Pattern Recognition,* pp. 252-254.

Winter, D. (1991)*The Biomechanics and Motor Control of Human Gait: Normal, Elderly and Pathological* (2nd edition),. Walterloo Press, ISBN:0-88898-105-8, Ontario- Canada.

4

The Detection Data of Mammary Carcinoma Processing Method Based on the Wavelet Transformation

Meng Yao,
Zhifu Tao and Zhongling Han
East China Normal University
P. R. China

1. Introduction

At present, breast cancer is one of the major threaten diseases for female(Xu & Tang, 1996). Microwave breast tumor detection will be the future development direction of clinical diagnostic, as a safe, mobile and cost-effective method, and the hotspot of breast cancer non-invasive diagnosis for decades. Unlike traditional diagnostic method of soft X-ray mammography imaging, microwave near-field tomography is a safe method of imaging diagnosis of breast tumor which is very low damage when detect on the human body. From the perspective of breast tumors medical imaging diagnosis, it is required for the target location on the contour (line) information or hierarchical information. The imaging information of medical diagnosis of mammary tumor in early stage bases on the great different between the dielectric properties of the normal breast tissues and the malignant ones. Therefore, when the electromagnetic wave enters the malignant tumor tissues from the normal mammary tissues, it will interact with the media (for example, absorption, dispersion, emission, etc.), and then the electromagnetic wave transmission path is changed. The electric-field intensity and the magnetic field strength is enhanced in certain regions, while is weakened in other regions. Based on this characteristic, mammary gland tumors can be detected by surveying the electrical field which is produced by the reflection and the scattering generated by near-field microwave. This method can facilitate the early discovery of cancer, early identification of treatment programs and achieve early medical imaging diagnosis of breast cancer. The signals obtained by Breast Tumor Microwave Sensor System (BRATUMASS) are non-linear and non-steady. So the analysis of the detected signals is significantly important for detecting and diagnosing of the early-stage breast tumor. Delays of backscattering in cancerous tissue surfaces are different, for the dissimilar possible location of the cancerous tissue and the different distance from detecting point to cancerous tissue. The different delays can make the equivalent frequency emerge in anywhere of the effective frequency range of system. Moreover, backscattering occurs in other tissues surface. These aspects increase the frequency component and the difficulties of analysis. Fig. 1 shows the typical signal obtained from one detecting point.

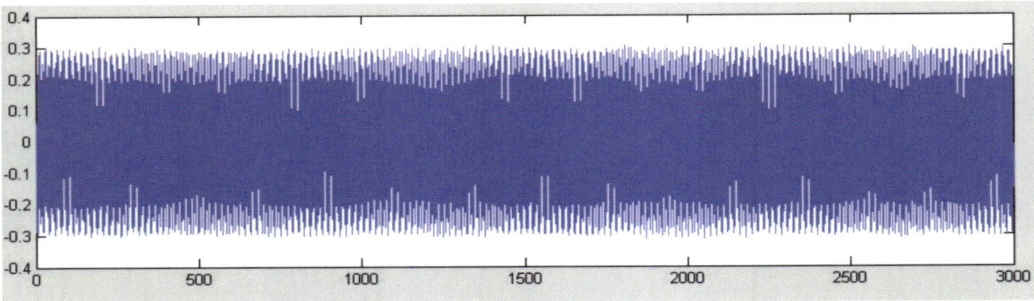

Fig. 1. Typical signal obtained from one detecting point. X-coordinate is the number of data, Y-coordinate is voltage, unit V. sampling interval is 0.002 s and sampling depth is 3000

2. The principle of detection system

Breast tumor microwave sensor system (BRATUMASS) consists of RF transceiver module, zero intermediate frequency (zero-IF) mixer. In the transmission, we use slot step frequency modulation method. In the receiver, we sample from zero IF output. In detection process, we arrayed the transceiver (R/T) antenna testing points around the test region and each sounding point of surface-wave (sagittal). There are 12 or 16 detection points of antenna devices, during the real detection process. The area enclosed by circular dashed line is the detected region. Figure 2 shows the block diagram of BRATUMASS.

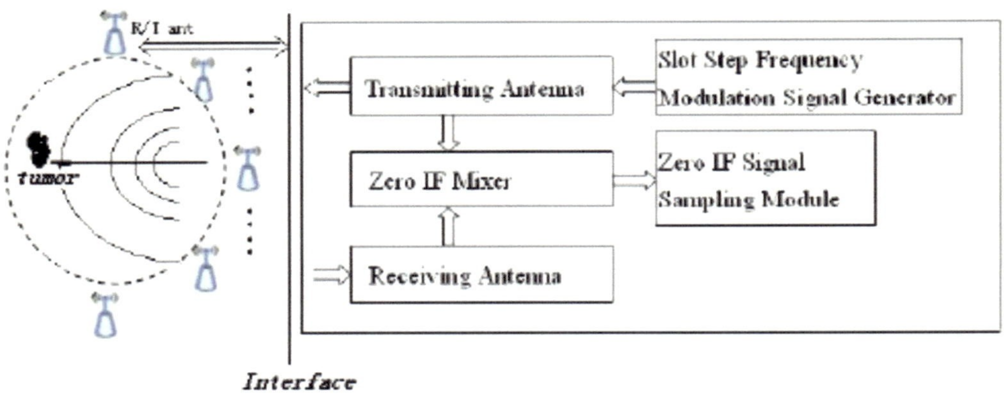

Fig. 2. Block diagram of BRATUMASS

For the large difference of dielectric properties between breast tissue and malignant tumor tissue, scattering will be happened in the interface of breast tissue and malignant tumor tissue. Figure 3 shows the relationship of reflection and transmission at the two different mediums interface. Where Pi is incident power, Pr is reflected power, Pt is transmitted power.

In two non-magnetic medium interface, the relationship between reflection coefficient $\Gamma i,j$ and dielectric properties ε_1, ε_2 is:

$$\Gamma_{i,j} = \frac{\sqrt{\varepsilon_2} - \sqrt{\varepsilon_1}}{\sqrt{\varepsilon_2} + \sqrt{\varepsilon_1}} \quad (1)$$

The ratio of incident power to reflected power is:

$$\frac{P_r}{P_i} = \left|\Gamma_{i,j}\right|^2 \quad (2)$$

In each detecting, the location of the target and the transceiver are relatively fixed, so the frequency components of low-frequency signal output by zero-IF mixer are relatively fixed too. Extracted the power spectrum of zero-IF output low-frequency signal to decide the target corresponding reflection coefficient.(Fourier diffraction relationship)

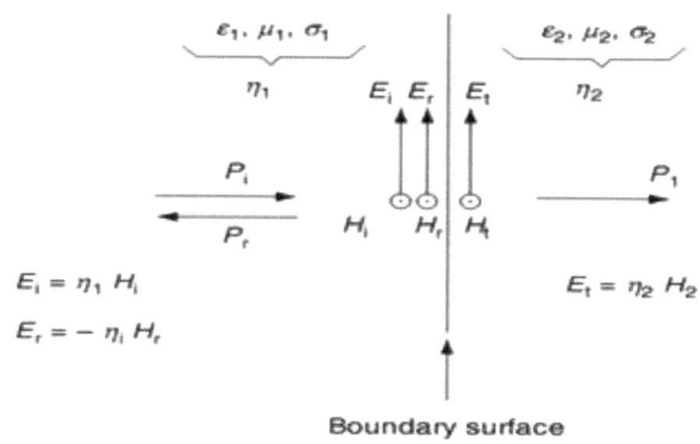

Fig. 3. Reflection and transmission of an microwave wave at the two different mediums interface. Pi = incident power, Pr = reflected power, Pt = transmitted power.

The collection reflection coefficient $\tilde{\Gamma}_{i,j}$ for the N layer media is:

$$\tilde{\Gamma}_{i,i+1} = R_{i,i+1} + \frac{T_{i,i+1}\tilde{\Gamma}_{i+1,i+2}T_{i+1,i}e^{2ik_{i+1,z}(d_{i+1}-d_i)}}{1 - R_{i+1,i}\tilde{\Gamma}_{i+1,i+2}e^{2ik_{i+1,z}(d_{i+1}-d_i)}} \quad (3)$$

Using our device parameters (transmitting frequency is 1.575 GHz and bandwidth is 200 MHz) to compute the parameter, we obtain dielectric properties , reflection coefficient and conductivity of the breast tissue as shown in table 1.

	Dielectric properties ε_r	Reflection coefficient Γ	Conductivity σ (S/m)
skin	36		2.64
normal breast tissue	10		0.24
malignant tumor tissue	50	0.49	2.8
mammary duct	11~14	0.2018~0.49	0.45
vascular		0.15~0.2018	

Table 1. Dielectric properties of typical breast tissues in 1.575GHz

The bandwidth of slot step frequency modulation emission signal is 200MHz, transmitting frequency 1.575GHz, scanning period 1ms. As shown in Figure 4(a), the real line stands for the transmitting signal and the dashed line for the theoretical receiving signal. The frequency of the scanning signal rises step by step. Figure 4(b) shows the zero-IF signal output from the mixer. IF contains the information for breast tumor location. The transmitting signal reaches the receiving antenna and mixer after being reflected by breast tumor, while the other part directly gets to the mixer through transmitting antenna coupling networks. The course difference of two parts, which can be deduced from zero-IF, the output of the mixer, is used to determine the distance between the antenna and the tumor.

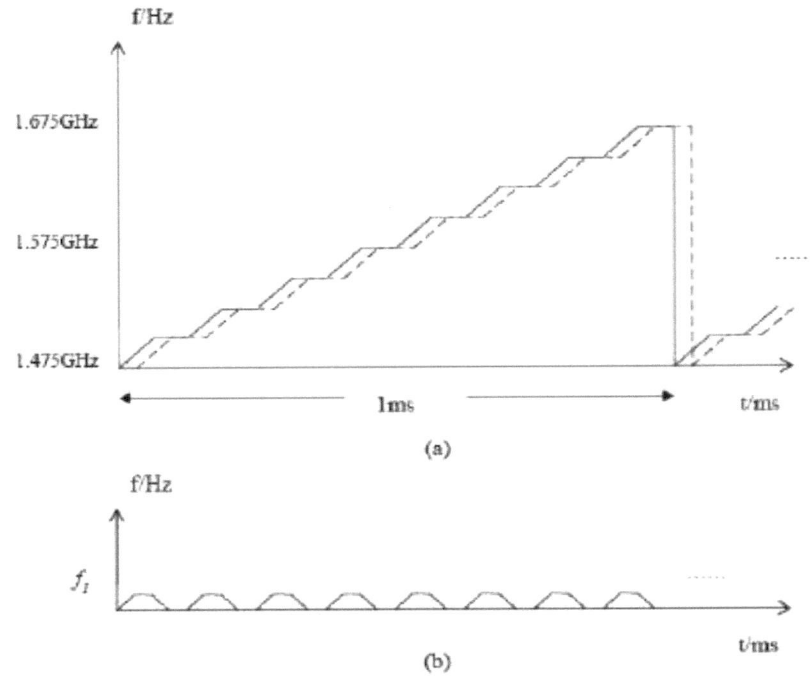

Fig. 4. Slot step frequency modulation signal.
(a) Real line stands for the transmitting signal, dashed line for the theoretical receiving signal (frequency --time)
(b) Zero-IF signal output (frequency --time)

Consider wave equation of the scattering field $u_s(r)$:

$$(\nabla^2 + k^2)u_s(r) = -k^2 f(r)u(r) \tag{4}$$

Where, $u(r)$ is the wave function of incident wave, k is the wave number, and $f(r)$ is the characteristic function of space scattering. The right of the equation is the source function of homogeneous medium in the spread of the scattering field $u_s(r)$ with wave number k. this type equation solving can use Green's function. Considering the two-dimensional situation, the source of the radiation field of free space is located in the $r_0 = (x_0, y_0)$, Green's function is given by:

$$g(r_1 \mid r_0) = \frac{1}{4} H_0(k|r_1 - r_0|) \tag{5}$$

Where, H_0 is the first zero-order Hankel function (that is the third type of Bessel function). By the principle of superposition, we can obtain the scattered field:

$$u_s(r) = \frac{ik^2}{4} \iint_S f(r_0) u(r_0) H_0(k|r - r_0|) dr_0 \tag{6}$$

Where S is the arbitrary area of surrounded objects cross-section in (x,y) plane. Also has:

$$H_0(k|r - r_0|) = \frac{1}{\pi} \int_{-\infty}^{\infty} \frac{1}{\beta} \exp\{i[\alpha(x - x_0) + \beta(y - y_0)]\} \tag{7}$$

In this equation, $u(r)$ includes $u_s(r)$, so we can only obtain approximate solution. In the first-order Born approximation, scattering is weak, and we can use the incident field to instead of the total field. Then we can get:

$$u_s(r) = \frac{ik^2 u_0}{4\pi} \iint_s f(r_0) \exp(iks_0 \bullet r_0) \int_{-k}^{k} \frac{1}{\beta} \exp\{i[\alpha(x - x_0) + \beta(y - y_0)]\} d\alpha dr_0 \tag{8}$$

When the detector is located in y = l, the detector receiving scattering field is:

$$u_s(x,l) = \frac{ik^2 u_0}{4\pi} \int_{-k}^{k} d\alpha \frac{1}{\beta} \exp(i(\alpha x + \beta l))] \iint_s f(x_0, y_0) \exp\{-i[\alpha(x_0) + (\beta - k) y_0)]\} dx_0 dy_0 \tag{9}$$

And

$$F(\omega_1, \omega_2)\Big|_{\substack{\omega_1 = \alpha \\ \omega_2 = \beta - k}} = F(\alpha, \beta - k) \tag{10}$$

Its also can be written as follows:

$$u_s(x,l) = \frac{ik^2 u_0}{4\pi} \int_{-k}^{k} \frac{1}{\beta} \exp(i(\alpha x + \beta l)] F(\alpha, \beta - k) d\alpha \tag{11}$$

Us(ω) is the Fourier transform of one-dimensional function $u_s(x, l)$ for the variable x, then:

$$U_s(\omega,l) = \frac{ik^2 u_0}{2} \frac{1}{\sqrt{k^2 - \omega^2}} \exp\left(i\sqrt{k^2 - \omega^2} l\right) F(\omega_1, \omega_2)\Big|_{\substack{\omega_1 = \omega \\ \omega_2 = \sqrt{k^2 - \omega^2} - k}}, \quad |\omega| < k \tag{12}$$

$F(\omega_1, \omega_2)$ is the Fourier transform of the space characteristics function $f(x, y)$. When ω_1 and ω_2 satisfied the relationship:

$$\omega_2 = \sqrt{k^2 - \omega_1^{\,2}} - k \,, \tag{13}$$

we can obtain the one-dimensional Fourier transform Us(ω, ·) of scattering measured data after properly weighted.

3. Description and mathematical model

We can obtain the sagittal distribution of the breast tissues dielectric properties from receiving signal. The sagittal distribution of the dielectric properties refers to the accumulated value of the dielectric properties along the arc circle whose center is the antenna. The sagittal of the arc circle varies from 0 to D, where D is the diameter of the target region. We have known the values of the dielectric properties along each arc circle. In the detection model, there are totally 12 or 16 antennas. For each antenna, the value of the dielectric properties along the arc circle can be calculated by the signal received from detection system. Now the problem is how to deduce the dielectric properties value of each point by the sagittal distribution.

As shown in Figure 5, dash-dotted frame is target region. We assume that the microwave reflection function in target region is Γ(x, y) which is bounded. We assume that there are N targets, recorded as M_i (i = 1, 2, ..., n) and P_j is the jth microwave detecting point. Where O is the coordinate center; L_i is the arc with center P_j, and radius $r_i = P_jM_i$. S_i is proportional to the integral of Γ(x, y) along the curve L_i, that is:

$$S_i(r_i) = \int_{Li} \Gamma(x,y)dl \,. \tag{14}$$

Receiving signal is the sequence of $S_1, S_2, \ldots, S_n$. BRATUMASS use the polar coordinate with P_j as the pole, P_jO as the polar. The target region used the Cartesian coordinate system. In practical calculation process, we need coordinate transformation.

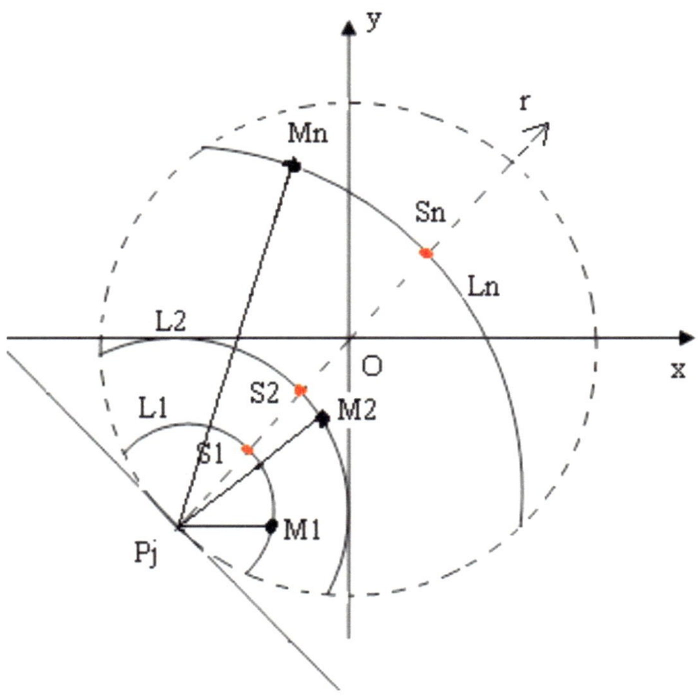

Fig. 5. Detection points and reflection interfaces

By Fourier diffraction theorem in microwave near field conditions, wave front projection is the system Fourier transforms. So back signal analysis is transformed the separation of projection data on different detecting arc. The detected target is manifested in the different duration of microwave transmission. In BRATUMASS, different time difference registers as the different frequencies outputting zero-IF signal. Sampling signal blend the reflections of all organization interfaces. Namely, the frequency components of sampling signal represent projection of all different detection arcs, and the back-wave's energy distribution relate to each reflection coefficient of detection arc.

All frequency components of back wave are decomposed by using Hilbert-Huang transform (HHT). This method based on empirical mode decomposition (EMD) and Hilbert transform (HT) is more suitable for analyzing nonlinear and non-stationary data than Fourier and wavelet method which depend on the priori function (Huang, N. E., etc.1998).

The high and low frequency components of zero-IF signal are output by HHT decomposition. Figure 6 is the EMD of each frequency component of one detection point outputted by zero-IF, Figure 7 is the corresponding FFT transformation.

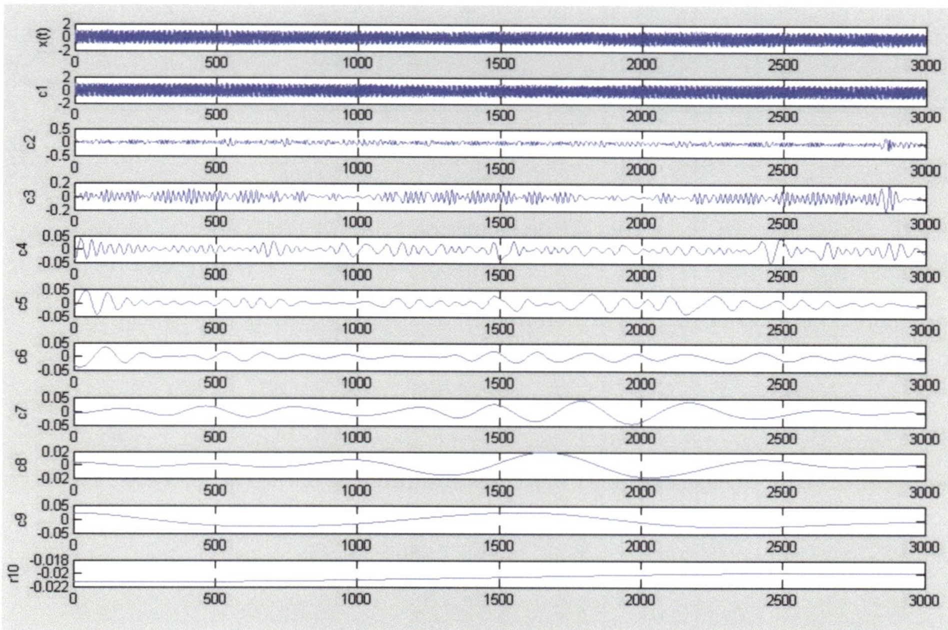

Fig. 6. The original signal and its IMF of the second detection point x(t) is the original signal, c1~c9 is the 9 IMFs, r10 is the surplus

From the Figure 6, we can see obviously each frequency component of output zero-IF of each target layer. We take out segregative frequency component from every layer and reconstruct the lesions information, combining with amplitude feature of back signal.

Microwave spreading in a homogeneous medium, power density attenuate in accordance with the law $1/R^2$. In application of BRATUMASS, detected target region is filled with uniform medium of dielectric properties ε_0, and then add two reflection units of dielectric properties ε_1 and ε_2 into the detected target region. The distances from the two units to the R/T antenna are r_1 and r_2, respectively.

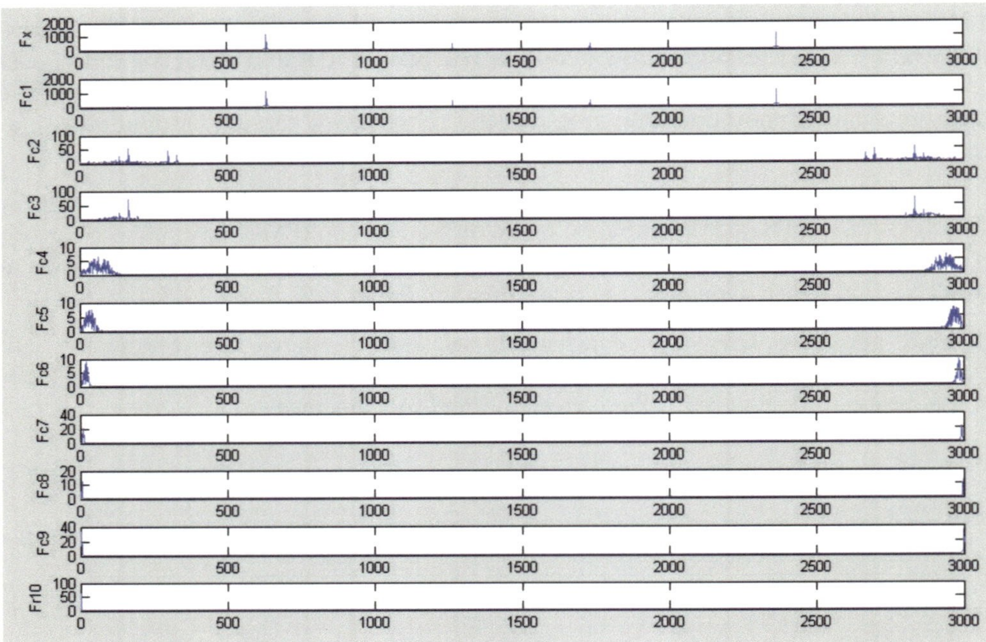

Fig. 7. The Fourier transform of the second detection point Fx is the Fourier transform of x(t); Fc1~Fc9 is the Fourier transform of the c1~c9; Fr10 is the Fourier transform of r10

The reflection coefficient from the first unit is:

$$\Gamma_1 = \frac{\sqrt{\varepsilon_1} - \sqrt{\varepsilon_2}}{\sqrt{\varepsilon_1} + \sqrt{\varepsilon_2}}, \tag{15}$$

Similarly, the reflection coefficient from the second unit is:

$$\Gamma_2 = \frac{\sqrt{\varepsilon_2} - \sqrt{\varepsilon_0}}{\sqrt{\varepsilon_2} + \sqrt{\varepsilon_0}} \tag{16}$$

When $|r_1| < |r_2|$ and the incident power is P_0, the incident power density of the first unit is:

$$S_1 = \frac{P_0}{4\pi r_1^2}, \tag{17}$$

and the incident power density of the second unit is:

$$S_2 = \frac{P_0}{4\pi r_2^2}. \tag{18}$$

The receiving power density from the first unit reflection is:

$$S_{11} = S_1 \times \Gamma_1 \times \frac{1}{4\pi r_1^2} = \Gamma_1 \times \frac{P_0}{16\pi^2 r_1^4} \tag{19}$$

and the receiving power density from the second unit reflection is:

$$S_{22} = S_2 \times \Gamma_2 \times \frac{1}{4\pi r_2^2} = \Gamma_2 \times \frac{P_0}{16\pi^2 r_2^4} \tag{20}$$

In general homogeneous delamination, the greater |r | is the farther spectrum midline leaves origin.

When $|r_1| \leq |r_2|$ and are close, meantime, the dielectric properties of two units are different greatly ($\varepsilon_2 > \varepsilon_1 > \varepsilon$), and unit is sorted according to the reflection power density, far distance unit is sorted at front and near distance unit at behind. That is spectrum centerline interleaving to appear. Whether use the malignant tumor tissue dielectric properties 55 (ε_2, $|r_2|$), background dielectric properties 10 (ε) and breast tissue (lobule, breast ducts and blood vessels) dielectric properties 30 (ε_1, $|r_1|$) to calculate, then there have:

$$\frac{\Gamma_1}{\Gamma_2} = \frac{r_1^4}{r_2^4} \tag{21}$$

And: $r_2 \geq \sqrt[4]{1.5} r_1 \approx 1.1 r_1$. Power spectrum appears overlap after sorted in terms of size of reflection power density, but the actual distances have 1.1 times relationship. Target dielectric properties, which $|r_2|$ corresponds, is 55 (ε_2), this is, the malignant tumor tissue boundary.

The Figure 8 is a typical example of spectrum centerline overlap. From the Figure8, we can see that there are a lot of positions to satisfy spectrum centerline overlap. However, the location satisfied $r_2 \geq \sqrt[4]{1.5} r_1 \approx 1.1 r_1$ is reduced greatly, as shown in Figure9. So this method can improve detection efficiency.

4. BRATUMASS system signal and data processing

The detection system (BRATUMASS) is re-illustrated in Figure10. BRATUMASS system is a time and distance measurement system, which can determine the distance from breast tumor to antenna by obtaining the time delay between directive wave and scattered wave.(Wang & Yao, 2006) The position of the breast tumor can locate by the distance.

Assume only one tissue interface in the effective detection space. The signal from transmitting antenna is:

$$v_1 = A_c \cos\left(2 * \pi * f_c + 2 * K_c * \int_t H(\tau) d\tau + \theta\right) \tag{22}$$

Where, H(τ)=sawtooth(τ) is triangular pulse.

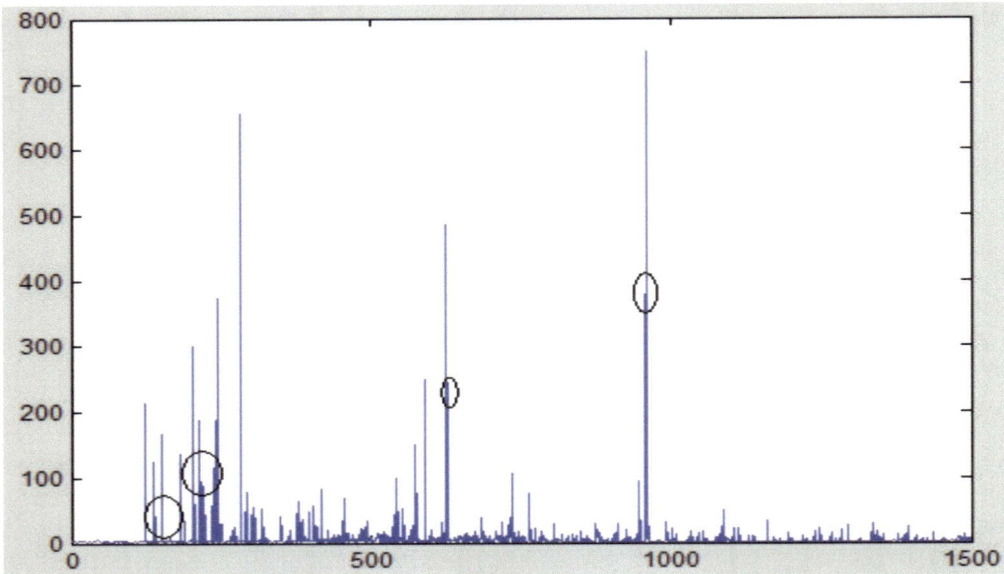

Fig. 8. The typical spectral chart according to the signal energy amplitude after the separation layer of the first patient fourth test signal echo. Abscissa is the signal ordinal number, the vertical axis is the size of the signal period, units are normalized, we can see from the chart a clear spectrum centerline overlap

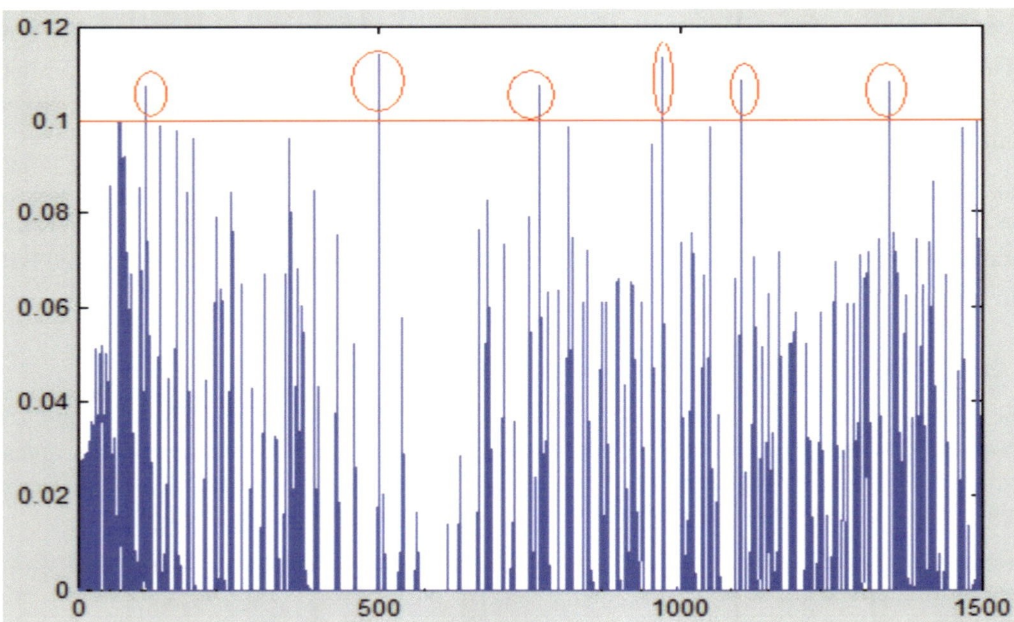

Fig. 9. In the energy spectrum, if the relative difference of energy between two-line is less than 20%, then the location probable is the distortion. Abscissa is the signal ordinal number, the vertical axis is the corresponding relative differences from the overlap line

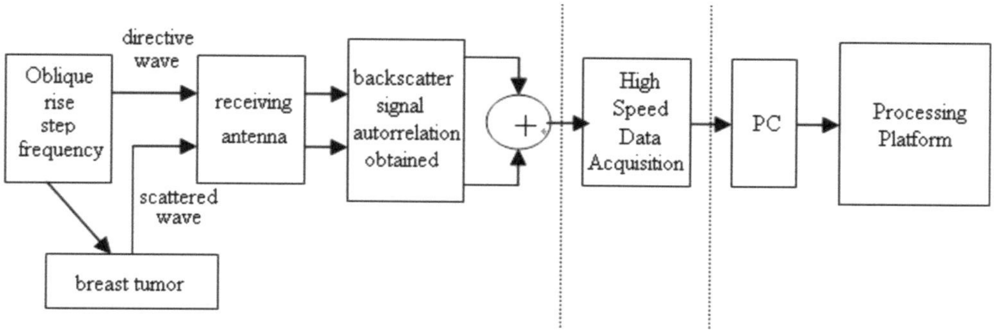

Fig. 10. The diagram of detection system

The received signal from antenna can divide into two parts: one is directive wave from the transmitting antenna; another is scattered wave which is reflected by breast tumor. Propagation distances and propagation times of directive wave and scattered wave are different, because the scattered wave has a time delay Δt relatived to directive wave. The system can estimate the distance from breast tumor to antenna by obtaining the time delay, then determine the position of the breast tumor by establishing the trajectory equation.(Zheng & Yao, 2005)

BRATUMASS is constructed based on the above-mentioned principle. In the detecting environment of BRATUMASS, the characteristics and delay of single back wave signal is satisfied: $|\Pi(\tau_i,\omega)| = 2\pi\kappa_i A^2\delta(\omega-\mu\tau_i)$ after calculation. (Tao, 2011) Where, $|\Pi(\tau_i,\omega)|$ is the amplitude spectrum of BRATUMASS signal, κ_i is the scattering coefficient of target i, τ_i is the delay of back wave, μ is the FM slope and A is the amplitude of transmitting signal.

It can be observed that a target which is existed in detection space for single sampling point corresponds to a spectral line in amplitude spectrum of BRATUMASS, whose frequency and amplitude are related to delay (location) and target characteristics, respectively. There have many back wave faces on the rough malignant tissue surface. Then the number of spectral lines in the region of malignant tissue is larger than them in the region of normal tissue. The spectral line represented characteristics of tissue can be determined by analysis the characteristics of back wave.

As early as in 1996, Zhonghua Ma et. al. (Zhang & Xu, 1996) introduced the fractal theory to analyze the karyomorphism of breast cancer and then indicated that the fractal dimension quantificationally describe the degree of irregularity of tumor karyomorphism, which has certain significance in pathology identifying normal and malignant tissue. Overall, the malignancy level is higher, the cell differentiation is worse and the corresponding fractal dimension is larger. By literature,(Hou & Zhu,2001) we obtained that an object with fractal structure often has a fine structure and very irregular interface, the larger of fractal dimension, the more fine structure of the surface and the more irregular and rougher.

In the actual detection, the back wave not only exists on the malignant tissue surface, but also on other tissues surface. And relatively speaking, the malignant tissue surface is rougher than normal tissues surface, so the back wave surfaces of malignant tissue are more and relative concentration. This makes the time delays, obtained by total back wave signals, large, which are corresponding to more frequency components after mixing. It is very difficult to separate back waves of malignant tissue from total back waves. However, back waves can be separated by using the characteristics of roughly layered distribution of organism tissue structure. That is to say, different tissue layers are corresponding to certain frequency band in effective frequency domain after mixing. We firstly extract layered structure and then separate and distinguish malignant tissue. That can decrease difficulty of distinguish.

5. Wavelet analysis of detection data

Multiresolution analysis is one applications of wavelet analysis. Then, the signals shown in Figure 1 are layered extraction by using wavelet analysis. In wavelet decomposition, the maximum frequency element is seen as one, so the wavelet decomposition of each layers are band-pass or low-pass filters. As close as possible to separate useful information and give detail information for rebuilding nidus, we use db3 wavelet and 11 layers in practice. Figure 11 shows separated results using the wavelet tool box of MATLAB.

Fig. 11. Separated results of signals of a detection point

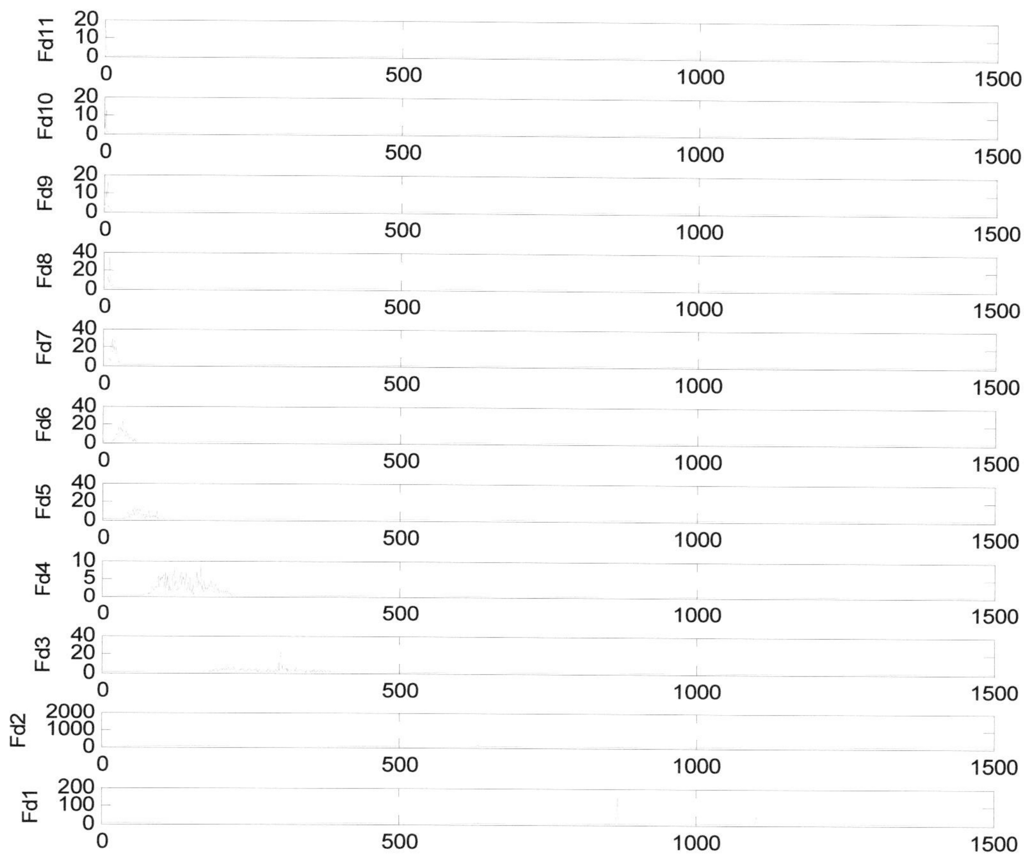

Fig. 12. FFT of each layer

The separated spectrums, which are the FFT of separated layer signals, are shown in Figure 12. The abscissa is corresponding to the spectrum ordinal number in order to remain consistent with latter.

From Figure 12, each layer separates out each frequency component in them own frequency band. Synthesized each layer separation results, spectral lines are ascendingly sequenced according its relative distances, which is the distances between the target and the transmit antenna. Then the information of tissue layers is demonstrated in Figure 13.

As Figure 13 shown, tissue layers are approximate uniformity and continuation in 30mm. There has a more than 2mm discontinuous tissue layer beyond 30mm, which is roughly fit with anatomical structure of breast. When a detection microwave enter into a breast, It firstly encounters a homogeneous fat layer about 1.5cm (3cm/2, microwave have forth and back wave), then it encounters other tissue layers after fat layer.

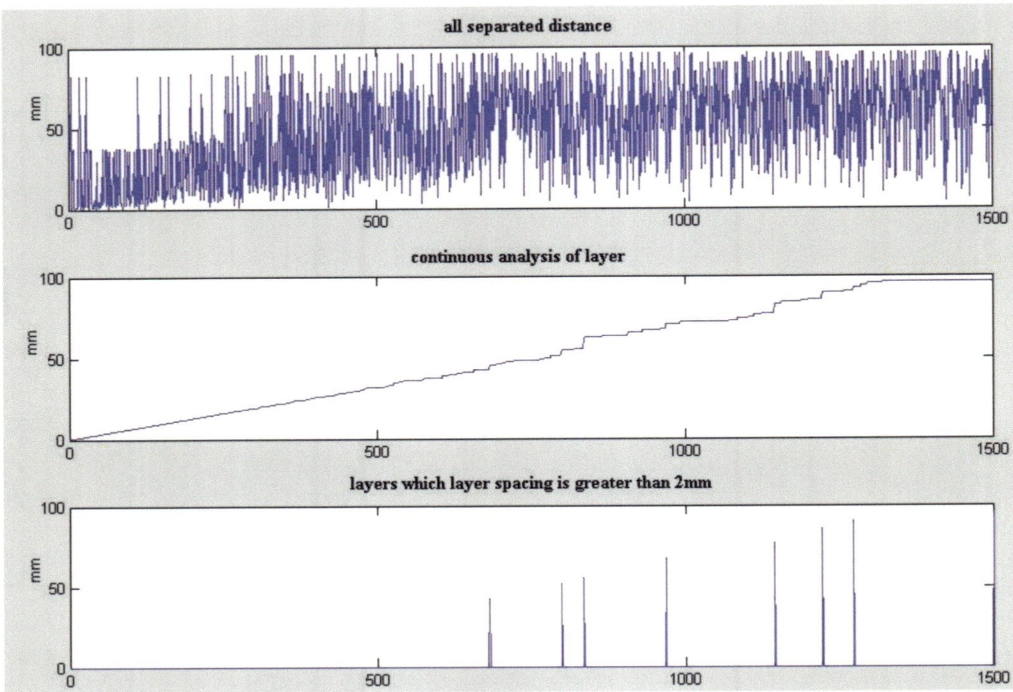

Fig. 13. Relative information of tissues layer

6. Discrimination model of characteristics of tissues

Discrimination model of same tissues: we assume that the distribution of same tissues is relatively gathered together as flocculent distribution in detection space. The results of detecting same tissues are basically same. So if the variance of data in detection space is relatively small (meet certain threshold), that space has only one tissue; conversely, that space has several different tissues. That is to say, it would be better of the differentiation of cell development in that space and more kinds of tissues. On the contrary, it would be fewer kinds of tissues.

Assumption of malignant discrimination: the rate of back wave can be used to discriminate the characteristics of tissues. We discriminate malignant tissues by using the adjacent degree between the mean of the spectral lines relatively gathered area and the rate of back wave of malignant tissues.

Consider the following mode function:

$$y(x) = \exp(-\frac{(\mu_d(x)\text{-Ld}(x))^2}{2\sigma^2}) \tag{23}$$

Where, σ is the variance of data window, μd is the mean of data window, d is the width of data window, x is the coordinate of center position of data window, Ld(x) is the ideal rate of back wave of malignant tissues in that position, y(x) is the total evaluation index of data window, which is the possibility of the existing malignant tissues in that window area.

Figure 14 illustrates analysis results of sampling data on a case of patient by using this algorithm. The malignant tissue is located the position confirmed by other medical methods as shown in Figure 14. The limit of malignant tissue is approximate 4cm. As shown in the right figure, the malignant tissue is emerged in the distance left antenna from 1cm to 5cm in the fourth sampling point. To the eighth sampling point, the analysis results reveal that the malignant tissue is most probably emerged in the distance left antenna from 5cm to 9cm. Because the sampling antenna is positioned using manual, excursion of mammary glands often occur in sampling. Those factors will increase the probability of error, but the calculation results of the limit of malignant tissue are basic right.

We also used the multi-resolution characteristics of wavelet to get the morphotype of malignant tissue under microwave range sounding signal. Figure 15 showed the processing results of same case of patient (figure 14) from first sample point to fourth sample point.

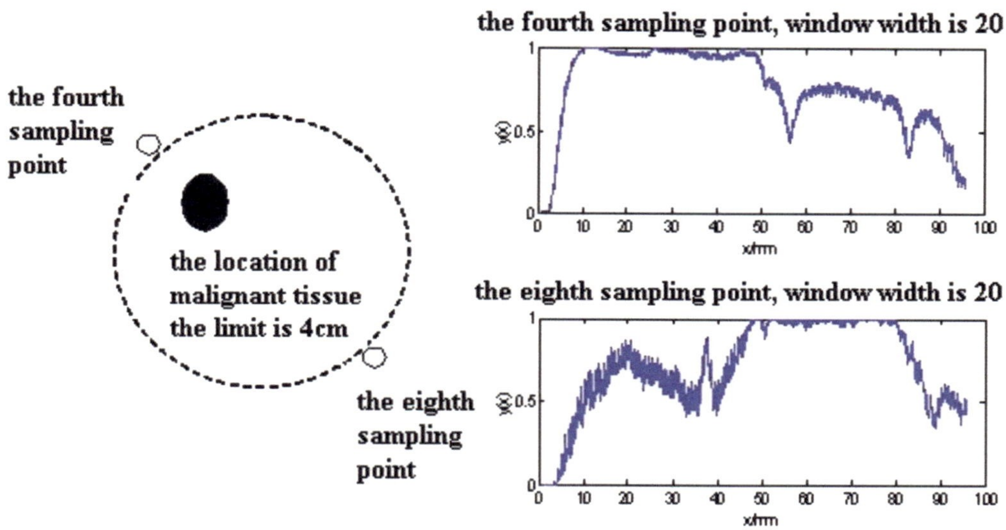

Fig. 14. The position of malignant and the calculation results of malignant degrees in sampling point.

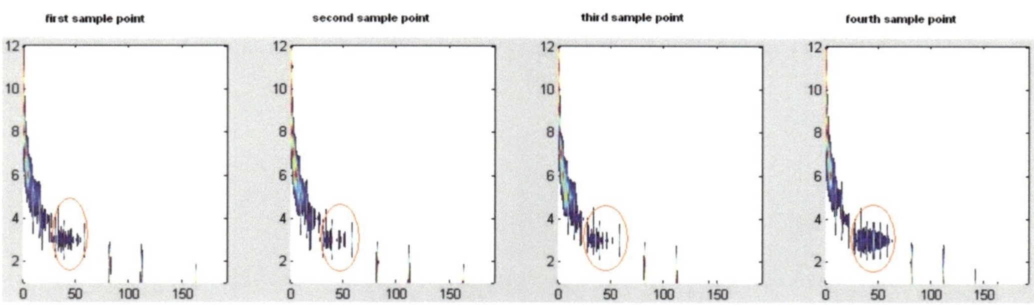

Fig. 15. The morphotype of malignant tissue under microwave range sounding signal

7. Conclusion

We can achieve the effective analysis to the reflected signals by interface of biological tissues using the multiresolution analysis characteristics of wavelet. At the same time, we can distinguish the malignant tissues and the normal tissues by using the relationship of frequency components which are separated and applying proper discrimination model of tissues. However, we don't detailedly discuss the discriminate threshold in the discrimination model of tissues. In fact, this threshold, in the same tissue and in the same microwave condition, is different at largest microwave response. That threshold needs to be further research.

8. Acknowledgments

This work has been performed while Prof. M. Yao was a visiting scholar in Michigan State University, thanks to a visiting research program from Prof. Erik D. Goodman. M. Yao would also like to acknowledge the support of Shanghai Science and Technology Development Foundation under the project grant numbers 03JC14026 and 08JC1409200, as well as the support of TI Co. Ltd through TI (China) Innovation Foundation.

9. References

Hou, Rong-tao; Zhu, Fei. J (2001) *Fractal theory and its significance*. China Computer & Communication 2001.1:196

Huang, N. E., etc R. (1998). *The empirical mode decomposition and the Hilbert spectrum for non-linear and non-stationary time series analysis.* pp. 903-995 Proc. R. Soc, London, 454 (A)

Tao, Zhi-fu. R.(2011) *Investigation on the methodologies of near-field microwave echo imaging integrity*. The Ph.D. Thesis of East China normal university, 2011.

Wang, Cui; Yao, Meng. J (2006) *Practical Near-field Microwave Sounding Image Method for Early-stage Breast Cancer* Chinese Journal of Scientific Instrument. 2006(S3)

Xu,Kaiye; Tang.Aorong; M. (1996) *The Imaging Diagnosis of Mammary Gland Disease.* Shanghai Science and Technology Press, Shanghai China

Zhang, Zhong-hua; Xu Yuan-ding. J(1996) *Fractal Analysis of Cell Nucleus of Breast Carcinoma.* Tumor. 1996,16(2):63

Zheng, Shijun; Yao, Meng. J (2006) *HFSS Simulation on the Microwave Sounding for Mammary Tumor* Chinese Journal of Scientific Instrument. 2005(S1)

5

Wavelet Transforms in Sport: Application to Biological Time Series

Juan Manuel Martín-González[1] and Juan Manuel García-Manso[2]
[1]*Department of Physics, University of Las Palmas de Gran Canaria*
[2]*Department of Physical Education, University of Las Palmas de Gran Canaria*
Spain

1. Introduction

In sports, biological signals are often used to control and design the sports activity. One of the most common used signals is heart rate (HR). Heart rate variability (HRV) refers to natural fluctuations in the interval between normal heartbeats that occurs while individuals rest or exercise. HRV results from the dynamic interplay between the multiple physiologic mechanisms that regulate HR, and it mainly reflects an expression of the interplay between the sympathetic and parasympathetic nervous systems (Task Force, 1996). Two main oscillatory processes interact with the heart as feedback and forward mechanisms, via autonomic pathways: the modulation of the heart rate by breathing, known as respiratory sinus arrhythmia (RSA), and the short-term blood pressure control, known as baroreflex. These main rhythms usually appear in the high and low frequency ranges of HRV, respectively; however, the dynamic interactions in the cardiovascular system may change this typical spectrum. However, the intrinsic properties of the complex autonomic regulation of cardiovascular function are difficult to measure since, even at rest, emotions and mental loading may affect it.

Usually, HRV is used as a non-invasive method to measure the cardiac autonomous input and it is analyzed from different viewpoints: time domains, frequency analysis and non-linear dynamics. According to the Task Force (1996), the power spectrum for time series, at rest, can be classified as follows: (i) power in the very low frequency range (VLF), *0.003–0.04 Hz*, (ii) power in the low frequency range (LF), *0.04 - 0.15 Hz*, and (iii) power in the high frequency range (HF), *0.15 - 0.4 Hz*. The HF normally reflects respiratory-related activity and appears to be mediated by vagal tone. LF is linked to the baroreceptor reflex and can be mediated by the vagus and cardiac sympathetic nerves. (Houle and Billman,1999) suggested that the LF component results from an interaction of the sympathetic and parasympathetic nervous systems and, as such, doesn't accurately reflect changes in the sympathetic activity. VLF zone is related, especially, by thermoregulation fluctuations and vasomotor tone (Task Force, 1996). With exercise, the very high frequency (VHF) should be taken into account (> *0.4 Hz*).

A difficulty in determining the value of these frequency bands during exercise is given by the lack of stationary in data series. When these time series are not stationary in frequency the problem is even more serious. The so-called "time-frequency analysis" provides the observer with a tool to detect changes both in time domain and in frequency domain, simultaneously. Several models have been proposed: The Short-Time Fourier Transform (STFT), the Gabor Transform (Windowed Fourier Transform), the Wigner-Ville Distribution (WVD) and its refinements, and finally the Wavelet Transform (see Daubechies, 1992; Chui, 1992; Cohen, 1992; Abry and Flandrin, 1996; Teich et al. 1996). In this paper, we used the Wavelet Transform as this method has proven a powerful tool suited for the analysis of the time–frequency localization and non-stationary behaviour of time series as for the previous treatment of the signal: detrending, smooth, filters, etc.

The Wavelet Transform (WT) decomposes a series into time-scale or time-frequency domains, which allows identification of temporal changes of dominant modes of variability, while the Fourier transform solely gives the spectral contents of the whole series. In this chapter, we will see some applications of this method to nonstationary data series very different among them, but obtained from the same system (cardiorespiratory system). Therefore, in this chapter we show the usefulness of methods based on the Wavelet Transform (WT) from two completely different viewpoints: during a cycloergometer test to exhaustation (elite cyclists) and when a subject is at rest in a mindfulness meditation state (MM) (Zen meditation or Zazen).

The Physical Education Department and the Physic Department of the University of Las Palmas (ULPGC) have been working jointly on this projects. All individuals tested to the present date were volunteers who signed their consent once informed on the aims of the experiments. The study was conducted according to the guidelines of the Helsinki Declaration adopted by Worldwide Medical Association on Research on Humans.

2. Practical aplications

2.1 Data analysis

The time-scale or scalogram of a signal is the squared modulus of its Continuous Wavelet Transform (CWT): $|W(a,b)|^2$, and is an average power spectrum for all the scales or frequencies, similar to a smoothed Fourier for each time. The scale parameter a and the localization parameter b assume continuous values, and the scalogram of a time series can be visually represented by an image or a field of isolines. In the present work, the Wavelet basic function as known as Morlet wavelet, was employed to analyse the temporal variation of the HRV. The Morlet wavelet is a modulated Gaussian function, which is well localised in time and frequency:

$$\psi_0(t) = \pi^{-1/4} e^{i\omega_0 t} e^{\frac{-1}{2}t^2} \tag{1}$$

Where ω_0 is the non-dimensional frequency which defines number of cycles of Morlet wavelet (Torrence and Compo, 1998). For large ω_0 the frequency resolution improves, though at the expense of decreased time resolution. For this reason, we chose various values for the parameter and found that $\omega_0 = 20$, is well suited for our purposes.

Integrating $|WT(a,b)|^2$ over a specific scale or frequency band provides the time dependent power of the signal in that frequency band:

$$S(t) = \int_{f_1}^{f_2} |WT(a,b)|^2 df \tag{2}$$

In the Discrete Wavelet Transform (DWT) scheme, the signal *f* of length *N* is decomposed into both approximation (*cAj*), and detail (*cDj*) coefficients by the use of two quadrature mirror filters (quadrature mirror filter bank). At each decomposition or reference level *J*, the approximation coefficients cA_J and detail coefficients $cD_1, cD_2, \ldots, cD_J$ are obtained, and we can reconstruct the approximation signal $A_J(t)$ and the details signal $D_j(t)$, $j=1 \ldots J$. Therefore, the signal *f(t)* may be expressed as the sum of a smooth part plus details as follows:

$$f(t) = A_J(t) + \sum_{j=1}^{J} D_j(t) \tag{3}$$

A discrete wavelet transform (DWT), with a Daubechies (Db8) type base function was initially used to obtain the mean HR signal and to eliminate the trend of the signal over time. Further mathematical details on this procedure can be found in Percival and Walden (2000).

First at all and in all cases analyzed, the measurement errors (artefacts and spurious data) and ectopic heart beats were manually checked and eliminated from the RR interval data. It has also been taken into account that the HRV time series produces an irregularly time-sampled signal. To recover an evenly sampled signal, a linear interpolation was applied to each time series.

3. Case 1

HRV analysis during exercise: All individuals performed an incremental ramp test cycloergometer exercise on a Monark-816, including breathing gas analyses and HR measurements. They performed a gradual effort starting at *100 W*, increasing *5 W* every *12* sec., until exhaustion. RR interval data (the time-interval between each heart beat) were collected by using the heart rate monitor *Polar S810i* (Polar Electro Oy, Kempele, Finland).

The HR interval time series is non-stationary during physical exercise, and therefore the time series include a low frequency baseline trend component. The lowest frequency components are useful for studies on long-term modulation, but they may affect the power spectra of the HRV signal used by us. Detrending is usually used to remove the effects of non-periodic low-frequency changes in the time series before doing further analysis, and in practice, linear or polynomial trend removal and high-pass filtering are the most commonly applied algorithms to do it. Nevertheless, here we propose the DWT method to remove extremely slowly oscillating components from HRV data, considering that it provides greater control of the content in low frequency to be removed, and it is important to understand the effect of detrending on the spectral properties of the time series. DWT was used to decompose the HRV signal into *J* wavelet scales, with Daubechies (Db8) wavelet filters (García-Manso et al., 2007). Then, the detrended signal is reconstructed using the wavelet coefficients at the first *J-1* scales.

In the Figure 1 we can observe that the values of HR (in milliseconds) increase as a response to the stress originated by an increase of the load. The Figure 1 also shows the results of applying the DWT to the signal.

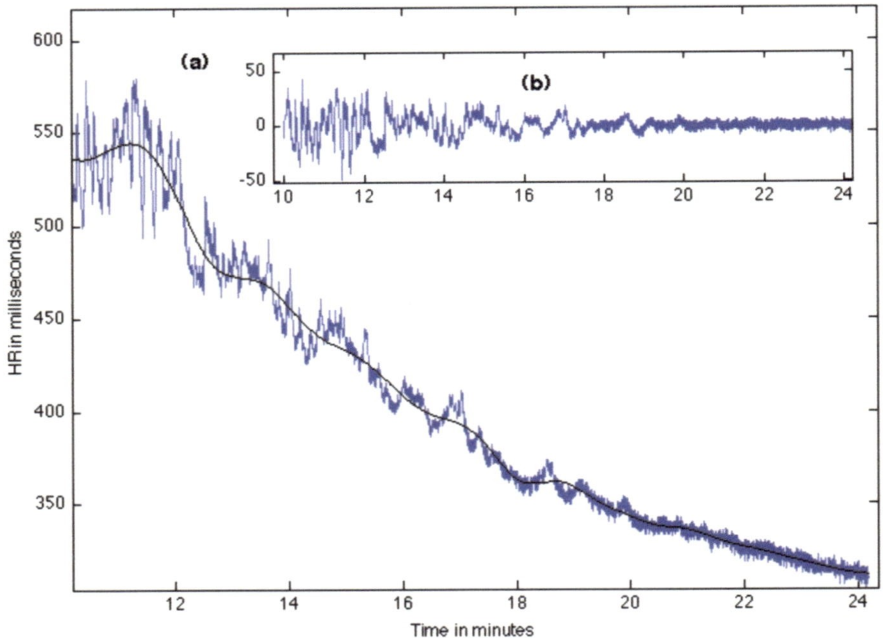

Fig. 1. a) The incremental exercise signal of subject 1 (HR in milliseconds). Also, the trend according to the DWT is shown (Daubechies 8, $J = 7$). b) The detrended incremental exercise signal.

Figure 2 shows the results of applying the CWT to the detrended signal. The method highlights three zones during the incremental test: *the activation, transitional* and *crisis or alarm zones.* It seems to represent three different functional mechanisms, which may be associated to characteristic metabolic processes (area of aerobic prevalence, transition aerobic-anaerobic and area of anaerobic prevalence). The *Transitional Zone* is characterized to be a phase of certain stabilization of HF and LF within the range of minimum values, which may be observed during the test (see Figure 2). Another relevant aspect of the transitional zone it is that the peaks of LF show a linear and positive increment as the load intensity increases.

In a *crisis or alarm zone,* in most of the analysed cases, if the effort is maintained despite the high fatigue caused by the exercise, the signal tends to concentrate on the high or very high frequency bands, indicating clearly that the subjects exert themselves at their highest potential limits. Some subjects give up executing their maximum effort, due to motivation reasons, local fatigue or low resistance to fatigue, instead of using their functional system to the utmost. For this reason, the use of traditional indicators as the stability of VO_2 max, RC over *1.2* or others, may not suffice for the detection of the entering in the crisis zone, while Continuous Wavelet Transform (CWT) proves to be a highly sensitive and useful technique to explain in detail the heart behaviour during incremental exercise.

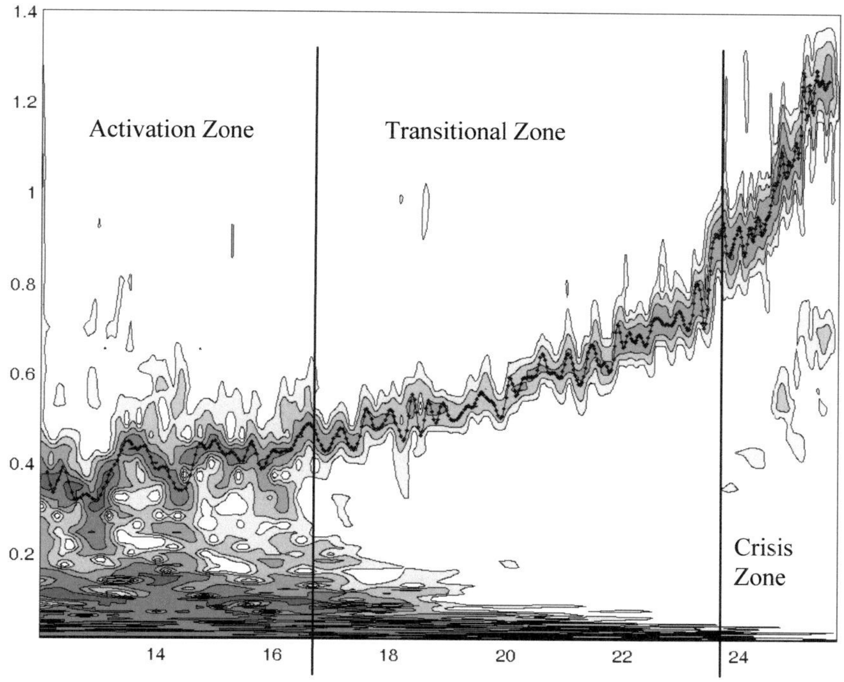

Fig. 2. Contour map of the wavelet coefficients of the heart rate variability shows the changes in the frequency components with time, obtained using a Morlet base function where $\omega_0 = 20$. The x-axis represents time in minutes, and the y-axis the frequency in Hz. Coefficients below a certain value have been eliminated in order to improve the readability of the spectrogram. The dotted lines show respiratory frequency values in Hz.

One of the most interesting applications is identifying the anaerobic threshold (level of exertion where your body must switch from aerobic metabolism to anaerobic metabolism). Some studies used WT for this purpose (Cottin et al., 2006; Cottin et al., 2007; García-Manso et al., 2008) with the following methodology:

- The evolution of frequency peaks (f_p) of the HRV HF-VHF band. The aerobic and anaerobic thresholds were determined on the basis of the changes in the values of f_p in the test. The aerobic threshold was taken as the first inflection point of the kinetics of the f_p slope, and the anaerobic threshold as the second inflection.
- The evolution of the product of the HF-VHF spectral energy value ($PS \cdot f_p$) by (f_p). This variable decreased with load until reaching a minimum value; the first inflection point representing the aerobic threshold; it then remained stable until it started to show a moderate increase, the second inflection point representing the anaerobic threshold.

4. Case 2

HRV analysis during Zen meditation: Zen Buddhist meditation was chosen because this practice does not involve voluntary efforts to concentrate in a single object; it has a mindfulness approach, which means that it includes the reflexive observation of the whole perceptual field and does not use any external method to reach a meditative state. The absence of voluntary efforts and the inexistence of any rule determining how the breathing pattern should be, makes this "technique" especially suitable to investigate the intrinsic properties of the autonomic nervous system (Lehrer et al., 1999; Phongsuphap et al., 2008). Differently than simple rest or sleeping, during mindfulness meditation the mental contents are observed with detachment, creating a delicate state of consciousness that involves both sustained attention and deep relaxation. Lutz et al. (2008) call this state "*open monitoring meditation*" and explain that it "involves nonreactive monitoring of the content of experience from moment to moment, primarily as a mean to recognize the nature of emotional and cognitive patterns".

Sample selection and data collection: A total of *19* subjects (*7* females and *13* males, mean age *43.65* ± *7.60* years, mean meditation experience *9.81* ± *8.82* years), with a consistent practice (at least three times a week) of Zen meditation, participated in this study. Exclusion criteria included the presence of cardiovascular disease or any disease that affects the autonomic balance and the use of any medication that could influence the results. The RR interval data were collected by using Polar S810i, which seems to be an adequate instrument to minimize the possible disturbance that data collection may cause in a meditation process. In any specific case we took simultaneous measures of breathing and HRV, using the multi-function monitoring equipment I-330-C2+. The subjects were instructed to sit quietly for *10* minutes before the beginning of the meditation, already in the position they customarily take during Zazen practice. They sat upright in a cross-legged position on a cushion (*zafu*), with the hands held together in front of the navel. The eyes are kept semi-open and the back upright to avoid drowsiness. After the baseline recording, a bell rings and the subjects start meditating, as they always do, for *40* minutes. After that, the bell rings again and an additional 5-minute recording was taken during quiet sitting. All data were collected in the meditation rooms of *Luz Serena* Temple (Valencia, Spain) and of the *Dojo Zanmai San* (Tenerife, Spain), which follow the same tradition and are systemically integrated.

In this study we found evidences that the evolution in Zen meditation can be characterized for specific patterns of cardiac variability and we found a tendency towards a shift in RSA to the LF range (Peresutti et al., 2010). So, for long-term practitioners the power in the HF ranges significantly decreases and it seems that there is a tendency towards a frequency coupling as regards the years the subject have meditated. In long-term practitioners the attention is opened to the diverse experience content; in the case of instructors with over *10* years of experience, the RSA frequency decreases and couples with LF oscillations, but there is no resonance. We found that even when the breathing frequency is higher than *0.15 Hz*, the RSA and the LF oscillations may coincide, but do not produce a resonant effect.

The Figure 3 represents the WCT of RR interval for a Zen instructor (ZI) with over 11 years of experience. The box *(a)* is the time series; *(b)* the scalogram and *(d)* is the power spectrum (FFT) of HRV signal. In box *(c)* shows the sum of wavelet coefficients (power) in the LF band for each time, obtained using equation (2). The LF band in the box (d) appears as a broad band. However, the wavelet analysis reveals that in reality is a narrow band that varies in time. We can see (horizontal arrow in Figure) how, at the beginning of the meditation, frequency drops to *0.065 Hz* and the power increases significantly. This time interval, about *5* minutes, is responsible for the presence of a sharp peak in the Fourier Spectrum around *0.065 Hz*.

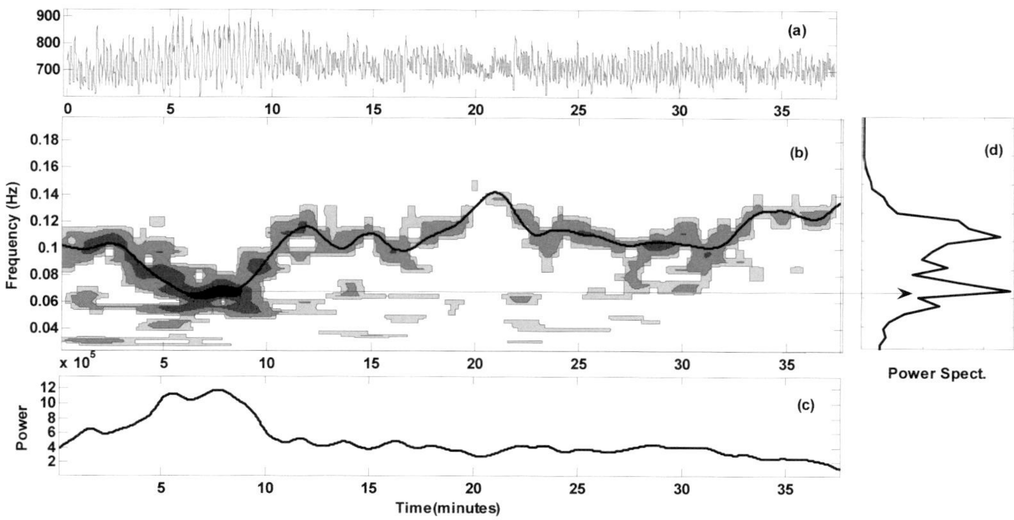

Fig. 3. Wavelet analysis of HRV. It shows the temporal evolution of the low frequency band LF (*0.04-0.15 Hz*) for Zen instructor (ZI), during mindfulness meditation. (*a*) RR interval time series. (*b*) Scalogram. The higher coefficient values are shown in greyscale (contour lines) and lower values were excluded. Darker areas represent higher values. The solid line corresponds to the highest values coefficients (peaks). (*c*) Sum of wavelet coefficients (power) in the LF band for each time. (*d*) FFT of the time series, where the axes have been invested, coinciding the axis *y* with the frequency (LF band only) and the axis *x* with the power. The axis *y* of (*b*) and (*d*) are the same.

In the case of more advanced meditators (Zen Masters with over *20* years of practice) breathing oscillates within the LF range, coupled with the other cardiovascular rhythms and producing resonance, although there may be variations in frequency, especially in the second half of meditation.

The WCT-based analysis also allows us to analyze the simultaneous action, but independent, of HRV and breathing during meditation in two different subjects: Zen instructor and the Zen Master. The Figures 3 and 4 show the wavelet analysis of both HRV and respiratory data, for the Zen instructor and the Zen Master (ZM), respectively.

It should be noted that in the first half of meditation there is less variation in the LF range. Nevertheless, in the second half there is much more variation of the LF probably due to the appearance of a breathing frequency higher than *0.15 Hz* although the RSA and the LF oscillations still coincides, which means that, although the frequencies vary, their relation remains stable, or in other words, the ratio remains almost constant. In long-term practitioners, the attention is opened to the diverse experience content; for the ZI the RSA frequency decreases and couples with LF oscillations, but there is no resonance. The resonance phenomenon described here occurs when the breathing oscillates within the LF range, coupled with the other cardiovascular rhythms. We found that even when the breathing frequency is higher than *0.15 Hz*, the RSA and the LF oscillations may coincide, but do not produce a resonant effect, note the presence of VLF oscillations (Figure 4). Further discussion on this point can be found in: Vaschillo et al. (2006), Cysarz et al. (2005), Peng et al. (1999).

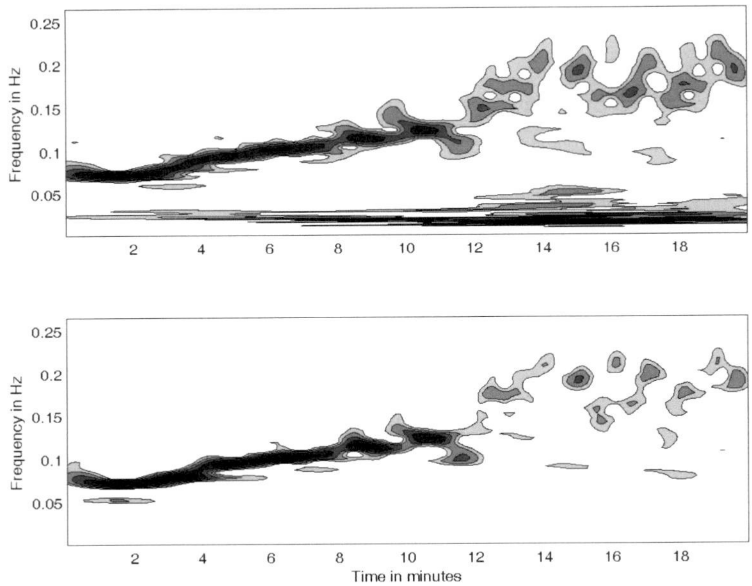

Fig. 4. Wavelet analysis of HRV (top) and respiration (bottom) showing the temporal evolution of the frequencies for ZI during mindfulness meditation. The higher coefficient values (power) are shown in greyscale (contour lines) and lower values were excluded. The darker areas represent more power.

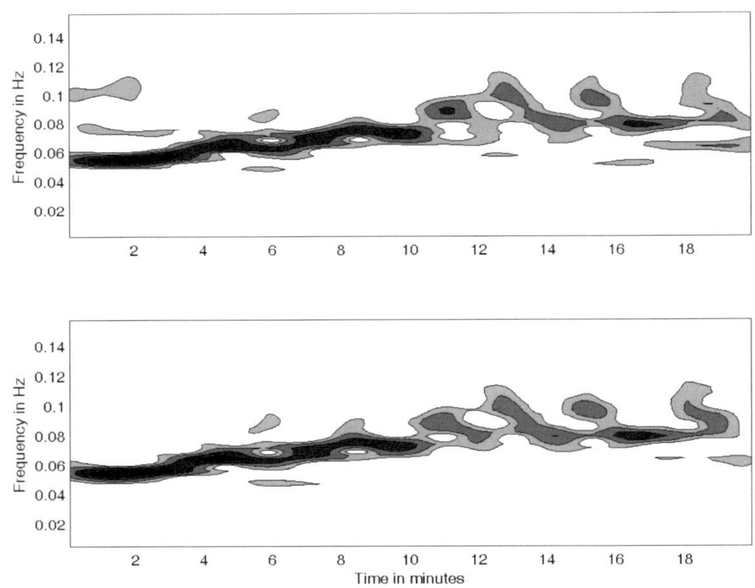

Fig. 5. Wavelet analysis of HRV (top) and respiration (bottom) showing the temporal evolution of the frequencies for ZM during mindfulness meditation. The higher coefficient values (power) are shown in greyscale (contour lines) and lower values were excluded. The darker areas represent more power.

For the Zen Master (Figure 5), the same pattern as shown for ZI is observed, with two different phases in the meditation; however, for ZM, even in the second half, the breathing oscillates exclusively in the LF range and the resonant effect never disappears. We believe that this more irregular pattern in the second half of Zazen, typical among long-term practitioners, could be related to an even less controlled state, when attention stability is well established and a greater depth in the meditation can emerge. If we consider that there is a tendency towards a coupling between HRV characteristic frequencies regarding the years of meditation practice, and also the previous works and the present results from the Zen Master, it is possible that the appearance of a resonant effect between long-term practitioners characterizes the pure state of mindfulness.

5. Conclusion

From the analysis of the experimental data obtained, it may be stated that the wavelet analysis proves as a subtle and precise tool for the detailed study of the cardiac response to physical exercise or coupling of biological rhythms in the absence of physical movement and mental processes dominated. The application of this method allows, on one hand, a global analysis of the behaviour and, on the other, a detailed study of concrete phases or zones and precise responses occurring during exercise.

6. References

Abry P., Flandrin P., 1996. Wavelets in Medicine and Biology (CRC Press, Boca Raton, FL), pp. 413-437.

Chui CK., 1992. An Introduction to Wavelets, Academic Press, San Diego.

Cohen L.1995. Time-frequency analysis. Prentice Hall PTR. Englewood Cliffs, NJ.

Cottin F, Lepretre PM, Lopes P, Papelier Y, Medigue C, Billat V. 2006.Assessment of ventilatory thresholds from heart rate variability in well-trained subjects during cycling. Int J Sports Med. 27:959-967.

Cottin F, Medigue C, Lopes P, Lepretre PM, Heubert R, Billat V. 2007. Ventilatory thresholds assessment from heart rate variability during an incremental exhaustive running test. Int J Sports Med. 28:287-294

Cysarz D, Büssing A. 2005. Cardiorespiratory synchronization during Zen meditation. Eur J Appl Physiol; 95(1):88–95.

Daubechies I., 1992. Ten Lectures on Wavelets, Number 61 in CBMS-NSF Series in Applied Mathematics, SIAM, Philadelphia.

García-Manso JM, Sarmienton S., Martín-González JM., Calderón FJ., da Silva-Grigoletto ME. 2008. Wavelet transform analysis of heart rate variability for determining ventilatory thresholds in cyclists Rev And Med Deporte, 3:90-97.

García-Manso JM., Martín JM., Sarmiento S., Calderón J., Benito P. 2007. Analysis of reply HRV in an incremental effort test: analysis time-frequency. Fit & Performance J. 6(3): 181-187.

Houle M. S., and Billman, G. E., 1999. Low-frequency component of the heart rate variability spectrum: a poor marker of sympathetic activity. Am. J. Physiol. 276: H215-H223.

Lehrer PM, Sasaki Y, Saito Y. 1999. Zazen and cardiac variability. Psychosom Med; 61:812–821.

Lutz A, Slagter HA, Dunne JD, Davidson RJ. 2008. Attention regulation and monitoring in meditation. Trends Cogn Sci; 12(4):163–169.

Malik M., Camm A.J., 1993. Components of heart rate variability - what they really mean and what we really measure. Am. J. Cardiol. 72(11): 821-822.

Peng CK, Mietus JE, Liu Y, Khalsa G, Douglas PS, Benson H, Goldberg AL. 1999. Exaggerated heart rate oscillations during two meditation techniques. Int J Cardiol; 70:101–107.

Percival D., & Walden A. 2000. *Wavelet Methods for Time Series Analysis.* Cambridge: Cambridge University Press.

Peresutti C.; Martín-González JM; García-Manso JM; Mes, D. 2010. Heart Rate dynamics in different levels of Zen meditation. International Journal of Cardiology. 145(1): 142-146.

Phongsuphap S, Pongsupap Y, Chandanamattha P, Lursinsap C. 2008. Changes in heart rate variability during concentration meditation. Int J Cardiol; 130(3):481–484.

Task Force of the European Society of Cardiology and the North American Society of Pacing and Electrophysiology, 1996. Heart rate variability: Standards of measurement, physiological interpretation, and clinical use. Eur. Heart J. 17:354-381.

Teich M.C., Heneghan C., Lowen S.B., Turcott R.G., 1996. Wavelets in Medicine and Biology. CRC Press, Boca Raton, FL, pp. 383-412.

Torrence C, Compo GP. 1998. A practical guide to wavelet analysis. Bull Am Met Soc; 79:61–78.

Vaschillo EG, Vaschillo B, Lehrer PM. 2006. Characteristics of resonance in heart rate variability stimulated by biofeedback. Appl. Psychophysiol Biofeedback; 31(2): 129–142.

6

2-D Discrete Wavelet Transform for Hand Palm Texture Biometric Identification and Verification

Juan José Fuertes Cebrián[1],
Carlos Manuel Travieso González[2] and Valery Naranjo Ornedo[1]
[1]Instituto Interuniversitario de Investigación en Bioingeniería y Tecnología Orientada al Ser Humano, Universitat Politècnica de València, I3BH/Labhuman
[2]Instituto Universitario para el Desarrollo Tecnológico y la Innovación en las Comunicaciones (IDETIC), Departamento de Señales y Comunicaciones, Universidad de Las Palmas de Gran Canaria
Spain

1. Introduction

In the competitive business world today, the need and demand for a biometric physical security solution has never been higher. Common biometric techniques include fingerprints, hand or palm geometry, retina, iris, or facial characteristics. Behavioural character includes signature, voice (which also has a physical component), keystroke pattern, and gait. Nowadays, most of the security systems developed into the society are based on hand image analysis (Masood et al., 2008; Pavesic et al., 2004), especially in the texture of the hand palm since they provide people with higher security in relation to authentication systems (Zhang, 2004 ; Yan & Long, 2008) and offer a good balance of performance characteristics; besides, they are relatively easy to use. Although other biometric techniques like fingerprint or hand and finger geometry let us achieve good results, a security and highly accurate biometric system can be built through the use of palm-print texture features. The user aligns the palm of his or her hand onto a metal surface with guidance pegs that read the hand attributes of that person. Then, the device records the users' hand information and sends it to its database for identification and verification. In short, biometric identification involves determining who a person is and biometric verification determines if a person is who he/she says.

Hand palm analysis offers many advantages similar to the other technologies such as ease of use, small data collection, resistant to attempt to fool a system and unlikely technology to emulate a fake hand. In this context, wavelet analysis plays an important role in biometric systems: the wavelet transform allows to extract the main hand users' features and to differentiate people. There are however several challenges to beat.

Besides high proprietary hardware costs and size, the aging of the people hands, the lack of technology accuracy and the biometrics inability to not recognize a fake hand pose a challenge. To overcome these drawbacks, Goh et al., 2006 presented a palm-print system made up of 75 individuals. It was based on wavelet transform and Gabor filter, with a verification result of 96.7% and an Equal Error Rate (EER) close to 4%. Liu et al., 2007, showed a research about the use of wavelet transform in palm-print. Classifying with the ISODATA algorithm got a 95% of identification accuracy with 180 palm-print of 80 people. Masood et al., 2008, developed a palm-print system using wavelet transforms. 50 people took part in the session (10 samples per person), reaching a 97.12% of accuracy with the combination of different wavelet families. Other authors have proposed different techniques of palm-print analysis: Guo et al., 2009, described a BOCV system, (Binary Orientation Co-Occurrence Vector) based on the linking of six Gabor features vectors. 7752 samples from 193 people were taken. The error rate was 0.0189%. Zhang et al., 2007, presented a novel 2D+3D palm-print biometric system made up to 108 individuals. The EER was 0.0022%. Badrinath & Gupta, 2009, proposed a prototype of robust biometric system for verification which uses features extracted using Speeded Up Robust Features (SURF) operator of human hand. The system was tested on IITK and PolyU database. It had FAR = 0.02%, FRR = 0.01% and an accuracy of 99.98% at original size. The system addressed the robustness in the context of scale, rotation and occlusion of palm-print.

This work will address a depth study about the use of 2-D discrete wavelet transform (DWT) in order to get a simple and robust biometric identification/verification system using the texture of the hand palm. Firstly, the hand palm image processing with scale, rotation and translation invariance (ROI) is obtained. Then, we will discuss and analyze several wavelet biometric families providing its advantages and disadvantages, showing identification and verification results, and concluding with the biometrics of the future. The general block diagram of the system is shown in Figure 1. After obtaining the ROI and applying different filter levels, the recognition rate will be reached with Support Vector Machine (SVM) (Steinwart & Christmann, 2008), a supervised classifier used to authenticate people.

Fig. 1. Functional block diagram of the system. After extracting the palmprint, it is processed with DWT algorithm in order to get a recognition rate by means of SVM classifier.

This paper is set up as follows: section 2 reminds some basics of wavelet transform and main features of the mathematical algorithm. In section 3, the image processing, feature extraction and classification system are detailed. Section 4 shows the methodology proposed and section 5 the experiments performed. A brief comparison with methods developed by other authors is presented in section 6 and finally, the conclusion is given in section 7 together with the future work.

2. Mathematical basis of wavelet transform

Similarity to Fourier analysis, continuous wavelet analysis splits a signal up into several waveforms which are scaled up/down by a factor "*a*" and displaced by a factor "*b*":

$$\psi_{a,b} = \frac{1}{\sqrt{a}}\psi(\frac{t-b}{a}) \tag{1}$$

The result of applying wavelet analysis indicates the wavelet coefficients CWT *(a,b)*, which are dependent on the scale and the position:

$$\mathrm{CWT}(a,b) = \mathrm{W} \cdot \mathrm{f}(a,b) =< \mathrm{f(t)}, \psi_{a,b} >= \int_{-\infty}^{+\infty} \mathrm{f(t)} \cdot \psi^*(\frac{t-b}{a})\, \mathrm{dt} \tag{2}$$

If they are multiplied by the scaled and displaced wavelets, the signals can be generated. In Figure 2 the factorization of a signal in different wavelet waveforms is shown. Small "*a*" values detect fast changes in the details, and high "*a*" values detect slow changes.

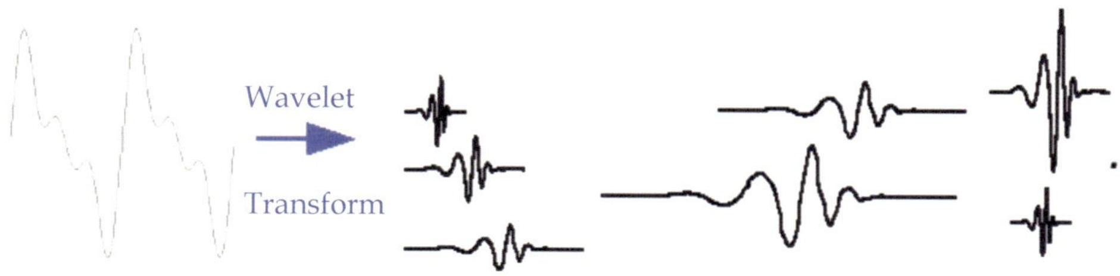

Fig. 2. An example of different wavelet waveforms.

But if we would have to calculate all the wavelet coefficients for all possible scale values, it would be costly. Therefore, in this paper the Discrete Wavelet Transform (DWT) is performed (Villegas & Pinto, 2006), sampling the CWT in values of scale and position which are a power of two $a = 2^{\mathrm{j}}, \mathrm{b} = \mathrm{k} \cdot a,\ \mathrm{j,k} \in \mathrm{Z}$

$$DWT(j,k) = \int_{-\infty}^{+\infty} f(t) \cdot \psi_k^j(t)dt \tag{3}$$

where $\psi_j^k(t) = 2^{-\frac{j}{2}} \cdot \psi(2^{-j} \cdot t - k)$.

In this work, the DWT is used to split a signal in several resolution levels (Gonzalez & Wood, 2008). The multi-resolution split can be performed projecting the signal in two orthogonal sub-spaces called approximations and details. An efficient way to implement the DWT is using a "bank of filters". It was developed in 1989 by Mallat, whose algorithm is the diagram which is known in the signal processing community as "two channels sub-band encoder" (see Figure 3). The signal is split into two parts: low frequencies (approximations) and high frequencies (details) (Mallat, 1989). The low pass filter *H(w)* is related to the high pass filter *G(w)* by means of the equation:

$$\mathrm{g}[\mathrm{L}-1-\mathrm{n}] = (-1)^{\mathrm{n}} \cdot \mathrm{h}[\mathrm{n}] \tag{4}$$

where L is the number of levels of the bank of filters. In section 4 we will explain the use of wavelet transform, but before that, in section 3 the process to extract the hand palm image and the classification system used to get a successful recognition rate will be presented.

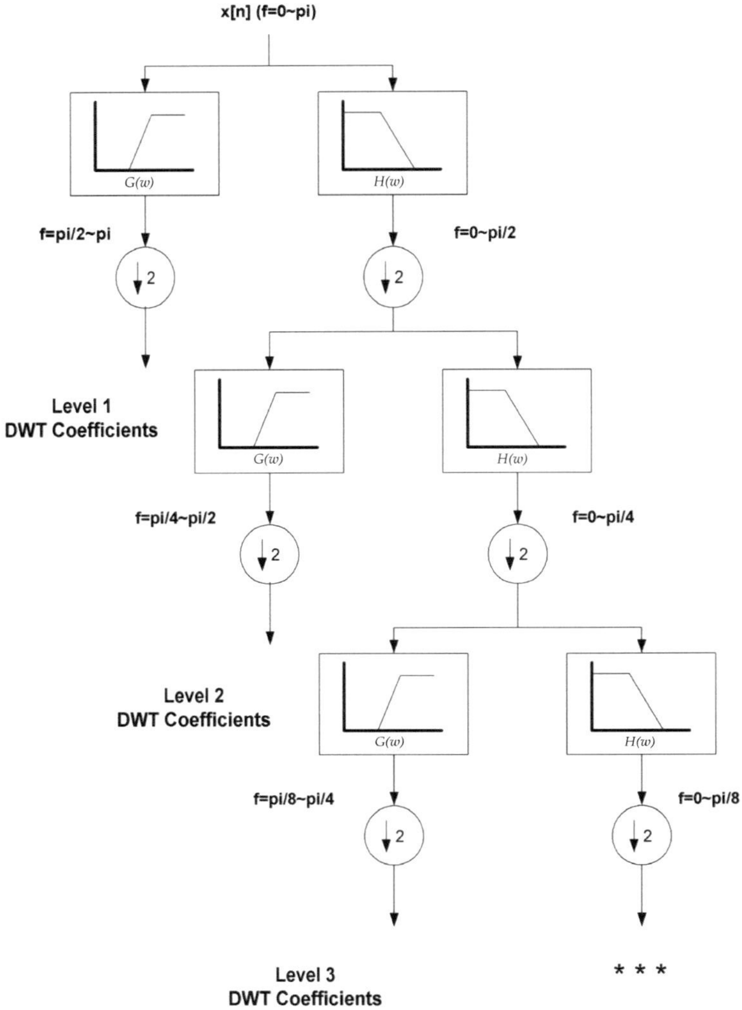

Fig. 3. Bank of filters proposed by Mallat to implement the Discrete Wavelet Transform.

3. Image pre-processing, features extraction and classification system

Once hand images have been acquired, each one is converted from 256 gray levels to a binary image, applying a thresholding algorithm where the threshold is obtained empirically from the training samples of the database. Then, a closing operation is performed in order to delete background noise and to select the user hand.

The goal of image pre-processing is to obtain a region of interest (ROI) with the main lines of the hand so as to extract the most important features of the user. In this way, the first step is to work out the tops and the valleys of the fingers, through the localization of the 4 fingers of the hand (through 8 initial points) with the exception of the thumb, tracing horizontal lines (see Fig. 4, Left). At this point, hand-contour is obtained subtracting the eroded from the dilated hand-image. Finding out the maximum of the contour between the 2 points of each finger the ends are obtained. Finding out the minimum between the 2 consecutive points of different fingers the valleys are calculated. Then, the ROI is obtained after lining up the valley of the little finger with the valley of the hearth-index finger (see Fig. 4, Right), in order to get an image with scale, rotation and translation invariance (in this work it will also be evaluated the system performance if the valleys are not aligned). The vertical size is fixed to 300 pixels and the horizontal size can vary some pixels depending on the fingers gap. Once the ROI is obtained, wavelet transform can be applied.

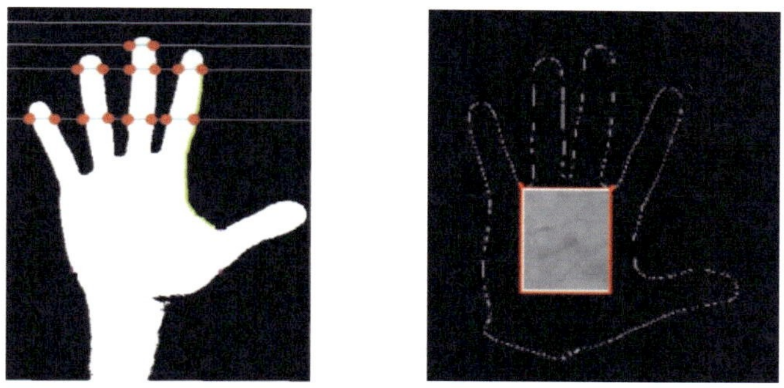

Fig. 4. Left, valleys and tops are obtained through the initial points. Right, the ROI is extracted after lining up the hand with the valleys.

A successive chain of low filters with cut-off *pi/2* is used in order to separate the thin details from the thick details of the palm-print image. It lets to emphasize the difference between the diverse gray levels. Specifically, taking one user palmprint approximation A_{k-1}, we can obtain the *k-th* level decomposition applying a successive chain of filters, low pass (H_L) or high pass (H_H), resulting in the approximation (A_k), horizontal (H_k), vertical (V_k) and diagonal (D_k) detail images (Figure 5). In this work, we have split the original palmprint in approximation and diagonal images, studying the response of the first one due to its high recognition rate.

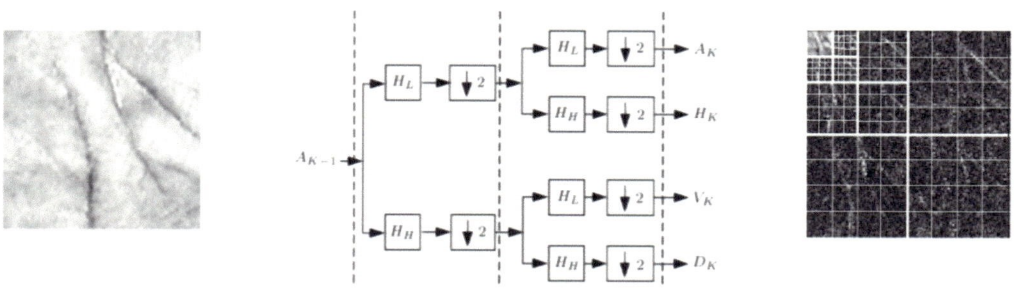

Fig. 5. Left, the original palmprint. Middle, the diagram to split de palmprint in the four images. Right, the decomposition of the palmprint in 3rd wavelet level.

Later, the size of the resulting palm-print image can be reduced to different sizes before introducing them to the classifier in order to optimize the system performance.

3.1 Support Vector Machine (SVM)

A general supervised classifier divided into two fundamental blocks is used, with the training and test block, just as it is shown in the Figure 6. In that figure, the classification process is shown after obtaining and extracting the data. With the training parameters, a score feature which is used in the test step is modelled.

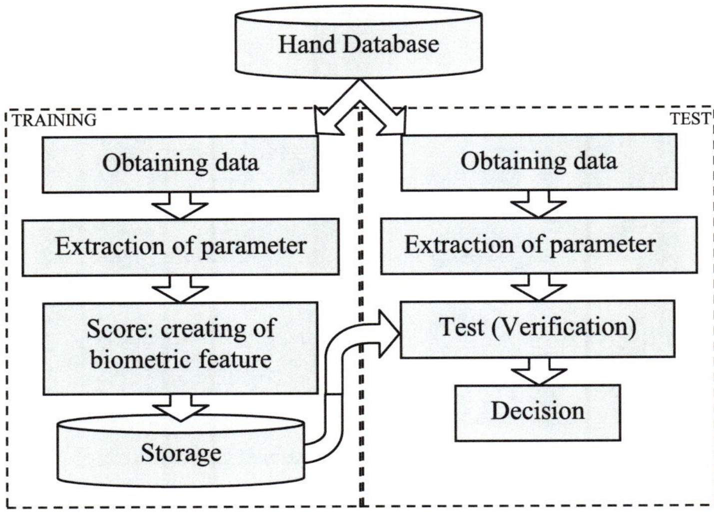

Fig. 6. Block diagram of the supervised identification/verification system. Left, training block; right, test block.

To evaluate the hand features, Support Vector Machines (SVM, Steinwart & Christmann, 2008) are used when they work as identifier or verifier, where only known users appear in the test data (closed mode). At that moment, linear (Figure 7) or RBF kernel (Figure 8) can be used to verify/identify users, if the data are or not linearly separable.

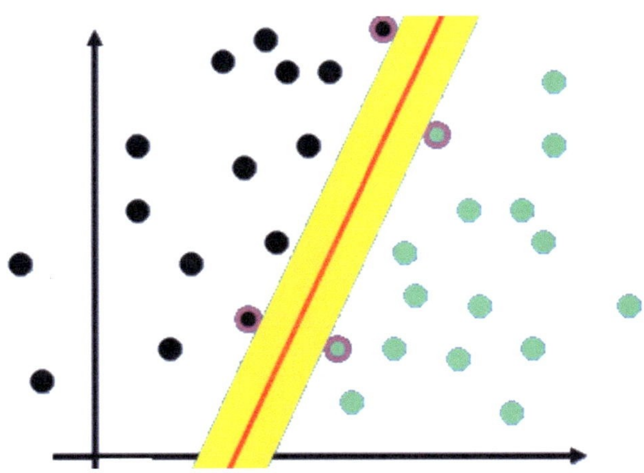

Fig. 7. Maximization of the hyperplane with linear kernel.

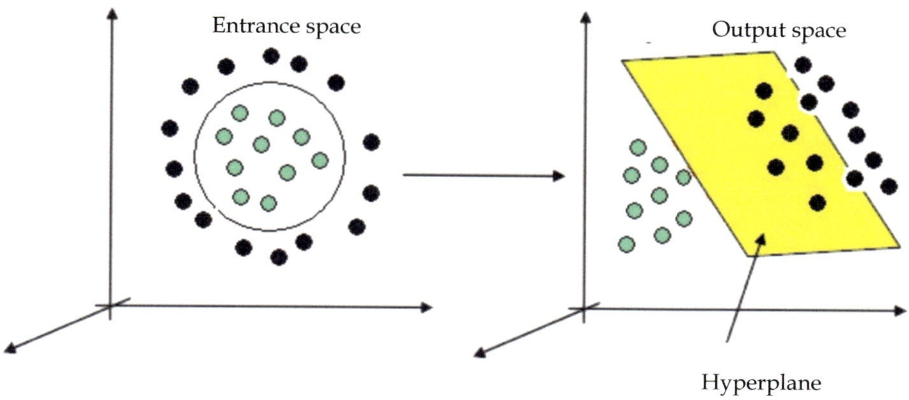

Fig. 8. Data transformation to higher space with RBF kernel.

Analyzing Figures 7 and 8, it would be possible to say that SVM builds a hyperplane of separation in the entrance space in two possible modes: in the first, the input space is converted into higher dimension characteristics space, by means of a (nucleus or kernel) (non)- linear transformation; in the second, the optimum hyperplane of separation (MMH, Maximal Margin Hyperplane) is built. This hyperplane maximizes the distance of the vectors which belong to different classes. Thus, if S is a set of N vectors $\boldsymbol{x}_i \in R^n$ where i=1...N, each vector $\boldsymbol{x}_i$ belongs to one of the two identifying classes as $y_i \in \{-1, 1\}$. If the two classes are linearly separable, a unique optimum hyperplane exists and it is defined by:

$$\boldsymbol{w} \cdot \boldsymbol{x} + b = 0 \tag{5}$$

where $\boldsymbol{w}$ denotes the normal vector to the hyperplane and $b/||\boldsymbol{w}||$ is the distance between the origin and the hyperplane. It provides a greater margin of separation among the classes, and it divides **S** leaving all the vectors of the same class in the same side of the hyperplane. This hyperplane can be obtained resolving the problem,

$$\begin{aligned} &\textit{minimize} && g(\boldsymbol{w}) = \tfrac{1}{2}||\boldsymbol{w}||^2 \\ &\textit{subject to} && y_i\,(\boldsymbol{w} \cdot \boldsymbol{x}_i + b) \geq 1, \quad i = 1, \ldots, N \end{aligned} \tag{6}$$

Therefore, it treats a problem of square programming whose solution is obtained applying the theorem of Karush-Kuhn-Tueker, which can be written as:

$$\boldsymbol{w} = \sum_{i=1}^{N} \alpha_i\, y_i \boldsymbol{x}_i \tag{7}$$

The associated vectors to the multiplicities of Lagrange (α_i) which are not null are called support vectors, and they are the unique vectors of S that are necessary to determine the optimum hyperplane. To extend the concept of SVM to non linear classifiers, it is carried out a transformation of the input space to other space of higher dimension, in which the data are linearly separable. Thus, each input vector is transformed into other vector $\mathbf{z} = \Phi(\mathbf{z})$ of greater dimension in the space of characteristics. Under certain conditions, given by the theorem of Mercer (Mercer et al., 1999), the scale product in the space of characteristics can be expressed by a certain nucleus $K(\boldsymbol{x},\boldsymbol{y})$. In this case, the problem is reduced to resolve;

$$\begin{aligned} &\textit{minimize} && g(\boldsymbol{w}) = \frac{1}{2}\|\boldsymbol{w}\|^2 + C\sum_{i=1}^{N}\xi_i \\ &\textit{subject to} && y_i\,(\boldsymbol{w}\cdot\boldsymbol{\Phi}(\boldsymbol{x}_i) + \mathrm{b}) \geq 1-\xi_i,\quad i = 1,\dots,N \end{aligned} \tag{8}$$

where ξ_i are lack variables which measure the degree of misclassification of the datum $\boldsymbol{x}_i$, and the vector $\boldsymbol{w}$ can be obtained again, as combination of the transformed support vectors. Support Vector Machines (SVM) are based on a bi-class system, where only two classes are considered. In this work, we have worked on identification and verification system, and for this reason, a one-versus-all and one-versus-one strategy are built respectively. To express the similarity between biometric patterns, in our case palm-print modality, the Recognition Rate (RR) in identifier and the Equal Error Rate (EER) in verifier mode are used. The higher the RR, the better the identifier performance.

In verifier mode, the EER of a system can be used to give a threshold independent performance measure (or TEER – Threshold of Equal Error Rate). The lower the EER, the better the verifier performance, since the total error rate which is the sum of the FAR (False Acceptance Rate) and the FRR (False Rejection Rate) at the point of the EER decreases. In theory, this work is fine if the EER of the system is calculated using an infinite and representative test set, but it is not possible under real world conditions. Therefore, to get comparable results it is necessary that the EERs, which are compared, are calculated on the same test data using the same test protocol. In our experiments, ERR and its TEER have been calculated.

4. Experimental methodology

The 1440 hand-images belonging to the 144 users of the database are acquired thanks to a 150 dots per inch general scanner putting the users' hand on its surface. Then, they are stored with 256 gray levels, 8 bits per pixel. The size of these images is set to 1403x1021 pixels after scaling them by a factor of 20% to facilitate later computation (See Table 1).

Properties of the hand images contained in the database	
Size	80% original size
Resolution	150 dpi
Colour	256 grey levels
File size	1405 Kbytes
Data matrix dimension	1403×1021

Table 1. Image Specifications.

Four experiments are performed with the database hand images.

1. In the first experiment, the recognition rate (RR) for 50 users is shown, after applying from 1 to 4 filter levels for each wavelet family above straight hand palm image (ROI obtained without aligning the valleys).
2. In the second experiment, the wavelet algorithm is evaluated above the sloping ROI (after aligning the valleys of the hand), under the same conditions in order to compare both methods.

3. In the third experiment, the recognition rate for straight and sloping ROI above the 144 users of the database to observe the dependence they have on the system is presented.
4. Finally, in the last experiment the best algorithm performance when the classifier works as verifier is shown. Besides, the False Acceptance Rate, the False Rejection Rate and the Equal Error Rate are detailed.

In order to optimize the results of these experiments, the research covers from the data analysis to classifier selection, studying several hand image sizes, DWT family, filter levels, RBF gamma values and the system performance when it works as identifier and verifier. In Figure 9 the different parameters studied in order are shown.

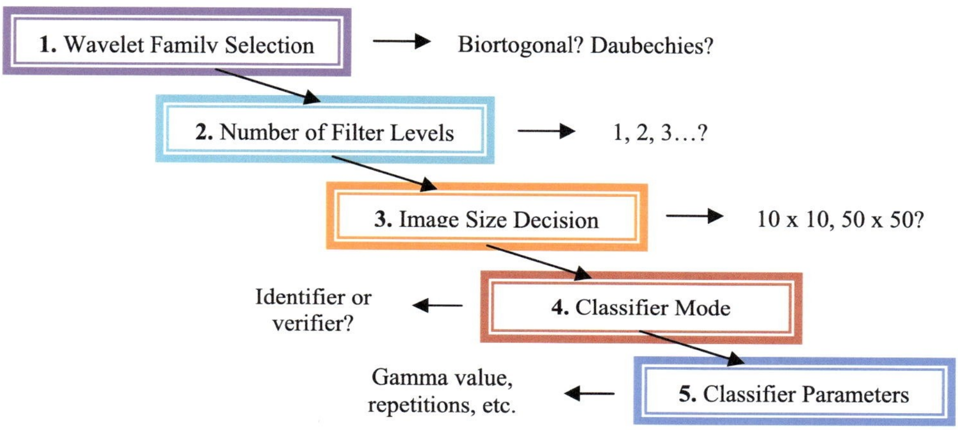

Fig. 9. The blocks and parameters studied to evaluate system accuracy.

The first step consists of the kind of filter and the number of levels to apply over the images. In this paper three wavelet families have been compared: *'haar'* or *'db1'*, *'daubechies5'* and *'biortogonal5.5'* (see Figure 10).

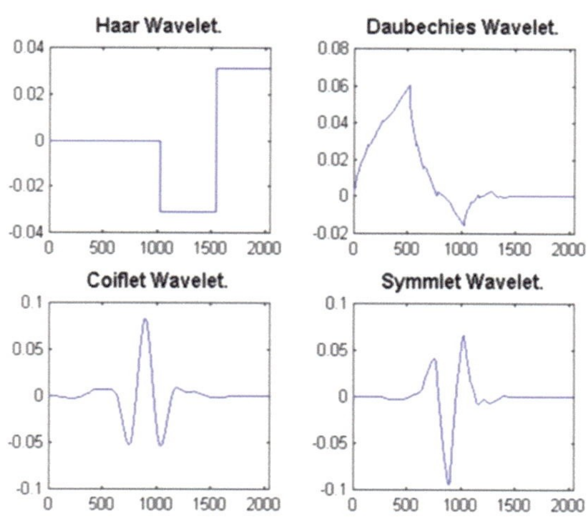

Fig. 10. Several Wavelet families: haar, daubechies, coiflet and symmlet.

The number of filter levels is varied from 1 to 4, but when 1 or 4 filters are applied the recognition rate decreases considerably. To facilitate the image filtering, it has been used a successive chain of low filters with cut-off pi/2 in order to separate the thin details from the thick details of the palmprint image, as it has been explained in section 2.

After the hand image is filtered, it can be rescaled to a particular size. However, best results are obtained after applying three consecutive low filters to the ROI, without reducing the hand-palm size (original size after the filters) as it will show in results section.

At this point, the classifier works as identifier or verifier. When it works as identifier, the system recognizes the identity of the person whose distance to the hyperplane is largest. If the system works as verifier, it accepts the identity of the person when the distance to the hyperplane is greater than the optimum threshold calculated in training step (Adán et al., 2008). Finally, the gamma value of the classifier is changed to get the best recognition rate.

5. Identification/verification results for biortogonal, daubechies and haar wavelet families

As it was introduced in the previous section, we have collected 1440 samples from 144 users, 10 samples per user (see Table I to know image specifications). Four images of each user are chosen randomly as training samples and the remaining six images are used to test the system. At this time, the extracted and filtered palmprint is introduced to the classifier in order to get the recognition rate. The results are shown in average (%) and typical deviation (std).

Each test has been done 10 times using linear and RBF (radial basis functions) SVM kernel, finding the final result through a cross-validation strategy. The supervised classification has been carried out with the library SVM_light (Cortes & Vapnik, 1995). In all experiments, we have evaluated the system in closed mode (only known users appear in the test data), where 50 or 144 users have been used as test and training data.

In Figures 11, 12 and 13, the recognition rates (along x-axis) of wavelet haar, daubechies5 and biortogonal5 families (experiment number 1) for different size palm image reductions (along y-axis) are shown. The number of filter levels is varied from 1 to 4 when the system works as identifier for 50 users (RBF classifier).

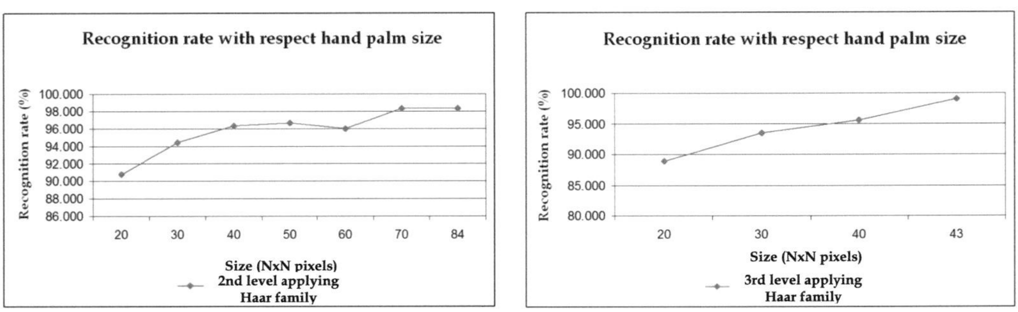

Fig. 11. Recognition rate with Haar Wavelet filter.

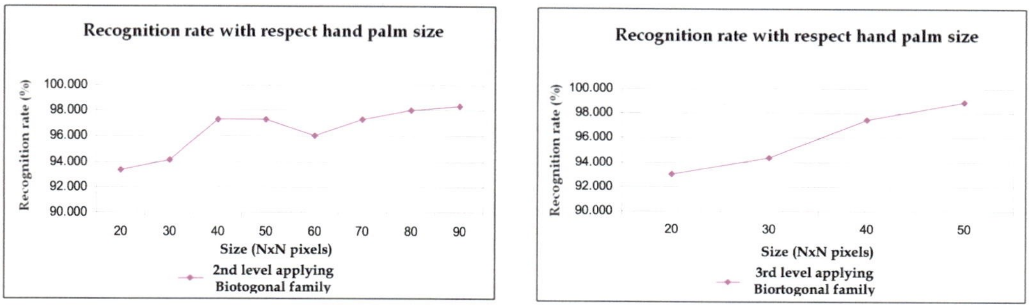

Fig. 12. Recognition rate with Biortogonal Wavelet filter.

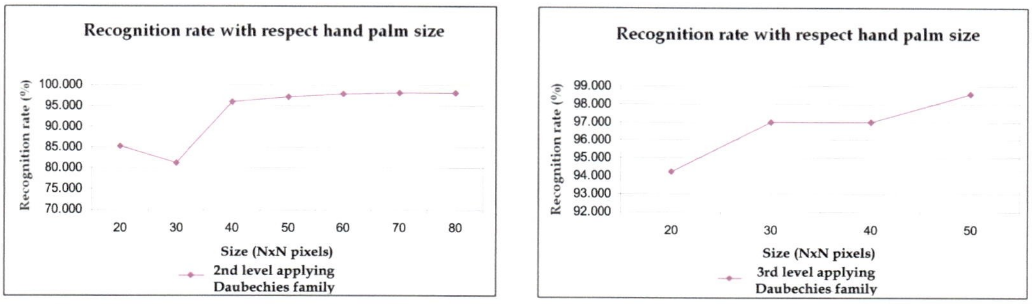

Fig. 13. Recognition rate with Daubechies Wavelet filter.

When the number of levels of the filter was 1 or 4, the recognition rate decreased drastically up to 85% approximately, with both linear and RBF kernels. In all cases, the best recognition rate is obtained when a 3rd filter level is applied without reducing the resulting images size. This happens because in the second and specifically in the third level, the information available to maximize the interclass relationship is highest.

Analyzing the figures, the higher the palmprint size, the higher the recognition rate. This happens because important discriminative features are ruled out when the image is reduced to a small size. The size is not increased either, since it supposes a higher computational cost and the recognition rate keeps constant. Moreover, when the wavelet family is the Biortogonal5.5, the recognition rate is higher than with the other signal families. When the applied filter is daubechies5 or haar, the recognition rate progress is similar to biortogonal5.5 filter.

Before testing the wavelet filter for the full database, it was applied the algorithm over the SLOPING ROI for 50 users (experiment number 2) in order to evaluate the influence of pre-processing steps (aligning the valleys).

As Table 2 shows, the results obtained for the three wavelet families are similar either linear or RBF kernel. For this reason, the three wavelet families must be evaluated for 144 people. In Table 3 the highest recognition rate with linear and RBF kernels for 144 users in identifier mode is shown (experiment number 3). In this case, the results demonstrate the difference when the ROI is aligned with its valleys (SLOPING ROI) or not (STRAIGHT ROI).

IDENTIFIER - 50 USERS		
Features	Linear recognition rate	RBF recognition rate
Wavelet Haar 3 filters	99.33% ± 0.00	99.17% ± 0.06
Wavelet db5 3 filters	99.66% ± 0.22	99.50% ± 0.06
Wavelet bior5.5 3 filters	99.83% ± 0.06	99.67% ± 0.22

Table 2. Wavelet identification results for 50 users using the *sloping* ROI.

STRAIGHT ROI - IDENTIFIER			SLOPING ROI - IDENTIFIER	
Features	Linear recognition	RBF recognition	Linear recognition	RBF recognition
haar 3 filter levels	99.11% ± 0.59	99.00% ± 0.77	99.54% ± 0.01	99.73% ± 0.03
db5 3 filter levels	98.44% ±1.59	98.55% ± 1.15	99.54% ± 0.01	99.31% ± 0.12
bior5.5 3 filter levels	98.77% ± 0.26	98.88% ± 0.26	99.76% ± 0.01	99.65% ± 0.02

Table 3. Wavelet identification results for 144 users using the *straight* and *sloping* ROI.

According to tables 2 and 3, the recognition rate is similar regardless of database users' number, unlike other texture algorithms such as derivative method or Gabor filter (Fuertes et al., 2010). The best recognition rate for 144 users (99.76% ± 0.01) is reached with linear kernel for biortogonal5.5 wavelet. This result is similar to 99.83% reached for 50 users. The third level of the low pass filter provides the suitable value of the texture to maximize the interclass relationship in order to get the large discriminative information. This is possible because it makes better use of the frequency-space resolution, and consequently the classifier is able to differentiate the users properly.

In Table 4, the EER for the best method (biortogonal wavelet family) is shown. The experiment is performed after applying 3 filters when the system works as verifier (experiment number 4). For a threshold of -0.5400, the EER is 0.60%. This result agrees with others proposed in the literature, although a higher accuracy system must be built with the combination of other techniques.

VERIFIER - 144 USERS			
Features	Threshold	Error average (percentage)	EER
Wavelet 3 filter levels (bior5.5)	-0.6120	0.70%	0.60%
	-0.5760	0.65%	
	-0.5400	0.60%	
	-0.5040	0.58%	
	-0.4680	0.59%	
	-0.5760	0.65%	
	-0.5400	0.66%	
	-0.5040	0.64%	
	-0.4680	0.65%	

Table 4. Wavelet verification results for 144 users using the *sloping* ROI.

In Figure 14 it is also shown how the FAR (False Acceptance Rate) and the FRR (False Rejecting Rate) change depending on the threshold.

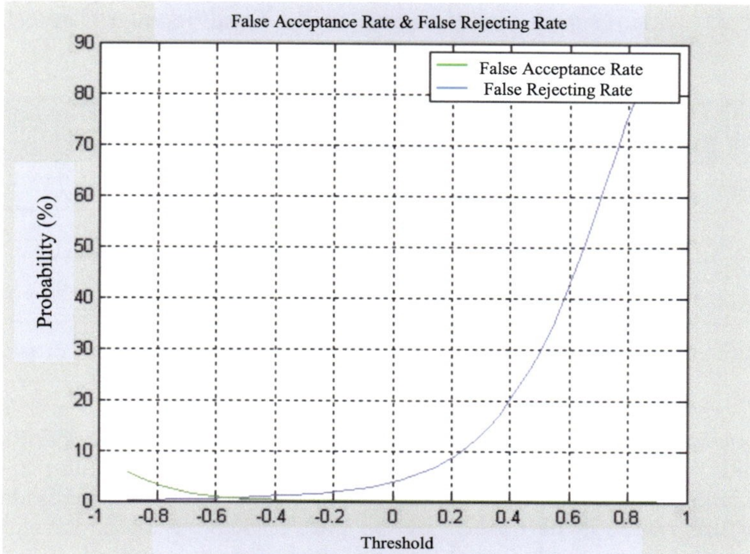

Fig. 14. FAR, FRR and EER for 144 database users.

As Figure 14 shows, the FAR and FRR curves decrease depending on threshold value. The point where both curves intersect other define the best accuracy of the verifier system. In next section we will discuss about these results and the system performance. It will be also introduced the future work and possible improvements in our work.

6. Comparison with other methods and discussion

In this section a brief comparison with wavelet experiments performed by other authors is detailed. Table 5 shows the recognition rate when the system works as identifier for several wavelet analysis which have been introduced in section 1.

Authors	Features	Users	Samples by user	Recognition rate
This work	Wavelet bior5.5 3 filter levels	144	10	99.76%
Xian-Qian et al., 2002	Wavelet 3 filter levels Haar	200	5	99%
Masood et al., 2008	Wavelet, biortogonal, symlet y meyer	50	10	97.12%
Liu et al., 2007	Wavelet (ISODATA)	80	180 (altogether)	95%

Table 5. Identification results published by other authors about wavelet transform.

A comparison with other texture-geometry techniques performed in the past by the authors of this work is also shown in Table 6 (Fuertes et al., 2010). In this case, a recognition rate of 99.73% was obtained with the Gabor algorithm and a 99.46% with the derivative method. In addition, a 99.90% was reached with the study of the hand geometry, especially with the finger widths.

Methods (Fuertes et al. 2010)	Linear recognition	RBF recognition
2D Gabor	99.73% ± 0.03	99.73% ± 0.03
Derivative Method	99.46% ± 0.09	99.46% ± 0.13
4 fingers(20 widths each finger)	99.38% ± 0.02	99.90% ± 0.01

Table 6. Identification results published by the authors of this work about other hand image techniques.

The results obtained in this work agree with others proposed in the literature. The best texture method, the biortogonal (bior5.5) wavelet, reaches a value similar to the obtained with the haar or db5 wavelet families. In Table 6, the results of the best methods of geometry and other texture methods performed in the past also agree with wavelet analysis. A combination of all these methods would improve the system recognition.

The conclusions arising from the experiments and its results will be explained in section 7, considering the future work and several improvements.

7. Conclusions and future work

In this paper, the performance of 2-D wavelet transform for hand palm biometric system has been presented with the intention of contributing to biometric field.

The best result of the identification system is obtained when the biortogonal5.5 wavelet is performed, with a recognition rate of 99.76% for the *sloping* ROI. These algorithms were applied to 50 and 144 users, and the recognition rate was similar independently of the number of users. It lets the system be useful for different number of people. In addition, the hand palm texture extraction algorithm is intuitive, simple and quick, with a computational load similar to geometrical parameter extraction.

It is crucial to emphasize the importance when the ROI is obtained. It has to be extracted after aligning the valleys of the hand: in this way, the area to study covers the main hand lines, minimizing the zones where the noise is higher. This issue has been proved with the experiments performed.

If the system works as verifier, the EER is 0.60%. There are other algorithms with best performance, but the possibility to combine it with other techniques makes wavelet algorithm suitable for verify people. We propose the combination of the wavelet method, proposed in this work, with other palm-print algorithms as derivative or Gabor analysis or with other hand recognition algorithms applying several fusion levels (Ferrer et al., 2007, Morales et al., 2008).

The proposed biometric feature is well adapted to the SVM classifier, which identify the feature degree of simplification necessary for the best performance. One drawback we have to mention is the necessity of operating with clean hands. Painted or dirty hands would cause an identification mistake. In this case a combined geometric/palm biometric feature would be more advisable.

Specifically for this research line, it will be interesting a deep study about the stability of the wavelet algorithm for many users. It is not the same to test a system with 100 or 1000 users in order to get a system regardless of number of users. If the number of users is long and unknown, wavelet analysis can be an excellent technique to recognize people.

The future of hand scanning appears to be static. Public acceptance of biometrics is growing slowly, until issues about privacy and security bring up to an acceptable level by majority of people. From a cost standpoint, hand image analysis is more expensive than fingerprint technology and just as effective. If it is used in conjunction with other techniques, passwords, and smart card and tokens then, it could prove to be a viable method. "Recent surveys suggest people are feeling more comfortable using biometric security, which could result in a $3 billion spending increase in biometrics over the next years" (Security world news, 2010).

One way to overcome this scepticism is to use new multimodal techniques (using more than one biometric system simultaneously to confirm identification), as vein impression or 3D facial recognition. Besides, a new method as tongue scanning identification is being tested at the Hong Kong Polytechnic University's Biometrics Research Centre. The tongue shapes of different people are different, and thus the tongue can be used to tell different subjects.

Future experiments about the use of free-contact systems and the combination of palmprint analysis with emergent techniques will be performed in order to contribute to scientific spreading.

8. Acknowledgements

This work has been partially supported by "Catedra Telefónica - ULPGC 2009/10" (Spanish Company), and partially supported by Spanish Government under funds from MCINN TEC2009-14123-C04-01. Authors want to thanks to Processing Digital Signal Division from Institute for Technological Development and Innovation in Communications (PDSD-IDETIC) for the work which all the Division has done on the building of the Palm-print database.

9. References

Adán, M.; Adán, A.; Vázquez, A.S. & Torres, R. (2008). Biometric verification/identification based on hands natural layout. *Image and Vision Computing*, Vol. 26, No. 4, pp. 451–465, April 2008.

Badrinath, G. & Gupta, P. (2009). Robust Biometric System Using Palmprint for Personal Verification. *Advances in Biometrics, Lecture Notes in Computer Science*, Vol. 5558, pp. 554-565, 2009.

Cortes, C. & Vapnik, V. (1995). Support Vector Networks. Machine Learning, Vol. 20, pp. 273-297, 1995.

Ferrer, M.A.; Morales, A.; Travieso, C.M. & Alonso, J.B. (2007). Low Cost Multimodal Biometric identification System Based on Hand Geometry, Palm and Finger Print Texture. *41st Annual IEEE International Carnahan Conference on Security Technology*, pp. 52-58, 2007.

Fuertes, J.J.; Travieso, C.M.; Ferrer, M.A. & Alonso, J.B. (2010). *Intra-modal biometric system using hand-geometry and palmprint texture. Security Technology (ICCST), 2010 IEEE International Carnahan Conference on,* pp. 318-322, October 2010.

Goh, M.K.; Connie, T.; Teoh, A.B.J. & Ngo, D.C.L. (2006). A Fast Palm Print Verification System. *Proceedings of the International Conference on Computer Graphics, Imaging and Visualization*, pp. 168-172, July 2006.

Gonzalez, R.C. & Wood, R.E. (2008). *Digital Image Processing*, 3rd ed., Prentice Hall, ISBN 9780-131687288 Upper Saddle River, NJ, 2008.

Guo, Z.; Zhang, D., Zhang, L. & Zuo, W. (2009). Palmprint verification using binary orientation co-occurrence vector. *Pattern Recognition Letters*, Vol. 30, No. 13, pp. 1219-1227, May 2009.

Liu, F.; Lin, C.X.; Cui, P.Y. & Dong, T. (2007). Palmprint recognition based on ISODATA clustering algorithm. *Proceedings of the 2007 International Conference on Wavelet Analysis and Pattern Recognition*, Vol.3, pp. 1129-1133, Beijing, China, 2-4 Nov. 2007.

Mallat, S.G. (1989). A Theory for Multiresolution Signal Decomposition: The Wavelet Representation, *IEEE Transactions on Pattern Analysis and Machine Intelligence*, Vol.11, No. 7, pp. 674-693, ISSN 0162-8828, 1989.

Masood, H. ; Mumtaz, M.; Butt, M.A.A. ; Mansoor, A.B. & Khan, S.A. (2008). Wavelet Based Palmprint Authentication System, *Proceedings of IEEE International Symposium of Biometric and Security Technologies*, 1-7, Islamabad, Pakistan, April 2008.

Mercer Schöfkopf, B.; Burges, C. & Smola, A. (1999). *Pairwise Classification and Support Vector Machines*, The MIT Press, Cambridge, Massachusetts, pp. 255-268, 1999.

Morales, A.;Ferrer, M.; Díaz, F.; Alonso, J. & Travieso, C. (2008). Contact-free hand biometric system for real environments. *Proceedings of the 16th European Signal Processing Conference (EUSIPCO)*, Laussane, Switzerland, September 2008.

Pavesic, N. ; Ribaric, S. & Ribaric, D. (2004). Personal Authentication Using Hand-Geometry and Palmprint Features, *Proceedings of the Workshop on Biometrics at ICPR04*, Cambridge, UK, 2004.

Security World News. (2010). *http://www.securityworldnews.com/2010/02/07/the-future-of-biometrics/* Last visit: 01-09-2011.

Steinwart, I. & Christmann, A. (2008). *Support Vector Machines*, Information Science and Statistics, Springer, ISBN 978-0-387-77241-7, New York, 2008.

Villegas, O.O.V. & Pinto R.E. (2006). *Procesamiento Digital de Imágenes en el Dominio Wavelet*, IEEE Looking Forward, Vol. 13, pp. 13-16, IEEE Computer Society, 2006.

Xian Qian W.; Zhang, D. & Wang, K. (2002). Wavelet based palmprint recognition. *Proceedings of the First International Conference on Machine Learning and Cybernetics*, Vol. 3, pp. 1253-1257, Beijing, 4-5 November 2002.

Yan, H. & Long, D. (2008). A novel bimodal identification approach based on hand-print. *Proceedings of the 2008 Congress on Image and Signal Processing*, Vol.4, 506-510, 2008.

Zhang, D.D. (2004). Palmprint Authentication, International Series on Biometrics, Vol.3, 256 p., 2004.

Zhang, L.; Guo, Z.; Wang, Z. & Zhang, D. (2007). Palmprint Verification using Complex Wavelet Transform. *IEEE International Conference on Image Processing*, Vol.2, pp. 417-420, 2007.

Zhang, D.; Kanhangad, V.; Luo, N. & Kumar, A. (2010). Robust palm print verification using 2D and 3D features. *Journal on Pattern Recognition*, Vol. 43, No. 1, pp. 358-368, January 2010.

7

Brain Computer Interface with Wavelets and Genetic Algorithms

Abdolreza Asadi Ghanbari[1], Mir Mohsen Pedram[2], Ali Ahmadi[3], Hamidreza Navidi[4], Ali Broumandnia[5] and Seyyed Reza Aleaghil[5]
[1]Young Researchers Club, Boroujerd Branch, Islamic Azad University, Boroujerd,
[2]Electrical Engineering Department, Tarbiat Moallem University, Tehran,
[3]Electrical and Computer Department, Khajeh Nasir Toosi University of Technology, Tehran,
[4]Applied Mathematics and Computer Sciences Department, Shahed University, Tehran,
[5]Islamic Azad University-South Tehran Branch, Tehran,
Iran

1. Introduction

For many years people have speculated that electroencephalographic activity or other electrophysiological measures of brain function might provide a new non-muscular channel for sending messages and commands to the external world - a brain-computer interface (BCI). BCI is a rapidly growing field of research combining neurophysiological insights, statistical signal analysis, and machine learning (Blankertz et al., 2007). The goal of BCI research is to build a communication channel from the brain to computers bypassing peripheral nerves and muscle activity. Among different techniques for the noninvasive measurement of the human brain, the electroencephalography (EEG) is commercially affordable and has excellent temporal resolution which enables real-time interaction through BCI.

Most popular and many scientific speculations about BCIs start from the 'mind-reading' or 'wire-tapping' analogy, the assumption that the goal is simply to listen in on brain activity as reflected in electrophysiological signals and thereby determine a person's wishes. This analogy ignores the essential and central fact of BCI development and operation. A BCI changes electrophysiological signals from mere reflections of central nervous system (CNS) activity into the intended products of that activity: messages and commands that act on the world. It changes a signal such as an EEG rhythm or a neuronal firing rate from a reflection of brain function into the end product of that function: an output that, like output in conventional neuromuscular channels, accomplishes the person's intent. A BCI replaces nerves and muscles and the movements they produce with electrophysiological signals and the hardware and software that translate those signals into actions.

As a replacement for the brain's normal neuromuscular output channels, a BCI also depends on feedback and on adaptation of brain activity based on that feedback. Thus, a BCI system must provide feedback and must interact in a productive fashion with the adaptations the brain makes in response to that feedback. This means that BCI operation depends on the interaction of two adaptive controllers: the user's brain, which produces the signals measured by the BCI; and the BCI itself, which translates these signals into specific commands.

Successful BCI operation requires that the user develop and maintain a new skill, a skill that consists not of proper muscle control but rather of proper control of specific electrophysiological signals; and it also requires that the BCI translate that control into output that accomplishes the user's intent. This requirement can be expected to remain even when the skill does not require initial training.

Present-day BCIs determine the intent of the user from a variety of different electrophysiological signals. These signals include slow cortical potentials, P300 potentials, and mu or beta rhythms recorded from the scalp, and cortical neuronal activity recorded by implanted electrodes. They are translated in real-time into commands that operate a computer display or other device. Successful operation requires that the user encode commands in these signals and that the BCI derive the commands from the signals. Thus, the user and the BCI system need to adapt to each other both initially and continually so as to ensure stable performance. Current BCIs have maximum information transfer rates up to 10–25 bits/min. This limited capacity can be valuable for people whose severe disabilities prevent them from using conventional augmentative communication methods.

At the same time, many possible applications of BCI technology, such as neuroprosthesis control, may require higher information transfer rates. Future progress will depend on: recognition that BCI research and development is an interdisciplinary problem, involving neurobiology, psychology, engineering, mathematics, and computer science; identification of those signals, whether evoked potentials, spontaneous rhythms, or neuronal firing rates, that users are best able to control independent of activity in conventional motor output pathways; development of training methods for helping users to gain and maintain that control; delineation of the best algorithms for translating these signals into device commands; attention to the identification and elimination of artifacts such as electromyographic and electro-oculographic activity; adoption of precise and objective procedures for evaluating BCI performance; recognition of the need for long-term as well as short-term assessment of BCI performance; identification of appropriate BCI applications and appropriate matching of applications and users; and attention to factors that affect user acceptance of augmentative technology, including ease of use, and provision of those communication and control capacities that are most important to the user.

Three major problems in this novel technology are identifies the brain signal features best suited for communication and artifacts that can occur during the signal acquisition.

Artifacts are undesired signals that can introduce significant changes in brain signals and ultimately affect the neurological phenomenon. Artifacts are attributed either to non-physiological sources (such as 50/60 Hz power-line noise, changes in electrode impedances, etc.) or physiological sources, such as potentials introduced by eye or body movements. Different methods for artifact removal are proposed in the literature. One of the most successful methods is Independent component analysis (ICA) (N. Xu et al., 2004). This method based on a common successful assumption in EEG research is that signals are generated by a linear mixing of independent sources in the brain and other external components and used for artifact removing of EEG signals (Hyvärinen et al., 2000). In this chapter, we use two artifacts removal methods: ICA and liner filtering.

The performance of a BCI, like that of other communication systems, depends on its signal-to-noise ratio. The goal is to recognize and execute the user's intent, and the signals are those aspects of the recorded electrophysiological activity that correlate with and thereby reveal that intent. The user's task is to maximize this correlation; and the system's first task is to measure the signal features accurately, i.e. to maximize the signal-to-noise ratio.

Feature extraction methods can greatly affect signal-to-noise ratio. Good methods enhance the signal and reduce central nervous system and non-CNS noise. This is most important and difficult when the noise is similar to the signal. For features extraction from the raw EEG data many methods such as time domain, frequency domain, and time-frequency domain are used. Since the EEG is non-stationary in general, it is most appropriate to use time-frequency domain methods like wavelet transform (WT) as a mean for feature extraction (Asadi Ghanbari et al., 2009). In this chapter, we used WT for feature extraction. The WT provides a more flexible way of time-frequency representation of a signal by allowing the use of variable sized windows.

Feature selection is one of the major tasks in classification problems. The main purpose of feature selection is to select a number of features used in the classification and at the same time to maintain acceptable classification accuracy. Various algorithms have been used for feature selection in the past decades. One of the best methods that can be used for features selection is GA (Te-Sheng et al., 2006). The GA plays the role of selector to select a subset of features that can best describe the classification. In this chapter, we employed this idea and used neural network classifier to compare the feature selection classification performance. The GA is a powerful feature selection tool, especially when the dimensions of the original feature set are large (Te-Sheng et al., 2006). Reducing the dimensions of the feature space not only reduces the computational complexity, but also increases estimated performance of the classifiers.

In this chapter, we show how we can convert EEG activity into cursor movement by a BCI using an appropriate feature extraction scheme. The proposed automated method for the classification of EEG activity is based on signal preprocessing, feature extraction and classification. The power spectrum, variance and mean of the Daubechies mother wavelet transform and Hilbert transform used for feature extraction. Finally, we implemented a feed-forward multi-layer perceptron (MLP) with a single hidden layer with five neurons, a probabilistic neural network (PNN).and support vector machine (SVM) classifier with Gaussian RBF kernel.

1.1 Definition and features of a BCI

1.1.1 Dependent and independent BCIs

A BCI is a communication system in which messages or commands that an individual sends to the external world do not pass through the brain's normal output pathways of peripheral nerves and muscles. For example, in an EEG-based BCI the messages are encoded in EEG activity. A BCI provides its user with an alternative method for acting on the world. BCIs fall into two classes: dependent and independent.

A dependent BCI does not use the brain's normal output pathways to carry the message, but activity in these pathways is needed to generate the brain activity (e.g. EEG) that does carry it. For example, one dependent BCI presents the user with a matrix of letters that flash one at a time, and the user selects a specific letter by looking directly at it so that the visual evoked potential (VEP) recorded from the scalp over visual cortex when that letter flashes is much larger that the VEPs produced when other letters flash (Sutter , 1992). In this case, the brain's output channel is EEG, but the generation of the EEG signal depends on gaze direction, and therefore on extraocular muscles and the cranial nerves that activate them. A dependent BCI is essentially an alternative method for detecting messages carried in the brain's normal output pathways: in the present example, gaze direction is detected by monitoring EEG rather than by monitoring eye position directly. While a dependent BCI does not give the brain a new communication channel that is independent of conventional channels, it can still be useful (Sutter , 1992).

In contrast, an independent BCI does not depend in any way on the brain's normal output pathways. The message is not carried by peripheral nerves and muscles, and, furthermore, activity in these pathways is not needed to generate the brain activity (e.g. EEG) that does carry the message. For example, one independent BCI presents the user with a matrix of letters that flash one at a time, and the user selects a specific letter by producing a P300 evoked potential when that letter flashes (Donchin et al., 2000). In this case, the brain's output channel is EEG, and the generation of the EEG signal depends mainly on the user's intent, not on the precise orientation of the eyes (Polich., 2000). The normal output pathways of peripheral nerves and muscles do not have an essential role in the operation of an independent BCI. Because independent BCIs provide the brain with wholly new output pathways, they are of greater theoretical interest than dependent BCIs. Furthermore, for people with the most severe neuromuscular disabilities, who may lack all normal output channels (including extraocular muscle control), independent BCIs are likely to be more useful.

1.1.2 BCI use is a skill

Most popular and many scientific speculations about BCIs start from the 'mind-reading' or 'wire-tapping' analogy, the assumption that the goal is simply to listen in on brain activity as reflected in electrophysiological signals and thereby determine a person's wishes. This analogy ignores the essential and central fact of BCI development and operation. A BCI changes electrophysiological signals from mere reflections of central nervous system (CNS) activity into the intended products of that activity: messages and commands that act on the world. It changes a signal such as an EEG rhythm or a neuronal firing rate from a reflection of brain function into the end product of that function: an output that, like output in conventional neuromuscular channels, accomplishes the person's intent. A BCI replaces nerves and muscles and the movements they produce with electrophysiological signals and the hardware and software that translate those signals into actions.

The brain's normal neuromuscular output channels depend for their successful operation on feedback. Both standard outputs such as speaking or walking and more specialized outputs such as singing or dancing require for their initial acquisition and subsequent maintenance continual adjustments based on oversight of intermediate and final outcomes (Ghez et al., 2000). When feedback is absent from the start, motor skills do not develop properly; and when feedback is lost later on, skills deteriorate.

As a replacement for the brain's normal neuromuscular output channels, a BCI also depends on feedback and on adaptation of brain activity based on that feedback. Thus, a BCI system must provide feedback and must interact in a productive fashion with the adaptations the brain makes in response to that feedback. This means that BCI operation depends on the interaction of two adaptive controllers: the user's brain, which produces the signals measured by the BCI; and the BCI itself, which translates these signals into specific commands.

Successful BCI operation requires that the user develop and maintain a new skill, a skill that consists not of proper muscle control but rather of proper control of specific electrophysiological signals; and it also requires that the BCI translate that control into output that accomplishes the user's intent. This requirement can be expected to remain even when the skill does not require initial training. In the independent BCI described above, the P300 generated in response to the desired letter occurs without training. Nevertheless, once this P300 is engaged as a communication channel, it is likely to undergo adaptive modification (Coles et al., 1995), and the recognition and productive engagement of this adaptation will be important for continued successful BCI operation.

That the brain's adaptive capacities extend to control of various electrophysiological signal features was initially suggested by studies exploring therapeutic applications of the EEG. They reported conditioning of the visual alpha rhythm, slow potentials, the mu rhythm, and other EEG features (Neidermeyer., 1999). These studies usually sought to produce an increase in the amplitude of a specific EEG feature. Because they had therapeutic goals, such as reduction in seizure frequency, they did not try to demonstrate rapid bidirectional control, that is, the ability to increase and decrease a specific feature quickly and accurately, which is important for communication. Nevertheless, they suggested that bidirectional control is possible, and thus justified and encouraged efforts to develop EEG-based communication. In addition, studies in monkeys showed that the firing rates of individual cortical neurons could be operantly conditioned, and thus suggested that cortical neuronal activity provides another option for non-muscular communication and control (Schmidt., 1980).

At the same time, these studies did not indicate to what extent the control that people or animals develop over these electrophysiological phenomena depends on activity in conventional neuromuscular output channels (Dewan., 1967). While studies indicated that conditioning of hippocampal activity did not require mediation by motor responses (Black., 1971), the issue was not resolved for other EEG features or for cortical neuronal activity.

1.1.3 The parts of a BCI

Like any communication or control system, a BCI has input (e.g. electrophysiological activity from the user), output (i.e. device commands), components that translate input into output, and a protocol that determines the onset, offset, and timing of operation. Fig. 1 shows these elements and their principal interactions.

1.1.3.1 Signal acquisition

In the BCIs discussed here, the input is EEG recorded from the scalp or the surface of the brain or neuronal activity recorded within the brain. Thus, in addition to the fundamental distinction between dependent and independent BCIs electrophysiological BCIs can be categorized by whether they use non-invasive (e.g. EEG) or invasive (e.g. intracortical) methodology. They can also be categorized by whether they use evoked or spontaneous inputs. Evoked inputs (e.g. EEG produced by flashing letters) result from stereotyped sensory stimulation provided by the BCI. Spontaneous inputs (e.g. EEG rhythms over sensorimotor cortex) do not depend for their generation on such stimulation. There is, presumably, no reason why a BCI could not combine non-invasive and invasive methods or evoked and spontaneous inputs. In the signal-acquisition part of BCI operation, the chosen input is acquired by the recording electrodes, amplified, and digitized.

1.1.3.2 Feature extraction

The digitized signals are then subjected to one or more of a variety of feature extraction procedures, such as spatial filtering, voltage amplitude measurements, spectral analyses, or single-neuron separation. This analysis extracts the signal features that (hopefully) encode the user's messages or commands. BCIs can use signal features that are in the time domain (e.g. evoked potential amplitudes or neuronal firing rates) or the frequency domain (e.g. mu or beta-rhythm amplitudes) (McFarland et al , 2000). A BCI could conceivably use both timedomain and frequency-domain signal features, and might thereby improve performance) (Schalk et al , 2000).

In general, the signal features used in present-day BCIs reflect identifiable brain events like the firing of a specific cortical neuron or the synchronized and rhythmic synaptic activation in sensorimotor cortex that produces a mu rhythm. Knowledge of these events can help guide BCI development. The location, size, and function of the cortical area generating a rhythm or an evoked potential can indicate how it should be recorded, how users might best learn to control its amplitude, and how to recognize and eliminate the effects of non-CNS artifacts. It is also possible for a BCI to use signal features, like sets of autoregressive parameters, that correlate with the user's intent but do not necessarily reflect specific brain events. In such cases, it is particularly important (and may be more difficult) to ensure that the chosen features are not contaminated by EMG, electrooculography (EOG), or other non-CNS artifacts.

1.1.3.3 The translation algorithm

The first part of signal processing simply extracts specific signal features. The next stage, the translation algorithm, translates these signal features into device commands orders that carry out the user's intent. This algorithm might use linear methods (e.g. classical statistical analyses (Jain et al, 2000) or nonlinear methods (e.g. neural networks). Whatever its nature, each algorithm changes independent variables (i.e. signal features) into dependent variables (i.e. device control commands).

Effective algorithms adapt to each user on 3 levels. First, when a new user first accesses the BCI the algorithm adapts to that user's signal features. If the signal feature is mu-rhythm amplitude, the algorithm adjusts to the user's range of mu-rhythm amplitudes; if the feature is P300 amplitude, it adjusts to the user's characteristic P300 amplitude; and if the feature is the firing rate of a single cortical neuron, it adjusts to the neuron's characteristic range of firing rates. A BCI that possesses only this first level of adaptation, i.e. that adjusts to the user initially and never again, will continue to be effective only if the user's performance is very stable. However, EEG and other electrophysiological signals typically display short- and long-term variations linked to time of day, hormonal levels, immediate environment, recent events, fatigue, illness, and other factors. Thus, effective BCIs need a second level of adaptation: periodic online adjustments to reduce the impact of such spontaneous variations. A good translation algorithm will adjust to these variations so as to match as closely as possible the user's current range of signal feature values to the available range of device command values.

While they are clearly important, neither of these first two levels of adaptation addresses the central fact of effective BCI operation: its dependence on the effective interaction of two adaptive controllers, the BCI and the user's brain. The third level of adaptation accommodates and engages the adaptive capacities of the brain. As discussed in Previous Sections, when an electrophysiological signal feature that is normally merely a reflection of brain function becomes the end product of that function, that is, when it becomes an output that carries the user's intent to the outside world, it engages the adaptive capacities of the brain. Like activity in the brain's conventional neuromuscular communication and control channels, BCI signal features will be affected by the device commands they are translated into: the results of BCI operation will affect future BCI input. In the most desirable (and hopefully typical) case, the brain will modify signal features so as to improve BCI operation. If, for example, the feature is mu-rhythm amplitude, the correlation between that amplitude and the user's intent will hopefully increase over time. An algorithm that incorporates the third level of adaptation could respond to this increase by rewarding the user with faster communication. It would thereby recognize and encourage the user's development of greater skill in this new form of communication. On the other hand, excessive or inappropriate adaptation could impair performance or discourage further skill development. Proper design of this third level of adaptation is likely to prove crucial for BCI development. Because this level involves the interaction of two adaptive controllers, the user's brain and the BCI system, its design is among the most difficult problems confronting BCI research.

1.1.3.4 The output device

For most current BCIs, the output device is a computer screen and the output is the selection of targets, letters, or icons presented on it (Wolpaw et al, 1991). Selection is indicated in various ways (e.g. the letter flashes). Some BCIs also provide additional, interim output, such as cursor movement toward the item prior to its selection (Pfurtscheller et al, 2000). In addition to being the intended product of BCI operation, this output is the feedback that the brain uses to maintain and improve the accuracy and speed of communication. Initial studies are also exploring BCI control of a neuroprosthesis or orthesis that provides hand closure to people with cervical spinal cord injuries (Lauer et al, 2000). In this prospective BCI application, the output device is the user's own hand.

1.1.3.5 The operating protocol

Each BCI has a protocol that guides its operation. This protocol defines how the system is turned on and off, whether communication is continuous or discontinuous, whether message transmission is triggered by the system (e.g. by the stimulus that evokes a P300) or by the user, the sequence and speed of interactions between user and system, and what feedback is provided to the user.

Most protocols used in BCI research are not completely suitable for BCI applications that serve the needs of people with disabilities. Most laboratory BCIs do not give the user on/off control: the investigator turns the system on and off. Because they need to measure communication speed and accuracy, laboratory BCIs usually tell their users what messages or commands to send. In real life the user picks the message. Such differences in protocol can complicate the transition from research to application.

2. Our BCI system: Brain computer interface with wavelets genetic algorithms

2.1 Materials and methods

In this research, EEG signal used as the basic data for classification. The EEG data is from an open EEG database of University of Tuebingen. Two types of the EEG database are employed as (BCI Competition, 2003).

2.1.1 Dataset I

The datasets were taken from a healthy subject. The subject was asked to move a cursor up and down on a computer screen, while his cortical potentials were taken. During the recording, the subject received visual feedback of his slow cortical potentials (Cz-Mastoids). Each trial lasted 6s. During every trial, the task was visually presented by a highlighted goal at either the top or bottom of the screen to indicate negativity or positivity from second 0.5 until the end of the trial. The visual feedback was presented from second 2 to second 5.5. Only this 3.5 second interval of every trial is provided for training and testing. The sampling rate of 256 Hz and the recording length of 3.5s results in 896 samples per channel for every trial. This dataset contain 266 trials that 70% of this dataset is considered as train dataset and the rest are considered as test.

2.1.2 Dataset II

The datasets were taken from an artificially respirated ALS patient. The subject was asked to move a cursor up and down on a computer screen, while his cortical potentials were taken. During the recording, the subject received auditory and visual feedback of his slow cortical potentials (Cz-Mastoids). Each trial lasted 8s. During every trial, the task was visually and auditorily presented by a highlighted goal at the top or bottom of the screen from second 0.5 until second 7.5 of every trial. In addition, the task ("up" or "down") was vocalised at second 0.5. The visual feedback was presented from second 2 to second 6.5. Only this 4.5 second interval of every trial is provided for training and testing. The sampling rate of 256 Hz and the recording length of 4.5s results in 1152 samples per channel for every trial. This dataset contain 200 trials that 70% of this dataset is considered as train dataset and the rest are considered as test.

2.1.3 Proposed methods

The block diagram of the proposed method for EEG signal classification is depicted in Fig.1. The method is divided into five steps: (1) EEG acquisition and sampling, (2) EEG preprocessing, (3) calculation of feature vector, (4) feature selection, (5) classification (Goncharova et al, 2003; Hyvärinen & Oja, 2000; McFarland et al, 2005).

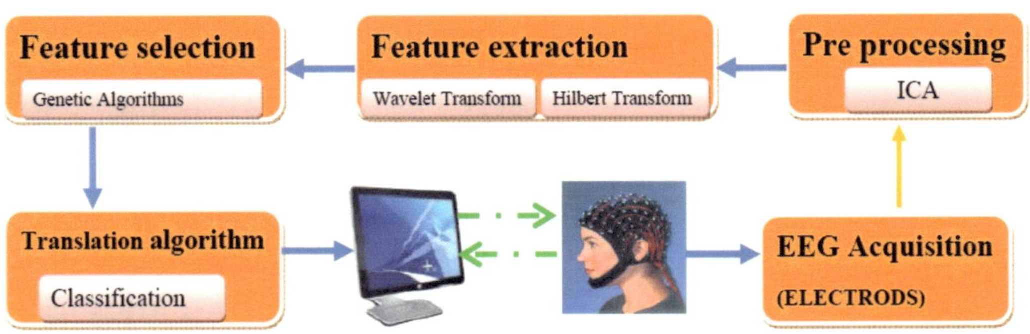

Fig. 1. Block diagram of the proposed method for EEG signal classification.

3. Artifacts in BCI systems

Artifacts are undesirable potentials that contaminate brain signals, and are mostly of non-cerebral origin. Unfortunately, they can modify the shape of a neurological phenomenon used to drive a BCI system. Thus, even cerebral potentials may sometimes be considered as artifacts. For example, in an MRP-based BCI system, a visual evoked potential (VEP) is considered as an artifact. Visual alpha rhythms can also appear as artifacts in a Mu-based BCI system (Goncharova et al, 2003). One problem with such artifacts is that they could mistakenly result in controlling the device (McFarland et al, 2005). Therefore, there is a need to avoid, reject or remove artifacts from recordings of brain signals.

Artifacts originate from non-physiological as well as physiological sources. Non-physiological artifacts originate from outside the human body (such as 50/60 Hz power-line noise or changes in electrode impedances), and are usually avoided by proper filtering, shielding, etc.

Physiological artifacts arise from a variety of bodily activities. Electrocardiography (ECG) artifacts are caused by heart beats and may introduce a rhythmic activity into the EEG signal. Respiration can also cause artifacts by introducing a rhythmic activity that is synchronized with the body's respiratory movements. Skin responses such as sweating may alter the impedance of electrodes and cause artifacts in the EEG signals.

Physiological artifacts such as ocular (EOG) and muscle (EMG) artifacts are much more challenging to handle than non-physiological ones. Moreover, controlling them during signal acquisition is not easy. There are different ways of handling these types of artifacts in BCI systems. In next Section, we examine the methods for handling Physiological artifacts in BCI systems.

3.1 Methods of handling artifacts

In this section, we briefly address methods of handling artifacts. Our focus throughout this section will be on Artifact removal in BCI systems.

3.1.1 Artifact avoidance

The first step in handling artifacts is to avoid their occurrence by issuing proper instructions to users. For example, users are instructed to avoid blinking or moving their body during the experiments. Instructing users to avoid generating artifacts during data collection has the advantage of being the least computationally demanding among the artifact handling methods, since it is assumed that no artifact is present in the signal (or that the presence of artifacts is minimal). However, it has several drawbacks. First, since many physiological signals, such as the heart beats, are involuntary, artifacts will always be present in brain signals. Even in the case of EOG and EMG activities, it is not easy to control eye and body movements during data recording. Second, the occurrence of ocular and muscle activity during an online operation of any BCI system is unavoidable. Third, the collection of a sufficient amount of data without artifacts may be difficult, especially in cases where a subject has a neurological disability. Finally, avoiding artifacts may introduce an additional cognitive task for the subject. For example, it has been shown that refraining from eye blinking results in changes in the amplitude of some evoked.

3.1.2 Artifact rejection

Artifact rejection refers to the process of rejecting the trials affected by artifacts. It is perhaps the simplest way of dealing with brain signals contaminated with artifacts. It has some important advantages over the artifact avoidance approach. For example, it would be easier for users to participate in the experiments and perform the required tasks, especially those individuals with motor disabilities. Also, the "secondary" cognitive task, resulting from a subject trying to avoid generating a particular artifact, will not be present in the EEG signal. Artifact rejection is usually done by visually inspecting the EEG or the artifact signals, or by using an automatic detection method.

3.1.3 Artifact removal

Artifact removal is the process of identifying and removing artifacts from brain signals. An artifact-removal method should be able to remove the artifacts as well as keeping the related neurological phenomenon intact. In this paper we use artifact removal methods that contain ICA and liner filtering.

3.1.3.1 Linear filtering

Linear filtering is useful for removing artifacts located in certain frequency bands that do not overlap with those of the neurological phenomena of interest (Barlow, 1984). For example, low-pass filtering can be used to remove EMG artifacts and high-pass filtering can be used to remove EOG artifacts. Linear filtering was commonly used in early clinical studies to remove artifacts in EEG signals (Zhou, 2005).

The advantage of using filtering is its simplicity. Also the information from the EOG signal is not needed to remove the artifacts. This method, however, fails when the neurological phenomenon of interest and the EMG or EOG artifacts overlap or lie in the same frequency band (de Beer et al, 1995). A look at the frequency range of neurological phenomena used in BCI systems unfortunately shows that this is usually the case. As a result, a simple filtering approach cannot remove EMG or EOG artifacts without removing a portion of the neurological phenomenon. More specifically, since EOG artifacts generally consist of low frequency components, using a high-pass filter will remove most of the artifacts. Such methods are successful to some extent in BCI systems that use features extracted from high-frequency components of the EEG (e.g., Mu and Beta rhythms). However, for BCI systems that depend on low frequency neurological phenomena (such as MRPs), this methods are not as desirable, since these neurological phenomena may lie in the same frequency range as that of the EOG artifacts.

In the case of removing EMG artifacts from EEG signals, filtering specific frequency bands of the EEG can be used to reduce the EMG activity. Since artifacts generated by EMG activity generally consist of high-frequency components, using a low-pass filter may remove most of these artifacts. Again, such methods may be successful to some extent for BCI systems that rely on low-frequency components (e.g., MRPs), but they cannot be effective for BCI systems that use a neurological phenomenon with high-frequency content (such as Beta rhythms).

3.1.3.2 Blind source separation (BSS)

BSS techniques separate the EEG signals into components that "build" them. They identify the components that are attributed to artifacts and reconstruct the EEG signal without these components (Choi et al, 2005). Among the BSS methods, Independent Component Analysis (ICA) is more widely used. ICA is a method that blindly separates mixtures of independent source signals, forcing the components to be independent. It has been widely applied to remove ocular artifacts from EEG signals (Jung et al, 2001). Preliminary studies have shown that ICA increases the strength of motor-related signal components in the Mu rhythms, and is thus useful for removing artifacts in BCI systems (Makeig et al, 2000).

One advantage of using BSS methods such as ICA is that they do not rely on the availability of reference artifacts for separating the artifacts from the EOG signals (Zhou, 2005). One disadvantage of ICA, along with other BSS techniques, is that they usually need prior visual inspection to identify artifact components (Jung et al, 2001). However, some automatic methods have been proposed (Joyce et al, 2004).

3.1.3.3 Independent component analysis

ICA was originally developed for blind source separation whose goal is to recover mutually independent but unknown source signals from their linear mixtures without knowing the mixing coefficients.

ICA is a computational technique for revealing hidden factors that underlie sets of measurements or signals. ICA assumes a statistical model whereby the observed multivariate data, typically given as a large database of samples, are assumed to be linear or nonlinear mixtures of some unknown latent variables. The mixing coefficients are also unknown. The latent variables are nongaussian and mutually independent and they are called the independent components of the observed data. By ICA, these independent components, also called sources or factors, can be found. Thus ICA can be seen as an extension to Principal Component Analysis and Factor Analysis. ICA is a much richer technique, however, capable of finding the sources when these classical methods fail completely.

In this chapter, we use a basic form of the FastICA algorithm is as follows (Hyvärinen & Oja, 2000):

$$\begin{aligned}
&1.\ \textit{Choose an initial (e.g random) weight vector } w.\\
&2.\ \textit{let } w^{+} = E\left\{xg\left(w^{T}x\right)\right\} - \left\{g'\left(w^{T}x\right)\right\}w\\
&3.\ w = w^{+}/\left\|w^{+}\right\|\\
&4.\ \textit{If not converged, go back to } 2.
\end{aligned} \tag{1}$$

Where $g = u\exp\left(-u^{2}/2\right)$, x observed data and w is a weight matrix that does ICA. Note that convergence means that the old and new values of w point in the same direction, i.e. their dot-product are (almost) equal to 1.

4. Feature extraction

For features extraction from the raw EEG data many methods such as time domain, frequency domain, and time-frequency domain are used. In this article we used Hilbert and Wavelet Transform for feature extraction.

4.1 Wavelet transforms

4.1.1 Wavelets

Wavelets are a recently developed signal processing tool enabling the analysis on several timescales of the local properties of complex signals that can present non-stationary zones. They lead to a huge number of applications in various fields, such as, for example, geophysics, astrophysics, telecommunications, imagery and video coding. They are the foundation for new techniques of signal analysis and synthesis and find beautiful applications to general problems such as compression and denoising.

The propagation of wavelets in the scientific community, academic as well as industrial, is surprising. First of all, it is linked to their capacity to constitute a tool adapted to a very broad spectrum of theoretical as well as practical questions. Let us try to make an analogy: the emergence of wavelets could become as important as that of Fourier analysis. A second element has to be noted: wavelets have benefited from an undoubtedly unprecedented trend in the history of applied mathematics. Indeed, very soon after the grounds of the mathematical theory had been laid in the middle of the 1980s (Meyer, 1990), the fast algorithm and the connection with signal processing (Mallat, 1989) appeared at the same time as Daubechies orthogonal wavelets (Daubechies, 1988). This body of knowledge, diffused through the Internet and relayed by the dynamism of the research community enabled a fast development in numerous applied mathematics domains, but also in vast fields of application.

Thus, in less than 20 years, wavelets have essentially been imposed as a fruitful mathematical theory and a tool for signal and image processing. They now therefore form part of the curriculum of many pure and applied mathematics courses, in universities as well as in engineering schools.

4.1.2 Wavelet transforms

For features extraction from the raw EEG data many methods such as time domain, frequency domain, and time-frequency domain are used. Since the EEG is non-stationary in general, it is most appropriate to use time-frequency domain methods like wavelet transform (WT) as a mean for feature extraction (Nazari et al, 2009). The WT provides a more flexible way of time-frequency representation of a signal by allowing the use of variable sized windows. In WT long time windows are used to get a finer low-frequency resolution and short time windows are used to get high-frequency information. Thus, WT gives precise frequency information at low frequencies and precise time information at high frequencies. This makes the WT suitable for the analysis of irregular data patterns, such as impulses occurring at various time instances. The EEG recordings were decomposed into various frequency bands through fourth-level wavelet packet decomposition (WPD). The decomposition filters are usually constructed from the Daubechies or other sharp mother wavelets, when the data has discontinuities. In this research, based on the analysis of the data, Daubechies mother wavelet was used in the decomposition. The power spectrum, variance and mean of the signal (each channel) are extracted as features. So the feature set for each subject in each trial consisted of 3*number of channels. As a result, the feature matrix was 266*18 and 200*21 for subject A and B respectively. Finally the feature matrix is normalized.

4.2 Hilbert transforms

Hilbert transforms are essential in understanding many modern modulation methods. These transforms effectively phase shift a function by 90 degrees independent of frequency. Of course practical implementations have limitations (Huang et al, 1998). For example, the phase shifting of a low frequency implies a long delay, which in turn implies a computational process that maintains a long history of the signal. Hilbert transforms are useful in creating signals with one sided Fourier transforms. Also the concepts of analytic functions and analytic signals will be shown to be related through Hilbert transforms.

The Hilbert transform is a convenient tool to use in dealing with band-pass signals (Huang, 2005). The Hilbert transform of a signal $x(t)$ will be denoted by $\hat{x}(t)$ and is obtained by passing $x(t)$ through a filter with the transfer function:

$$H(\omega) = -jsign\,\omega = \begin{cases} -j & for \quad \omega > 0 \\ 0 & for \quad \omega = 0 \\ j & for \quad \omega < 0 \end{cases} \tag{2}$$

This is illustrated in "Fig. 2". The Hilbert transform filter is an ideal 90° phase shifter. In the frequency domain:

$$\hat{X}(\omega) = H(\omega)X(\omega) = -jsign(\omega)X(\omega) \tag{3}$$

It can be shown that the inverse Fourier transform of $H(\omega)$ is $h(t) = \dfrac{1}{\pi t}$ Therefore, a signal and its Hilbert transform are related by the convolution integral:

$$\hat{X} = x(t) * \frac{1}{\pi t} = \frac{1}{\pi}\int_{-\infty}^{\infty} \frac{x(\tau)}{t-\tau} d\tau \tag{4}$$

Where * represents convolution.

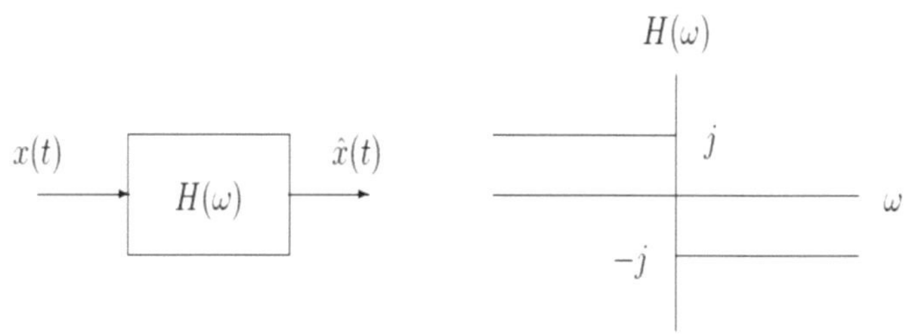

Fig. 2. System for Forming Hilbert Transforms.

5. Feature selection

Feature selection is one of the major tasks in classification problems. The main purpose of feature selection is to select a number of features used in the classification and at the same time to maintain acceptable classification accuracy. Besides deciding which types of features to use, the weighting of features also plays an important role in classification. Emphasizing features that have better discriminative power will usually boost classification. Feature selection can be seen as a special case of feature weighting, in which features that are eliminated are assigned zero weight. Feature selection reduces the dimensionality of the feature space, which leads to a reduction in computational complexity. Furthermore, in some cases, classification can be more accurate in the reduced space. Various algorithms have been used for feature selection in the past decades. One of the best methods that can be used for features selection is Genetic Algorithms (Sheng, 2006).

5.1 Genetic algorithms

Genetic Algorithms are adaptive heuristic search algorithm premised on the evolutionary ideas of natural selection and genetic (Andries, 2007). The basic concept of Gas is designed to simulate processes in natural system necessary for evolution. The main operator of GA to search in pool of possible solutions is Crossover, Mutation and selection.

The genetic search process is iterative: evaluating, selection and recombining string in the population during each one of iterations (generation) until reaching some termination condition. Evaluation of each string is based on a fitness function that is problem-dependent. It determines which of the candidate solutions are better. This corresponds to the environmental determination of survivability in national selection. Selection of a string, which represents a point in the search space, depends on the string's fitness relative to those of other strings in the population, those points that have relatively low fitness.

Mutation, as in natural systems, is a very low probability operator and just flips bit. The aim of mutation is to introduce new genetic material into an existing individual; that is, to add diversity to the genetic characteristics of the population. Mutation is used in support of crossover to ensure that the full range of allele is accessible for each gene.

Crossover in contrast is applied with high probability. It is a randomized yet structured operator that allows information exchange between points. Its goal is to preserve the fittest individual without introducing any new value.

The proposed approach to the use of Gas for Feature selection involves encoding a set of d, Feature s as a binary string of d elements, in which a 0 in the string indicates that the corresponding Feature is to be omitted, and a 1 that it is to be included. This coding scheme represents the presence or absence of a particular Feature from the Feature space (see Fig. 3). The length of chromosome equal to Feature space dimensions.

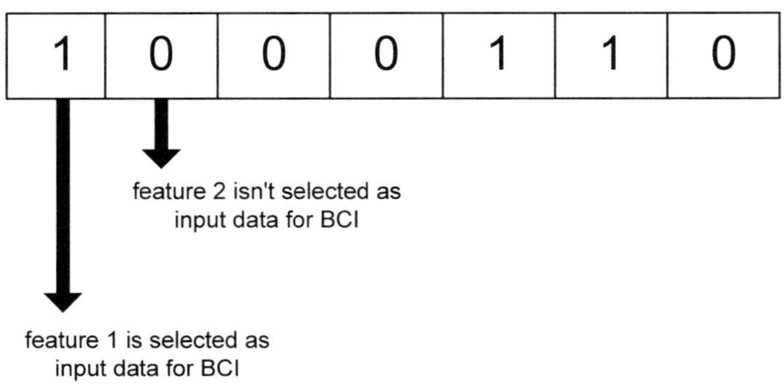

Fig. 3. Schema of the proposed GA-based feature selection approach

6. Translation algorithm

6.1 Neural Networks

An artificial neural network (ANN) is an interconnected group of artificial neurons simulating the thinking process of human brain. One can consider an ANN as a "magical" black box trained to achieve expected intelligent process, against the input and output information stream. ANN are useful in application areas such as pattern recognition, classification etc.

6.1.1 Multilayered Perceptron Neural Networks

The decision making process of the ANN is holistic, based on the features of input patterns, and is suitable for classification of biomedical data. Typically, multilayer feed forward neural networks can be trained as non-linear classifiers using the generalized back-propagation (BP) algorithm.

Our network has one hidden layer with five neurons and output layer with one neuron. Generalized BP algorithm with momentum used as training procedure. Momentum is a standard training technique which is used to speed up convergence and maintain generalization performance (Hagan, 1995). For hidden and output layers, we used bipolar and unipolar sigmoid functions respectively as decision function on the other hand we normalized weights and inputs. With these methods we achieved a NN classifier that is the most suitable classifier for the task at hand. We determined the most effective set as well as the optimum vector length for high accuracy classification. This NN classifier was trained and tested by using the feature sets described above.

By means of minimizing error optimized the number of neurons in hidden layer to five with tansig functions and sigmoid function for output layer.

6.1.2 Probabilistic Neural Network

The probabilistic approach to neural networks has been developed in the framework of statistical pattern recognition. Probabilistic neural network (PNN) is derived from radial basis function (RBF) network which is an ANN using RBF. RBF is a bell shape function that scales the variable nonlinearly. PNN is adopted for it has many advantages (Kim & Chang, 2007). Its training speed is many times faster than a BP network. PNN can approach a Bayes optimal result under certain easily met conditions. Additionally, it is robust to noise examples. We choose it also for its simple structure and training manner. The most important advantage of PNN is that training is easy and instantaneous. Weights are not "trained" but assigned. Existing weights will never be alternated but only new vectors are inserted into weight matrices when training. So it can be used in real-time. Since the training and running procedure can be implemented by matrix manipulation, the speed of PNN is very fast.

6.2 Support Vector Machine

The SVM is a relatively new classification technique developed by Vapnik (Avidan, 2004) which has shown to perform strongly in a number of real-world problems, including BCI.

The invention of SVM was driven by underlying statistical learning theory, i.e., following the principle of structural risk minimization that is rooted in VC dimension theory, which makes its derivation even more profound. The SVMs have been a topic of extensive research with wide applications in machine learning and engineering. The output of a binary SVM classifier can be computed by the following expression:

$$y = \operatorname{sgn}\left(\sum_{i=1}^{N} \alpha_i y_i k(x_i, x) + b\right) \tag{5}$$

where $\{x_i, y_i\}_{i=1}^{N}$ are training samples with input vectors $x_i \in R^d$, and class labels $y_i = \{-1, 1\}, \alpha_i \geq 0$, are Lagrangian multipliers obtained by solving a quadratic optimization problem, b is the bias, and $k(x_i, x_j)$ is called kernel function in SVM. The most commonly used kernel function is the Gaussian RBF as:

$$k(x_i, x_j) = \exp\left(\frac{-\|x_i - x_j\|^2}{2\sigma^2}\right) \tag{6}$$

The SVM for the linearly separable case find optimal separating hyper plane, as shown in "Fig. 4" (Chandaka et al, 2008).

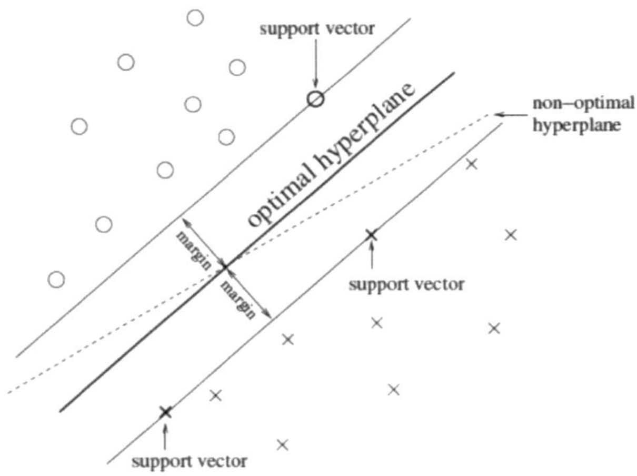

Fig. 4. SVM classification with a hyperplane that maximizes the separating margin between the two classes (indicated by data points marked by "X"s and "O"s). Support vectors are elements of the training set that lie on the boundary hyperplanes of the two classes.

7. Simulation results

In this paper, we proposed a scheme to combine liner filtering, Genetic Algorithm and neural network classifiers for EEG signal classification. Liner filtering and independent component analysis is used to artifact removal from EEG signals. The GA select essential EEG features then selected features serve as input feature vector for the following classifiers. Two neural networks, including probabilistic neural network (PNN), Multilayered

Perceptron (MLP) and support vector machine (SVM) were employed in the study and their effects were compared. In neural network structure, the output layer unit has sigmoid function, which makes network capable of nonlinearly mapping and capturing dynamics of signals. In SVM classifier different values for σ which is a very essential parameter in designing a SVM classifier with Gaussian RBF kernel examined and the best one selected.

To classify cursor movements two types of the EEG database are used, 70% of each dataset used for training and the rest for test classifiers. Both neural network classifiers and SVM demonstrated high classification accuracies with relatively small number of features. Between the three classifiers, SVM shows slightly better performance than MLP and PNN in terms of classification accuracy and robustness to different number of features. The results prove that the proposed scheme a promising model for the discrimination of clinical EEG signals. The performance of a classifier is not just measured as the accuracy achieved by the network, but aspects such as computational complexity and convergence characteristics are just as important. To reduce complexity, the GA used to select essential EEG features. This approach to BCI helps to reduce the computational complexity of the Classification process, and helps to improve transfer rate in real-time BCI systems.

Generally, the classification accuracy over files, which were included in training, is higher than the accuracy for the testing set. Tables I and II indicate the results of classification accuracy during training and test stages for both datasets. In comparison with the neural network classifier, SVM has a better training and test accuracy rate of neural network classifier, because of the nature of SVM classifier, this classifier is more general than neural network and this specification is very important in the use of classifiers. The most important advantage of PNN is that training is easy and instantaneous in comparison with SVM and MLP classifiers.

Features	**Wavelet transform**		**Hilbert transform**	
classifier	Training	Test	Training	**Test**
MLP	99.56%	86.75%	99.76%	**88.75%**
PNN	99.98%	88.75%	99.97%	**89.75%**
SVM	99.95%	90.25%	99.98%	**92.25%**

Table 1. Results of the dataset type I

Features	**Wavelet transform**		**Hilbert transform**	
classifier	Training	Test	Training	**Test**
MLP	99.46%	87.75%	99.56%	**88.25%**
PNN	99.95%	88.55%	99.97%	**88.75%**
SVM	99.92%	90.05%	99.95%	**91.25%**

Table 2. Results of the Dataset type II

8. Conclusion

Wavelets are a recently developed signal processing tool enabling the analysis on several timescales of the local properties of complex signals that can present non-stationary zones. They lead to a huge number of applications in various fields, such as, geophysics, astrophysics, telecommunications, imagery and video coding. They are the foundation for new techniques of signal analysis and synthesis and find beautiful applications to general problems such as compression and denoising. In order to extract the most suitable features from the raw EEG data different methods in time or frequency domain can be used. Since the EEG is non-stationary in general, it is most appropriate to use time-frequency domain methods like wavelet transform (WT) as a mean for feature extraction. The simulation results confirm this fact. The Genetic algorithm is applied in order to choose the best features from the feature space. GA is an evolutionary algorithm which its optimality has been proved in other fields, the computation complexity is low and it is an appropriate method in real time problems.

9. Acknowledgment

This study was supported by Islamic Azad University, Borujerd Branch, Iran. The authors would like to acknowledge staff of university.

10. References

A. Asadi Ghanbari, M. R. Nazari Kousarrizi, M. Teshnehlab, and M. Aliyari, "Wavelet and Hilbert Transform-based Brain Computer Interface", *IEEE International Conference on advances tools for engineering application*. Notre dame university- Lebanon, 2009.

Andries P. Engelblrecht, "Computational Intelligence An Introduction Second Edition," *John Wiley & Sons*, Ltd. 2007.

Barlow JS. EMG artifact minimization during clinical EEG recordings by special analog filtering. *Electroencephalogr Clin Neurophysiol*, 1984; 58:161–74.

BCI Competition 2003. http://ida.first.fraunhofer.de/projects/bci/competition.

Black AH. The direct control of neural processes by reward and punishment. *Am Sci* 1971; 59:236–245.

Blankertz B, Dornhege G, Krauledat M, Muller KR, Curio G. The non-invasive Berlin Brain–Computer interface: fast acquisition of effective performance in untrained subjects. *NeuroImage* 2007;37:539–50.

Choi S, Cichocki A, Park HM, Lee SY. Blind source separation and independent component analysis: *a review. Neural Inf Process-Lett Rev*, 2005; 6:1–57.

Coles MGH, Rugg MD. Event-related potentials: an introduction. In: Rugg MD, Coles MGH, editors. *Electrophysiology of the mind: event-related brain potentials and cognition*, New York, NY: Oxford University Press, 1995.

D. K. Kim and S. K. Chang, "Advanced Probabilistic Neural Network for the prediction of concrete Strength", *ICCES*, vol. 2, pp. 29-34, 2007.

Daubechies I., "Orthonormal basis of compactly supported wavelets", Comm. *Pure Appl. Math.*, vol. XLI, p. 909-996, 1988.

De Beer NA, van de Velde M, Cluitmans PJ. Clinical evaluation of a method for automatic detection and removal of artifacts in auditory evoked potential monitoring. *J Clin Monit* 1995;11:381–91.

Dewan AJ. Occipital alpha rhythm, eye position and lens accommodation. *Nature* 1967; 214:975–977.

Donchin E, Spencer KM, Wijesinghe R. The mental prosthesis: assessing the speed of a P300-based brain–computer interface. *IEEE Trans Rehabil Eng* 2000;8:174–179.

Ghez C, Krakauer J. Voluntary movement. In: Kandel ER, Schwartz JH, Jessell TM, editors. *Principles of neural science, 4th ed.* New York, NY: McGraw-Hill, 2000. pp. 653–674.

Goncharova II, McFarland DJ, Vaughan TM, Wolpaw JR. "EMG contamination of EEG: spectral and topographical characteristics," *Clin Neurophysiol*, 2003.

Hyvärinen, A. and Oja, E, "Independent Component Analysis: Algorithms and Applications," *Neural Computation*, vol 9, 1483-1492, 2000.

Jain AK, Duin PW, Mao J. Statistical pattern recognition: a review. *IEEE Trans Pattern Anal Machine Intell* 2000;22:4–37.

Joyce CA, Gorodnitsky IF, Kutas M. Automatic removal of eye movement and blink artifacts from EEG data using blind component separation. *Psychophysiology*, 2004;41:313–25.

Jung TP, Makeig S, Westerfield M, Townsend J, Courchesne E, Sejnowski TJ. Analysis and visualization of single-trial event-related potentials. *Hum Brain Mapp* 2001;14:166–85.

Lauer RT, Peckham PH, Kilgore KL, Heetderks WJ. Applications of cortical signals to neuroprosthetic control: a critical review. *IEEE Trans Rehabil Eng* 2000;8:205–208.

M. Hagan, H. Demuth, and M. Beale, "Neural Network Design," *PWS Publishing Company*, 1995.

M. R. Nazari Kousarrizi, A. Asadi Ghanbari, M. Teshnehlab, M. Aliyari, A. Gharaviri, "Feature Extraction and Classification of EEG Signals using Wavelet Transform, SVM and Artificial Neural Networks for Brain Computer Interfaces," *IEEE International Joint Conferences on Bioinformatics, Systems Biology and Intelligent Computing*, 2009.

Makeig S, Enghoff S, Jung TP, Sejnowski TJ. A natural basis for efficient brain-actuated control. *IEEE Trans Rehabil Eng*, 2000;8:208–11.

Mallat S., "A theory for multiresolution signal decomposition: the wavelet representation", *IEEE Trans.* on PAMI, vol. 2, no. 7, p. 674-693, 1989.

McFarland DJ, Sarnacki WA, Vaughan TM, Wolpaw JR, "Brain- computer interface (BCI) operation: signal and noise during early training sessions," *Clin Neurophysiol*; vol. 116, pp. 56–62, 2005.

McFarland DJ, Sarnacki WA, Vaughan TM, Wolpaw JR. EEG-based brain-computer interface communication effect of target number and trial length on information transfer rate. *Soc Neurosci Abstr* 2000;26:1228.

Meyer Y., "Ondelettes et opérateurs", vol. 1, Hermann, *Actualités* mathématiques, 1990.

N. E. Huang, Z. Shen, S. R. Long, M. L. Wu, H. H. Shih, Q. Zheng, N. C. Yen and H. H. Liu, "The empirical mode decomposition and Hilbert spectrum for nonlinear and nonstationary time series analysis," *Proc. Roy. Soc. London*, vol. 454, pp. 903-995, 1998.

N. Huang, "The Hilbert–Huang Transform in Engineering,", *CRC Press*, Boca Raton, FL, 2005.

N. Xu, X. Gao, B. Hong, X. Miao, S. Gao, and F.Yang, BCI Competition 2003—Data set IIb: Enhancing P300 wave detection using ICA-based subspace projections for BCI applications, *in IEEE Trans. Biomed. Eng.*, vol. 51, pp. 1067–1072, June 2004.

Neidermeyer E. The normal EEG of the waking adult. In: Niedermeyer E, Lopes da Silva FH, editors. Electroencephalography: basic principles, clinical applications and related fields, 4th ed. Baltimore, MD: *Williams and Wilkins*, 1999. pp. 149–173.

Pfurtscheller G, Neuper N, Guger C, Harkam W, Ramoser H, Schlo¨gl A, Obermaier B, Pregenzer M. Current trends in Graz Brain-Computer Interface (BCI) research. *IEEE Trans Rehabil Eng* 2000;8:216–219.

Polich J. P300 in clinical applications. In: Niedermeyer E, Lopes da Silva FH, editors. *Electroencephalography: basic principles, clinical applications and related fields*, 4th ed. Baltimore, MD: Williams and Wilkins, 1999. pp. 1073–1091.

S. Avidan, "Support Vector Tracking," *IEEE Trans. On Pattern Analysis and Machine Intelligence*, vol. 26, no. 8, pp.1064-1072, Aug. 2004.

S. Chandaka, A. Chatterjee, S. Munshi, "Cross-correlation aided support vector machine classifier for classification of EEG signals," *Expert Systems with Applications*, 2008.

Schalk G, Wolpaw JR, McFarland DJ, Pfurtscheller G. EEG-based communication and control: presence of error potentials. *Clin Neurophysiol* 2000;111:2138–2144.

Schmidt EM. Single neuron recording from motor cortex as a possible source of signals for control of external devices. *Ann Biomed Eng* 1980;8:339–349.

Sutter EE. The brain response interface: communication through visuallyinduced electrical brain responses. *J Microcomput Appl* 1992; 15:31–45.

Te-Sheng Li, " Feature Selection For Classificatin By Using a GA-Based Neural Network Approach", *Journal of the Chinese Institute of Industrial Engineers*, Vol. 23, No. 1, pp. 55-64, 2006.

Wolpaw JR, McFarland DJ, Neat GW, Forneris CA. An EEG-based brain- computer interface for cursor control. *Electroenceph clin Neurophysiol* 1991;78:252–259.

Zhou W, Gotman J. Removing eye-movement artifacts from the eeg during the intracarotid amobarbital procedure. *Epilepsia* 2005;46:409–14.

8

Improvement of Shimmer Parameter of Oesophageal Voices Using Wavelet Transform

Ibon Ruiz and Begoña García Zapirain
Deusto Institute of Technology, Deustotech-LIFE Unit, University of Deusto, Bilbao Spain

1. Introduction

This chapter presents an oesophageal speech enhancement algorithm. Such an exceptionally special type of voice is due to the laryngectomy undergone by those persons with larynx cancer. An oesophageal voice has extremely low intelligibility. The parameter values characterising the voice go beyond normal levels. This chapter proposes a method to improve its quality, which consists in improving Shimmer parameter using Wavelet transform and stabilizing the transfer function poles of the vocal tract model so as to improve a signal's formants. With this aim, the joint use of two techniques has been applied: on the one hand, Digital Wavelet Transform technique to normalise Shimmer and, on the other hand, an algorithm that transforms the modulus and phase of vocal tract's poles technique. The final speech improvement has been measured with the help of Multidemensional Voice Program (MDVP) (Deliyski, 1993) tools and the Shimmer and Harmonic to Noise Ratio (HNR) parameters.

Communication ability of human beings can be extremely influenced by voice disorders. When any problem in the larynx or changes in the voice pitch appear, it could be important to go to the specialist's office to examine the vocal folds movements.

Specialists use computational tools in the objective diagnosis of vocal folds pathologies by means of a set of acoustic parameters among others. There are some patients with severe degradations of speech, as they are the oesophageal voice of laryngectomees.

Patients who have undergone a laryngectomy as a result of larynx cancer have exceptionally low intelligibility. This is due to the removal of their vocal folds, which forces them to use the air flowing through the oesophagus: this is known as oesophageal speech. The characterization parameters for these kinds of oesophageal voices go beyond normal ranges, due to the low quality of the sound itself and its intelligibility.

The cancer of the vocal folds needs to pay special attention in its diagnosis, treatment, rehabilitation and monitoring mainly because it can cause death. Once the cancer has been detected, the otolaryngology (ORL) arranges the vocal folds removal. This implies that patients in such situation will not be able to produce laryngeal voice and hence, they lose the speaking ability. The second most common type of cancer is larynx cancer with a rate of

95%. Every year approximately 136,000 new cases of larynx cancer are diagnosed in the world, with an average survival rate of 5 years in 68% of the cases.

After the operation and during the rehabilitation, the patient will begin the learning stage of oesophageal speech: the voice produced due to the modulation of the air by means of the oesophagus. This will allow the patient to use oesophageal speech which has a degraded quality but it makes possible to maintain a fluid oral communication.

Low intelligibility is the main problem in both oral and telephone communications with other people. In addition, the noise of this kind of speech signal is especially high. This fact has an extremely negative effect on the objective voice parameters, such as pitch, jitter, shimmer and HNR (Harmonic to Noise Ratio). Thus, it is necessary to process voice signal in order to increase intelligibility. The voice enhancement will be measured by those objective parameters. Therefore, the main aim of this work is to recover the normal range of those parameters, to facilitate the laryngectomized collective communication.

Thus, it is necessary to process voice signals in order to increase intelligibility. The voice enhancement will be measured by the objective parameters. Therefore, the main aim of this work is to recover the normal range of the parameters, to facilitate the laryngectomized people communication.

The general objective of this work is to develop an algorithm to enhance and the voice for people who have voice disorders.

2. Methods and system design

2.1 Acoustic parameters

The voice enhancement will be measured by the objective parameters. Therefore, the main aim of this work is to recover the normal range of the parameters, to facilitate the laryngectomized collective communication.

The pitch (Baken & Orlikoff, 2000) is the property of a sound or musical tone measured by its perceived frequency. Due to de pseudo-periodic nature of the voiced speech, there are variations in the instantaneous frequency f_i so the pitch can be defined as

$$Pitch(Hz) = \frac{\sum_{i=1}^{N} f_i}{N} \qquad (1)$$

being N the number of extracted pitch periods.

Fundamental frequency estimation has consistently been a difficult topic in audio signal processing because is so difficult to define the time instants which define the voice cycles used to obtain their related instantaneous frequency, f_i.

Furthermore, in acoustical parameterization it is of capital importance to calculate those instants because they are basic features used in this kind of characterization.

Jitter (Baken & Orlikoff, 2000) is a parameter that represents the variation of the fundamental frequency:

Name	Notation	Definition	Units	id
Absolute Jitter	Jita	$Jitter(Hz) = \frac{\sum_{i=1}^{N-1} \left\| T^{(i)} - T^{(i+1)} \right\|}{N-1}$ T.- time period N.- number of extracted pitch periods	Hz	(2)
Jitter Percent	Jit	$Jitter(\%) = \frac{\frac{\sum_{i=1}^{N-1} \left\| T^{(i)} - T^{(i+1)} \right\|}{N-1}}{\sum_{i=1}^{N} \frac{T^{(i)}}{N}}$	%	(3)
Relative Average Perturbation	RAP	$RAP(\%) = 100 \times \frac{\frac{\sum_{i=2}^{i=N-1} \left\| \frac{T^{(i-1)} + T^{(i)} + T^{(i+1)}}{3} - T^{(i)} \right\|}{N-2}}{\sum_{i=1}^{N} \frac{T^{(i)}}{N}}$	%	(4)
Pitch Perturbation Quotient	PPQ	$PPQ(\%) = \frac{\frac{1}{N-4} \sum_{i=1}^{N-4} \left\| \frac{1}{5} \sum_{r=0}^{4} T^{(i+r)} - T^{(i+2)} \right\|}{\frac{1}{N} \sum_{i=1}^{N} T^{(i)}}$	%	(5)
Smoothed Pitch Perturbation Quotient	sPPQ	$sPPQ(\%) = \frac{\frac{1}{N-sf+1} \sum_{i=1}^{N-sf+1} \left\| \frac{1}{sf} \sum_{r=0}^{sf-1} T^{(i+r)} - T^{(i+m)} \right\|}{\frac{1}{N} \sum_{i=1}^{N} T^{(i)}}$ sf.- smoothing factor (typically odd) m.- ½*(sf-1)	%	(6)

Table 1. Jitter definition formulae

In the other hand, specialists also use the reference of shimmer (Baken & Orlikoff, 2000) which is the parameter that represents the amplitude perturbation of the voice signal. The voice produced in vocal folds is supposed to have the ability to maintain its amplitude almost constant, thus an increased value of shimmer may imply a symptom of a voice disorder.

The possible mathematical definitions of shimmer are the following:

Name	Notation	Definition	Units	id
Shimmer Percentage	Shim	$Shim = \dfrac{\frac{1}{N-1}\sum_{i=1}^{N-1}\left\|A^{(i)} - A^{(i+1)}\right\|}{\frac{1}{N}\sum_{i=1}^{N} A^{(i)}}$ $A^{(i)}$ - Extracted peak-to-peak amplitude data, N - Number of extracted impulses	%	(7)
Shimmer	ShdB	$ShdB = \frac{1}{N-1}\sum_{i=1}^{N-1}\left\|20\log\left(\frac{A^{(i+1)}}{A^{(i)}}\right)\right\|$	dB	(8)
Smoothed Amplitude Perturbation Quotient	sAPQ	$SAPQ = \dfrac{\frac{1}{N-sf+1}\sum_{i=1}^{N-sf+1}\left\|\frac{1}{sf}\sum_{r=0}^{sf-1} A^{(i+r)} - A^{(i+m)}\right\|}{\frac{1}{N}\sum_{i=1}^{N} A^{(i)}}$ sf - smoothing factor (typically odd) m.- ½*(sf-1)	%	(9)

Table 2. Shimmer definition formulae

Several authors have reported decent results using voice cycle detection (Chen & Kao, 2001) (Hagmüller & Kubina, 2006) and there are many techniques widely detailed in literature: time domain estimators (e.g. Zero Crossing Rate (Kedem, 1986)), fundamental frequency estimators (Dorken & Nawab, 1994) (Piszczalski & Galler, 1979), Autocorrelation methods (Yin Estimator, (Cheveigné & Kawahara, 2002)), Phase Space representation (Gibiat, 1988), Cepstrum (Flanagan, 1965) and Statistical Methods (Sano & Jenkins, 1989) (Doval & Rodet, Estimation of fundamental frequency of musical sound signals, 1991) (Doval & Rodet, 1993). Some of them define directly the voice cycles (Chen & Kao, 2001) while others are used to calculate a numerical approximation (Cheveigné & Kawahara, 2002) to the fundamental frequency value. In these ones, a further step is necessary in order to identify clearly which instants define the voice cycles.

The HNR is a general evaluation of noise present in the analyzed signal. It is defined as (10), $r_p(0)$ and $r_{ap}(0)$ being the respective energies of the periodic and aperiodic components:

$$HNR = \frac{r_p(0)}{r_{ap}(0)} \quad (10)$$

The measures have been made with the help of MDVP from Kay Electronics important software that gives good estimations of a signal's parameters. However, this software is not specialized in pathological speech. To fix this problem, the voiced period marks are needed to calculate the pitch and then the HNR have been manually introduced, one by one, on each signal.

The MDVP calculates the HNR as the average ratio of the harmonic spectral energy in the frequency range 70-4500 Hz and the enharmonic spectral energy in the frequency range 1500-4500 Hz (Deliyski, 1993) (Yumoto & Gould, 1982).

2.2 Wavelets

It is well known from Fourier theory that a signal can be expressed as the sum of a, possibly infinite, series of sines and cosines. This sum is also referred to as a Fourier expansion. The big disadvantage of a Fourier expansion however is that it has only frequency resolution and no time resolution. This means that although we might be able to determine all the frequencies present in a signal, we do not know when they are present. To overcome this problem in the past decades several solutions have been developed which are more or less able to represent a signal in the time and frequency domain at the same time.

The idea behind these time-frequency joint representations is to cut the signal of interest into several parts and then analyze the parts separately. It is clear that analyzing a signal this way will give more information about the when and where of different frequency components, but it leads to a fundamental problem as well: how to cut the signal? Suppose that we want to know exactly all the frequency components present at a certain moment in time. We cut out only this very short time window using a Dirac pulse1, transform it to the frequency domain and ... something is very wrong.

The problem here is that cutting the signal corresponds to a convolution between the signal and the cutting window. Since convolution in the time domain is identical to multiplication in the frequency domain and since the Fourier transform of a Dirac pulse contains all possible frequencies the frequency components of the signal will be smeared out all over the frequency axis. In fact this situation is the opposite of the standard Fourier transform since we now have time resolution but no frequency resolution whatsoever.

The underlying principle of the phenomena just described is Heisenberg's uncertainty principle, which, in signal processing terms, states that it is impossible to know the exact frequency and the exact time of occurrence of this frequency in a signal. In other words, a signal can simply not be represented as a point in the time-frequency space. The uncertainty principle shows that it is very important how one cuts the signal.

The *wavelet transform* or *wavelet analysis* is probably the most recent solution to overcome the shortcomings of the Fourier transform. In wavelet analysis the use of a fully scalable modulated window solves the signal-cutting problem. The window is shifted along the signal and for every position the spectrum is calculated. Then this process is repeated many times with a slightly shorter (or longer) window for every new cycle. In the end the result will be a collection of time-frequency representations of the signal, all with different resolutions. Because of this collection of representations we can speak of a multiresolution analysis. In the case of wavelets we normally do not speak about time-frequency representations but about time-scale representations, scale being in a way the opposite of frequency, because the term frequency is reserved for the Fourier transform.

2.2.1 The Continuous Wavelet Transform (CWT)

The wavelet analysis described in the introduction is known as the *continuous wavelet transform* or *CWT*. More formally it is written as:

$$\gamma(s,\tau) = \int f(t)\, \Psi^{*}_{s,\tau}(t)dt \quad (11)$$

where * denotes complex conjugation. This equation shows how a function $f(t)$ is decomposed into a set of basis functions $\Psi^{*}_{s,\tau}(t)$, called the wavelets. The variables s and - are the new dimensions, scale and translation, after the wavelet transform. For completeness sake equation (11) gives the inverse wavelet transform. I will not expand on this since we are not going to use it:

$$f(t) = \iint \gamma(s,\tau)\, \Psi^{*}_{s,\tau}(t)d\tau ds \quad (12)$$

The wavelets are generated from a single basic wavelet Ψ (t), the so-called *mother wavelet*, by scaling and translation:

$$\Psi^{*}_{s,\tau}(t) = \frac{1}{\sqrt{s}}\Psi\left(\frac{t-\tau}{s}\right) \quad (13)$$

In (13) s is the scale factor, τ is the translation factor and the factor $s^{-1/2}$ is for energy normalization across the different scales (Lió, 2003) (Ortolan, Mori, Pereira, Cabral, Pereira, & Cliquet, 2003).

It is important to note that in (11), (12) and (13) the wavelet basis functions are not specified. This is a difference between the wavelet transform and the Fourier transform, or other transforms. The theory of wavelet transforms deals with the general properties of the wavelets and wavelet transforms only. It defines a framework within one can design wavelets to taste and wishes.

2.2.2 Discrete wavelet

Now that we know what the wavelet transform is, we would like to make it practical. However, the wavelet transform as described so far still has three properties that make it difficult to use directly in the form of (11). The first is the redundancy of the CWT. In (11) the wavelet transform is calculated by continuously shifting a continuously scalable function over a signal and calculating the correlation between the two. It will be clear that these scaled functions will be nowhere near an orthogonal basis5 and the obtained wavelet coefficients will therefore be highly redundant. For most practical applications we would like to remove this redundancy.

Even without the redundancy of the CWT we still have an infinite number of wavelets in the wavelet transform and we would like to see this number reduced to a more manageable count. This is the second problem we have. The third problem is that for most functions the wavelet transforms have no analytical solutions and they can be calculated only numerically or by an optical analog computer. Fast algorithms are needed to be able to exploit the power of the wavelet transform and it is in fact the existence of these fast algorithms that have put wavelet transforms where they are today.

As mentioned before the CWT maps a one-dimensional signal to a two-dimensional time-scale joint representation that is highly redundant. The time-bandwidth product of the CWT is the square of that of the signal and for most applications, which seek a signal description with as few components as possible, this is not efficient. To overcome this problem *discrete wavelets* have been introduced. Discrete wavelets are not continuously scalable and translatable but can only be scaled and translated in discrete steps. This is achieved by modifying the wavelet representation (13) to create (Tohidypour, Seyyedsalehi, & Behbood, 2010) (Daubechies, 1992).

$$\Psi_{j,k}(t) = \frac{1}{\sqrt{s_0^j}}\Psi\left(\frac{t-k\tau_0 s_0^j}{s_0^j}\right) \tag{14}$$

Although it is called a discrete wavelet, it normally is a (piecewise) continuous function. In (14) j and k are integers and $s_0 > 1$ is a fixed dilation step. The translation factor τ_0 depends on the dilation step. The effect of discretizing the wavelet is that the time-scale space is now sampled at discrete intervals. We usually choose $s_0 = 2$ so that the sampling of the frequency axis corresponds to *dyadic sampling*. This is a very natural choice for computers, the human ear and music for instance. For the translation factor we usually choose $\tau_0 = 1$ so that we also have dyadic sampling of the time axis.

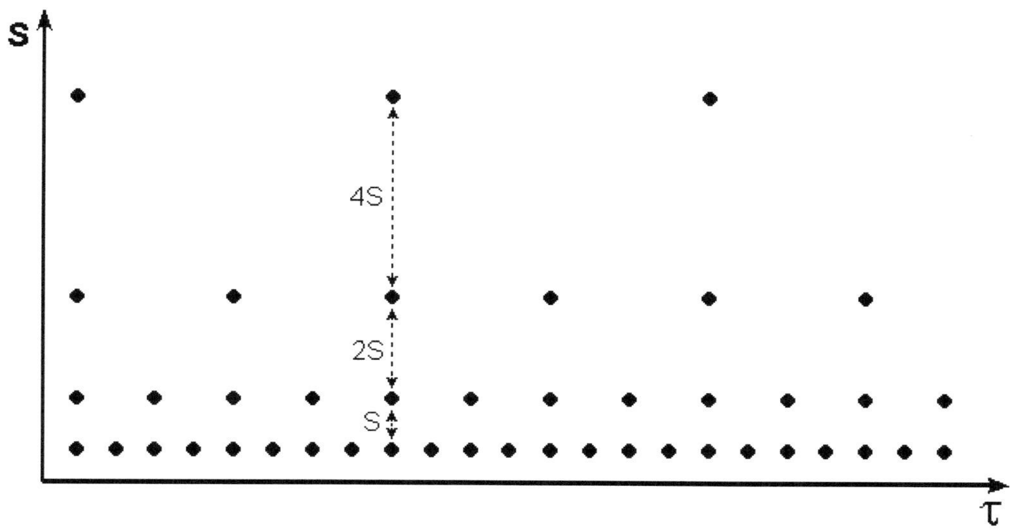

Fig. 1. Localization of the discrete wavelets in the time-scale space on a dyadic grid.

When discrete wavelets are used to transform a continuous signal the result will be a series of wavelet coefficients, and it is referred to as the *wavelet series decomposition*. An important issue in such a decomposition scheme is of course the question of reconstruction. It is all very well to sample the time-scale joint representation on a dyadic grid, but if it will not be possible to reconstruct the signal it will not be of great use. As it turns out, it is indeed possible to reconstruct a signal from its wavelet series decomposition. In (Daubechies, 1992) it is proven that the necessary and sufficient condition for stable reconstruction is that the energy of the wavelet coefficients must lie between two positive bounds, i.e.

$$A\|f\|^2 \leq \sum_{j,k}\left|\langle f, \Psi_{j,k}\rangle\right|^2 \leq B\,\|f\|^2 \tag{15}$$

where $\|f\|^2$ is the energy of $f(t)$, $A > 0$, $B < \infty$ and A, B are independent of $f(t)$. When equation ($A\|f\|^2 \leq \sum_{j,k}\left|\langle f, \Psi_{j,k}\rangle\right|^2 \leq B\,\|f\|^2$ (15)) is satisfied, the family of basis functions

$\Psi_{j,k}(t)$ with $j,k \in \mathbb{Z}$ is referred to as a *frame* with frame bounds A and B. When $A = B$ the frame is *tight* and the discrete wavelets behave exactly like an orthonormal basis. When $A \neq B$ exact reconstruction is still possible at the expense of a *dual frame*. In a dual frame discrete wavelet transform the decomposition wavelet is different from the reconstruction wavelet.

We will now immediately forget the frames and continue with the removal of all redundancy from the wavelet transform. The last step we have to take is making the discrete wavelets orthonormal. This can be done only with discrete wavelets. The discrete wavelets can be made orthogonal to their own dilations and translations by special choices of the mother wavelet, which means:

$$\int \Psi_{j,k}(t)\Psi^*{}_{m,n}(t)\, dt = \begin{cases} 1 & if\ j = m\ and\ k = n \\ 0 & otherwise \end{cases} \tag{16}$$

An arbitrary signal can be reconstructed by summing the orthogonal wavelet basis functions, weighted by the wavelet transform coefficients (Sheng, 1996):

$$\mathrm{f(t)} = \sum_{\mathrm{j,k}} \gamma(\mathrm{j,k})\Psi_{\mathrm{j,k}}(\mathrm{t}) \tag{17}$$

Equation (17) shows the inverse wavelet transform for discrete wavelets, which we had not yet seen.

Orthogonality is not essential in the representation of signals. The wavelets need not be orthogonal and in some applications the redundancy can help to reduce the sensitivity to noise (Sheng, 1996) or improve the *shift invariance* of the transform (Burrus, Goinath, & Guo, 1998). This is a disadvantage of discrete wavelets: the resulting wavelet transform is no longer shift invariant, which means that the wavelet transforms of a signal and of a time-shifted version of the same signal are not simply shifted versions of each other.

2.2.3 A band-pass filter

With the redundancy removed, we still have two hurdles to take before we have the wavelet transform in a practical form. We continue by trying to reduce the number of wavelets needed in the wavelet transform and save the problem of the difficult analytical solutions for the end.

Even with discrete wavelets we still need an infinite number of scalings and translations to calculate the wavelet transform. The easiest way to tackle this problem is simply not to use an infinite number of discrete wavelets. Of course this poses the question of the quality of the transform. Is it possible to reduce the number of wavelets to analyze a signal and still have a useful result?

The translations of the wavelets are of course limited by the duration of the signal under investigation so that we have an upper boundary for the wavelets. This leaves us with the question of dilation: how many scales do we need to analyze our signal? How do we get a lower bound? It turns out that we can answer this question by looking at the wavelet transform in a different way.

If we look wavelets proprieties we see that the wavelet has a band-pass like spectrum. From Fourier theory we know that compression in time is equivalent to stretching the spectrum and shifting it upwards:

$$F\{f(at)\} = \frac{1}{a} F\left(\frac{\omega}{a}\right) \quad (18)$$

This means that a time compression of the wavelet by a factor of 2 will stretch the frequency spectrum of the wavelet by a factor of 2 and also shift all frequency components up by a factor of 2. Using this insight we can cover the finite spectrum of our signal with the spectra of dilated wavelets in the same way as that we covered our signal in the time domain with translated wavelets. To get a good coverage of the signal spectrum the stretched wavelet spectra should touch each other, as if they were standing hand in hand (see Fig. 2). This can be arranged by correctly designing the wavelets.

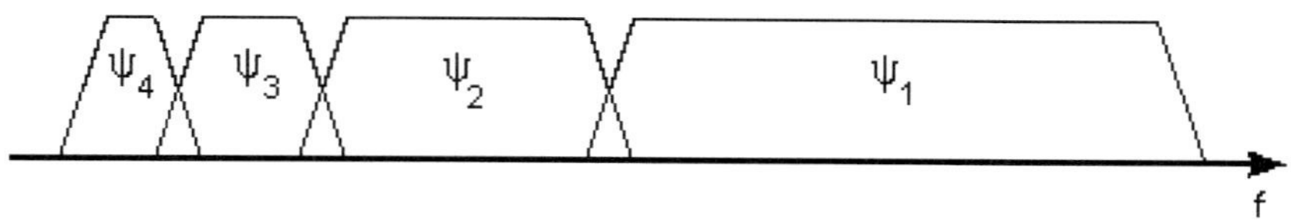

Fig. 2. Touching wavelet spectra resulting from scaling of the mother wavelet in the time domain

Summarizing, if one wavelet can be seen as a band-pass filter, then a series of dilated wavelets can be seen as a band-pass filter bank. If we look at the ratio between the centre frequency of a wavelet spectrum and the width of this spectrum we will see that it is the same for all wavelets. This ratio is normally referred to as the fidelity factor Q of a filter and in the case of wavelets one speaks therefore of a *constant-Q* filter bank.

2.2.4 The scaling function

The careful reader will now ask him- or herself the question how to cover the spectrum all the way down to zero? Because every time you stretch the wavelet in the time domain with a factor of 2, its bandwidth is halved. In other words, with every wavelet stretch you cover only half of the remaining spectrum, which means that you will need an infinite number of wavelets to get the job done.

The solution to this problem is simply not to try to cover the spectrum all the way down to zero with wavelet spectra, but to use a cork to plug the hole when it is small enough. This cork then is a low-pass spectrum and it belongs to the so-called *scaling function*. The scaling function was introduced by Mallat (Mallat, 1989). Because of the low-pass nature of the scaling function spectrum it is sometimes referred to as the *averaging filter*.

If we look at the scaling function as being just a signal with a low-pass spectrum, then we can decompose it in wavelet components and express it like (17):

$$\varphi(t) = \sum_{j,k} \gamma(j,k) \Psi_{j,k}(t) \quad (19)$$

Since we selected the scaling function $\varphi(t)$ in such a way that its spectrum neatly fitted in the space left open by the wavelets, the expression ($\varphi(t) = \sum_{j,k} \gamma(j,k) \Psi_{j,k}(t)$ (19) uses an infinite number of wavelets up to a certain scale j (see Fig. 3). This means that if we analyze a signal using the combination of scaling function and wavelets, the scaling function by itself takes care of the spectrum otherwise covered by all the wavelets up to scale j, while the rest is done by the wavelets. In this way we have limited the number of wavelets from an infinite number to a finite number.

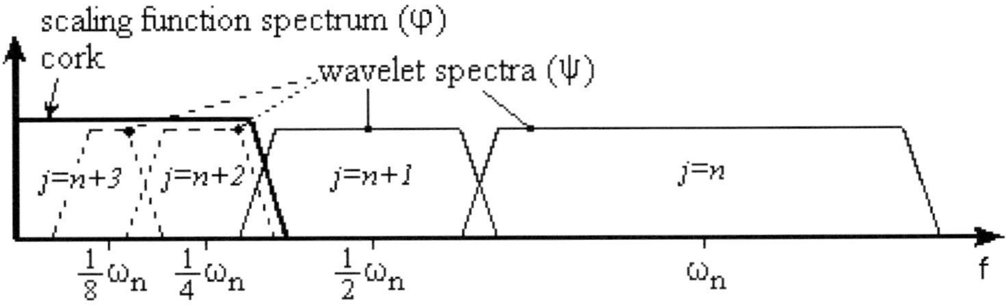

Fig. 3. How an infinite set of wavelets is replaced by one scaling function.

By introducing the scaling function we have circumvented the problem of the infinite number of wavelets and set a lower bound for the wavelets. Of course when we use a scaling function instead of wavelets we lose information. That is to say, from a signal representation view we do not lose any information, since it will still be possible to reconstruct the original signal, but from a wavelet-analysis point of view we discard possible valuable scale information. The width of the scaling function spectrum is therefore an important parameter in the wavelet transform design. The shorter its spectrum the more wavelet coefficients you will have and the more scale information. But, as always, there will be practical limitations on the number of wavelet coefficients you can handle. As we will see later on, in the discrete wavelet transform this problem is more or less automatically solved.

Summarizing once more, if one wavelet can be seen as a band-pass filter and a scaling function is a low-pass filter, then a series of dilated wavelets together with a scaling function can be seen as a filter bank.

2.2.5 Subband coding

Two of the three problems mentioned in section 4 have now been resolved, but we still do not know how to calculate the wavelet transform. Therefore we will continue our journey through multiresolution land.

If we regard the wavelet transform as a filter bank, then we can consider wavelet transforming a signal as passing the signal through this filter bank. The outputs of the different filter stages are the wavelet- and scaling function transform coefficients. Analyzing a signal by passing it through a filter bank is not a new idea and has been around for many years under the name *subband coding*. It is used for instance in computer vision applications.

The filter bank needed in subband coding can be built in several ways. One way is to build many band-pass filters to split the spectrum into frequency bands. The advantage is that the width of every band can be chosen freely, in such a way that the spectrum of the signal to analyze is covered in the places where it might be interesting. The disadvantage is that we will have to design every filter separately and this can be a time consuming process. Another way is to split the signal spectrum in two (equal) parts, a low-pass and a high-pass part. The high-pass part contains the smallest details we are interested in and we could stop here. We now have two bands. However, the low-pass part still contains some details and therefore we can split it again. And again, until we are satisfied with the number of bands we have created. In this way we have created an *iterated filter bank*. Usually the number of bands is limited by for instance the amount of data or computation power available. The process of splitting the spectrum is graphically displayed in figure 4. The advantage of this scheme is that we have to design only two filters; the disadvantage is that the signal spectrum coverage is fixed.

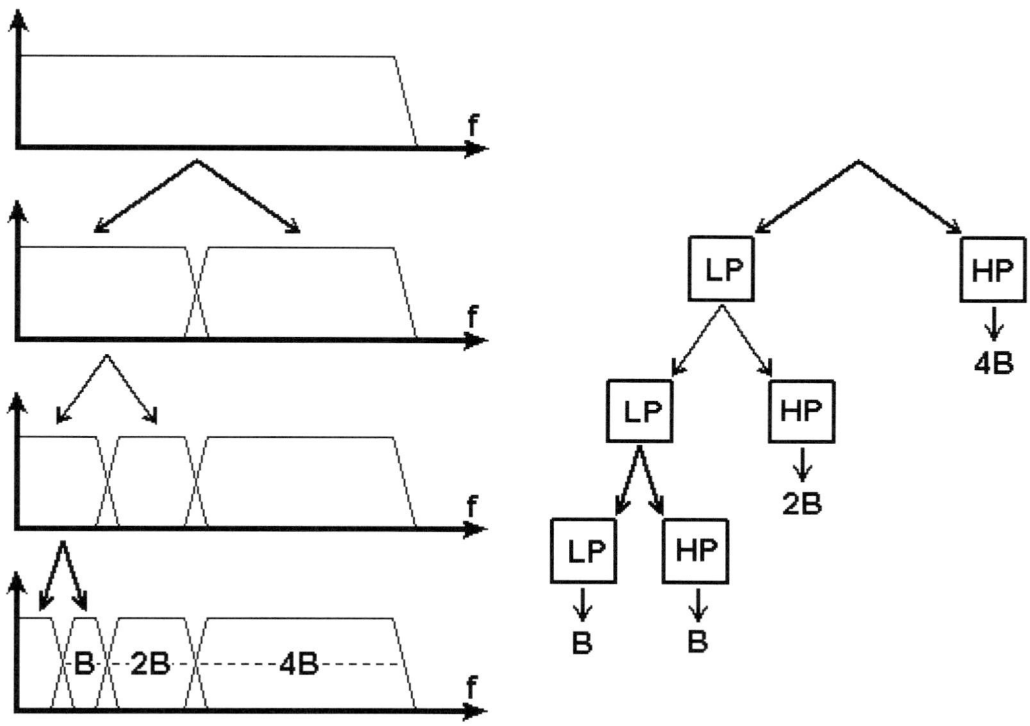

Fig. 4. Splitting the signal spectrum with an iterated filter bank.

Looking at figure 4 we see that what we are left with after the repeated spectrum splitting is a series of band-pass bands with doubling bandwidth and one low-pass band. (Although in theory the first split gave us a high-pass band and a low-pass band, in reality the high-pass band is a band-pass band due to the limited bandwidth of the signal.) In other words, we can perform the same subband analysis by feeding the signal into a bank of band-pass filters of which each filter has a bandwidth twice as wide as his left neighbour (the frequency axis runs to the right here) and a low-pass filter. At the beginning of this section we stated that this is the same as applying a wavelet transform to the signal. The wavelets give us the band-pass bands with doubling bandwidth and the scaling function provides us with the low-pass band. From this we can conclude that a wavelet transform is the same thing as a subband coding scheme using a constant-Q filter bank (Mallat, 1989). In general we will refer to this kind of analysis as a multiresolution analysis.

Summarizing, if we implement the wavelet transform as an iterated filter bank, we do not have to specify the wavelets explicitly! This sure is a remarkable result.

2.2.6 The Discrete Wavelet Transform (DWT)

In many practical applications and especially in the application described in this report the signal of interest is sampled. In order to use the results we have achieved so far with a discrete signal we have to make our wavelet transform discrete too. Remember that our discrete wavelets are not time-discrete, only the translation- and the scale step are discrete. Simply implementing the wavelet filter bank as a digital filter bank intuitively seems to do the job. But intuitively is not good enough, we have to be sure.

In (19) we stated that the scaling function could be expressed in wavelets from minus infinity up to a certain scale *j*. If we add a wavelet spectrum to the scaling function spectrum we will get a new scaling function, with a spectrum twice as wide as the first. The effect of this addition is that we can express the first scaling function in terms of the second, because all the information we need to do this is contained in the second scaling function. We can express this formally in the so-called multiresolution formulation (Burrus, Goinath, & Guo, 1998) or *two-scale relation* (Sheng, 1996):

$$\varphi(2^j t) = \sum_k h_{j+1}(k)\varphi(2^{j+1}t - k) \tag{20}$$

The two-scale relation states that the scaling function at a certain scale can be expressed in terms of translated scaling functions at the next smaller scale. Do not get confused here: smaller scale means more detail.

The first scaling function replaced a set of wavelets and therefore we can also express the wavelets in this set in terms of translated scaling functions at the next scale. More specifically we can write for the wavelet at level *j*:

$$\Psi(2^j t) = \sum_k g_{j+1}(k)\varphi(2^{j+1}t - k) \tag{21}$$

which is the two-scale relation between the scaling function and the wavelet.

Since our signal $f(t)$ could be expressed in terms of dilated and translated wavelets up to a scale *j*-1, this leads to the result that $f(t)$ can also be expressed in terms of dilated and translated scaling functions at a scale *j*:

$$f(t) = \sum_k \lambda_j(k)\varphi(2^j t - k) \tag{22}$$

To be consistent in our notation we should in this case speak of discrete scaling functions since only discrete dilations and translations are allowed. If in this equation we step up a scale to *j*-1, we have to add wavelets in order to keep the same level of detail. We can then express the signal $f(t)$ as

$$f(t) = \sum_k \lambda_{j-1}(k)\varphi(2^{j-1}t - k) + \sum_k \gamma_{j-1}(k)\Psi(2^{j-1}t - k) \tag{23}$$

If the scaling function $\varphi_{j,k}(t)$ and the wavelets $\Psi_{j,k}(t)$ are orthonormal or a tight frame, then the coefficients $\lambda_{j-1}(k)$ and $\gamma_{j-1}(k)$ are found by taking the inner products

If we now replace $\varphi_{j,k}(t)$ and $\Psi_{j,k}(t)$ in the inner products by suitably scaled and translated versions of (20) and (21) and manipulate a bit, keeping in mind that the inner product can also be written as an integration, we arrive at the important result (Burrus, Goinath, & Guo, 1998):

$$\lambda_{j-1}(k) = \sum_m h(m-2k)\,\lambda_j(m) \quad (24)$$

$$\gamma_{j-1}(k) = \sum_m g(m-2k)\,\gamma_j(m) \quad (25)$$

These two equations state that the wavelet- and scaling function coefficients on a certain scale can be found by calculating a weighted sum of the scaling function coefficients from the previous scale. Now recall from the section on the scaling function that the scaling function coefficients came from a low-pass filter and recall from the section on subband coding how we iterated a filter bank by repeatedly splitting the low-pass spectrum into a low-pass and a high-pass part. The filter bank iteration started with the signal spectrum, so if we imagine that the signal spectrum is the output of a low-pass filter at the previous (imaginary) scale, then we can regard our sampled signal as the scaling function coefficients from the previous (imaginary) scale. In other words, our sampled signal $f(k)$ is simply equal to $\lambda(k)$ at the largest scale!

As we know from signal processing theory a discrete weighted sum like the ones in (24) and (25) is the same as a digital filter and since we know that the coefficients $\lambda j(k)$ come from the low-pass part of the splitted signal spectrum, the weighting factors h(k) in ($\lambda j - 1(k) = \sum_m h(m-2k)\,\lambda_j(m)$ (24)) must form a low-pass filter. And since we know that the coefficients $\gamma j(k)$ come from the high-pass part of the splitted signal spectrum, the weighting factors g(k) in (24) must form a high-pass filter. This means that (24) and (25) together form one stage of an iterated digital filter bank and from now on we will refer to the coefficients h(k) as the scaling filter and the coefficients g(k) as the wavelet filter.

By now we have made certain that implementing the wavelet transform as an iterated digital filter bank is possible and from now on we can speak of the discrete wavelet transform or DWT. Our intuition turned out to be correct. Because of this we are rewarded with a useful bonus property of (24) and (25), the subsampling property. If we take one last look at these two equations we see that the scaling and wavelet filters have a step-size of 2 in the variable k. The effect of this is that only every other $\lambda j(k)$ is used in the convolution, with the result that the output data rate is equal to the input data rate. Although this is not a new idea, it has always been exploited in subband coding schemes, it is kind of nice to see it pop up here as part of the deal.

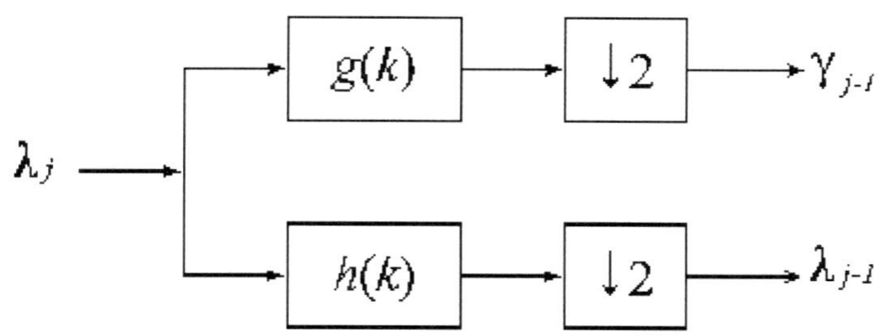

Fig. 5. Implementation of (23) and (24) as one stage of an iterated filter bank.

The subsampling property also solves our problem, which had come up at the end of the section on the scaling function, of how to choose the width of the scaling function spectrum. Because, every time we iterate the filter bank the number of samples for the next stage is halved so that in the end we are left with just one sample (in the extreme case). It will be clear that this is where the iteration definitely has to stop and this determines the width of the spectrum of the scaling function. Normally the iteration will stop at the point where the number of samples has become smaller than the length of the scaling filter or the wavelet filter, whichever is the longest, so the length of the longest filter determines the width of the spectrum of the scaling function.

2.3 Wavelets algorithm

The general goal of this investigation, and so of all the previous researches (García, Vicente, Ruiz, Angulo, & Aramendi, 2002) is the improvement of the oesophageal voices' quality (García, Vicente, Ruiz, Alonso, & Loyo, 2005). Certainly the specific aim of this research is the spectral and temporal correction of the shimmer and parameter of these voices by the Wavelet Transform.

One of the most important techniques applied in the spectral analysis is the Fourier Transform (STFT), which will allow recognizing the spectral components of speech signal, so it makes possible to distinguish pathological voices and process them. That transform has a resolution problem which is explained by Heisenberg Uncertainty Principle. The Wavelet Transform (WT) was developed to overcome some resolutions related problems of the STFT. It is possible to analyze any signal by using an alternative approach called the Multiresolution Analysis (MRA). MRA, as implied by its name, analyzes the signal at different frequencies with different resolutions. MRA is designed to give good time resolution and poor frequency resolution at high frequencies and good frequency resolution and poor time resolution at low frequencies.

As the signals used are digital, it is more useful to use Discrete Wavelet Transform (DWT) (Mallat, 1999). The DWT analyzes the signal at different frequency bands with different resolutions by decomposing the signal into a coarse approximation and detail information. The decomposition of the signal into different frequency bands is simply obtained by successive high pass and low pass filtering of the time domain signal. The original signal x[n] is first passed through a half band high pass filter g[n] and a low pass filter h[n]. This constitutes one level of decomposition and can mathematically be expressed as follows (Kadambe & Bourdreaux-Bartels, 1991):

$$y_{high}[k] = \sum_{n} x[n] \cdot g[2k-n] \tag{26}$$

$$y_{low}[k] = \sum_{n} x[n] \cdot h[2k-n] \tag{27}$$

Before applying the DWT, the signal is processed. That is, a resample of the original signal, x[n], at a sampling frequency of 12800 Hz. This is so done, as when applying the transformed DWT, the detail signals remain between the frequency bands that are suitable for pitch detection (Kadambe & Bourdreaux-Bartels, A Comparison of a Wavelet Functions for Pitch Detection of Speech Signals, 1991) (Kadambe & Bourdreaux-Bartels, 1992) (Wing-kei, Kwong-sak, & Kin-hong, 1995) (Nadeu, Pascual, & Herdondo, 1991). More specifically, the oesophageal voices have a pitch nearing 60 Hz. On doing the above-mentioned resample and the following transformed DWT, one of the details is found in the frequency band level of 50 Hz - 100 Hz. This means that the original pitch signal's information is located within this detail. Low-frequency noise present in oesophageal voices are found in the 0 Hz - 50 Hz level. We should eliminate this noise before modifying the pitch's peak amplitude.

In short, so as to control the high rates of the shimmer parameter in oesophageal voices, the following steps should be taken: carry out a resample of the original signal at Fs = 12800 Hz; after this the transformed DWT should be done, for which we have used "bior 6.8" as the mother wavelet. Trials with other mother wavelets were done and the results are quite similar to as regards shimmer measurements. Once the DWT transform has been done, the low-frequency noise in the 0 Hz - 50 Hz frequency band is eliminated. After this pre-processing, the amplitude of the maximums in the 50 Hz - 100 Hz frequency band are modified, as this is where the information on oesophageal voices is to be found.

Fig. 6 shows the frequency band tree when DWT is applied.

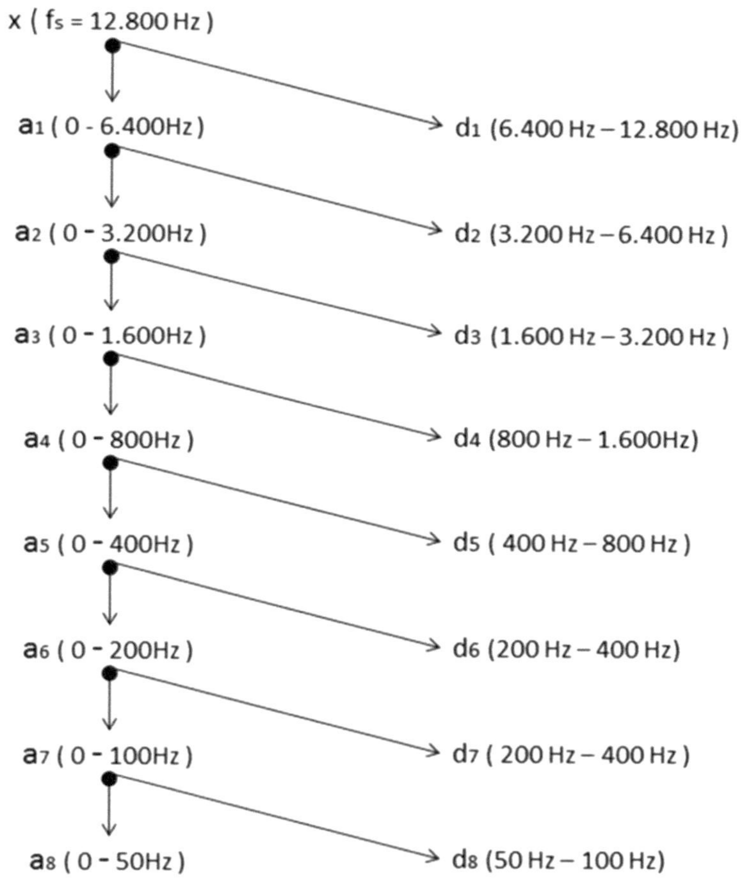

Fig. 6. Frequency band diagram

2.4 Poles stabilization algorithm

The stabilizations of poles, the second algorithm is responsible for analyzing and modifying the poles of the system modelized by the vocal tract. It works with an oesophageal voice signal from which the excitation has been separated from the tract, and it calculates the evolution of modulus and phase of each formant of the vowel modifying such poles.

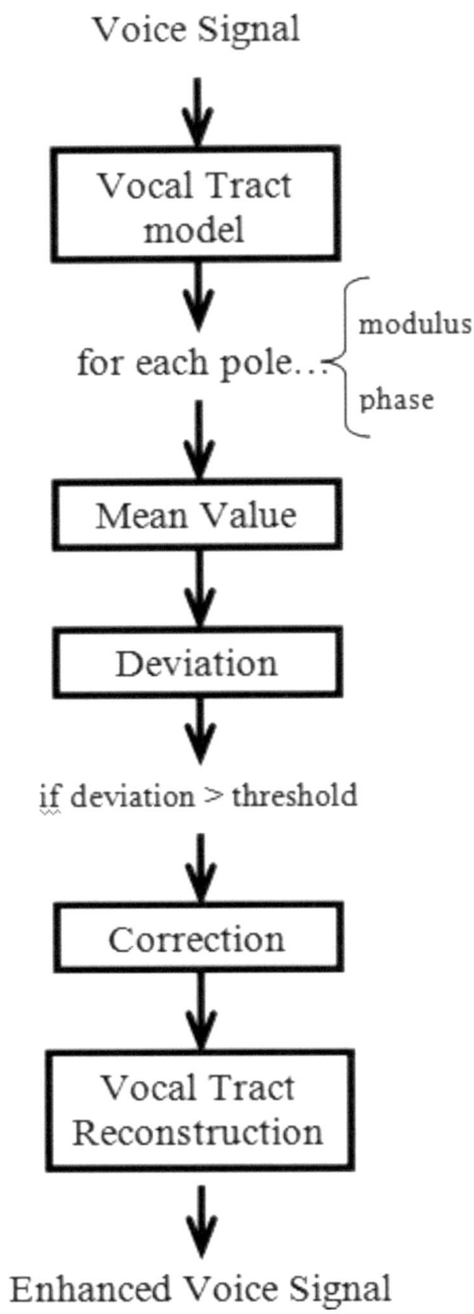

Fig. 7. Pole Stabilization Block Diagram

The stabilization of the first three formants is applied in those values of the vowel which is being enhanced by means of the modification of the first three poles, following these steps:

1. Calculation of the mean value of modulus and phase of each pole through the vocal signal.
2. Calculation of the maximum deviations relative to the mean modulus and phase of the first three poles.
3. Whether the deviations exceed a certain threshold is analyzed, if so the modulus correction is applied:
   a. ModulusModif=modulus+((1-modulus)*ConstMod);
   b. and the phase correction:
      AngModif=Angle-(ContPhase*(Angle+MeanPhase));
      being the correction implanted by means of "ConstMod" and "ConstPhase" parameters which can be adjusted for each voice.
4. Reconstruction of the filter that modelizes the vocal tract with the new poles of the system corrected and stabilized.

3. Results

On the one hand, in the DWT algorithm, the inputs of the developed algorithm are the samples of the oesophageal voice, which shimmer parameter have been previously evaluated. In the 100% of the studied cases obtained value for these parameters is out of the range of the normality. So it could be improved in order to increase the quality of the voice between the normality ranges specified by the scientific community. After the application of the algorithm based on the analysis and processing by Wavelet, the speech signal has been reconstructed. When measuring the shimmer in this reconstructed signal, the obtained results are the following:

Oesophageal Phoneme	Shimmer Real (%)	Transformed Shimmer (%)
a1	18,43	10,27
a2	17,27	10,82
a3	16,05	9,98
a4	19,62	12,79
a5	16,76	8,43
a6	18,58	12,99
a7	12,53	10,11
a8	16,98	9,99
a9	20,31	12,83
a10	18,22	10,61

Table 3. Table of shimmer values

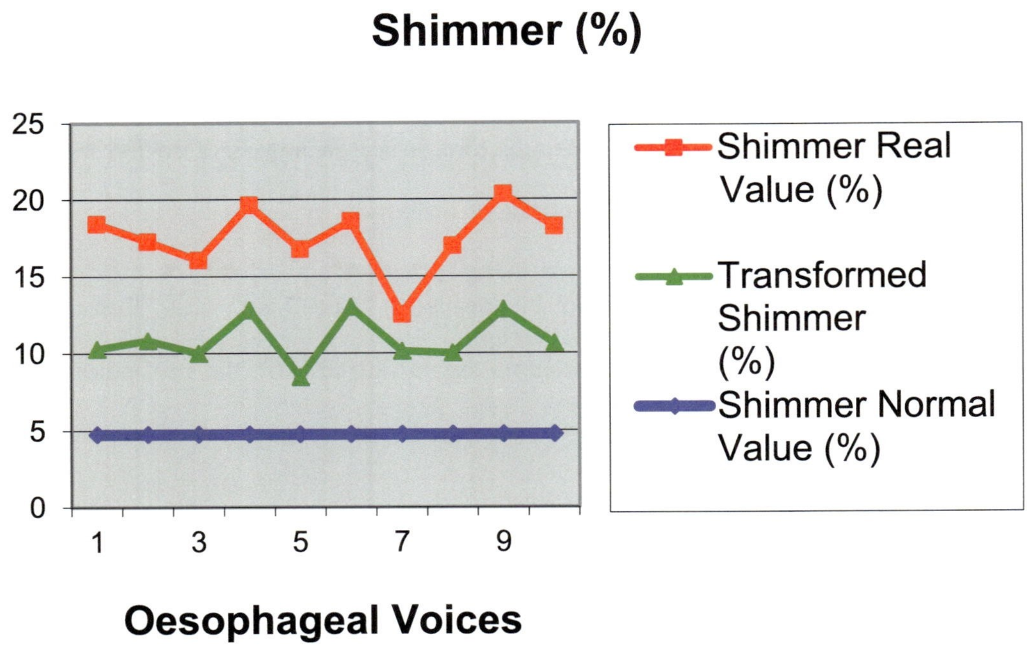

Fig. 8. Shimmer of different voices

Phoneme	Original HNR (dB)	Stabilization HNR (dB)	HNR (dB) increase
a1	-5.001	-1.701	3.300
a2	0.549	1.656	1.107
a3	-3.684	-2.219	1.465
a4	-4.901	-0.668	4.462
a5	-6.375	-2.631	3.744
a6	-6.803	-3.159	3.644
a7	-6.389	-4.451	1.938
a8	-8.724	-5.615	3.109
a9	-3.737	-0.040	3.697
a10	0.930	1.846	0.916
Average			2.941

Table 4. HNR measures with the /a/ phonemes.

As is shown in the table 3 the shimmer has improved in 9 of 10 cases. In four out of the ten voices researched a great goal has been reached. They are not only improved in terms of quality, moreover their values are situated nearest of the limits of normality stipulated in Fig. 8. On the other hand, in the pole's stabilization algorithm, in all cases an increase in harmonics to noise ratio has been achieved. For example, the value of "a1" signal is 5.001dB before processing and 1.701dB afterwards. The increase in improved oesophageal signal, in this case, has been 3.3dB. The fourth column in table 1 shows the enhancement of HNR (dB) before and after processing. It can be appreciated that the improvement in HNR (dB) ranges from 0.916dB, for "a10" signal, to 4.462dB, for "a4". Taking into account all the database, the average HNR improvement (dB) is 2.941dB.

As can be seen in Fig. 9 the increase of the HNR occurs in all voices of the database.

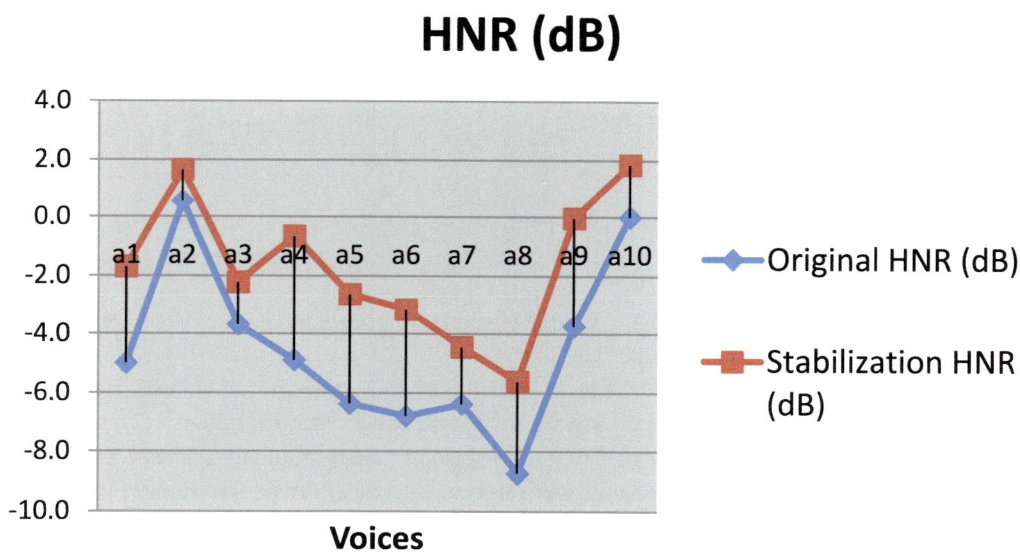

Fig. 9. HNR before and after algorithm

4. Conclusion

It can be concluded that the aimed objectives have been achieved because of the fact that the algorithms are very suitable.

The usage of the Wavelet Transform for the analysis and processing of oesophageal voices is successful in the improvement of the shimmer, which is the aim of the paper. In a extensive analysis its appreciable that it is also good for the improvement of other parameters such as the harmonics to noise ratio. Being a single wavelet detail, optimisation of the computational calculation when processing a simpler signal favours the application of the proposed algorithm to prototypes that process oesophageal signals in real time, in order to improve their quality. On the other hand, the close relationship between characterisation parameters, such as shimmer or jitter and the values of the signal situated in frequency intervals below 100Hz reinforces the suitability of working with bands inferior to the Wavelet Transform, which distinguish spectral components and enable the focusing on particular components.

Therefore, DWT and both pole stabilization improvement are suitable techniques in the speech enhancement context.

5. Acknowledgment

The authors wish to acknowledge the Deusto University which kindly lend infrastructures and material for this investigation. They would also like to thank to all the scholarships that so enthusiastically have collaborated with this project.

Especially it cannot be forgotten the help of the "Asociación Vizcaína de Laringectomizados" whose members, voluntarily lend his voices for this investigation, without their help it would not be possible to carry out this project.

This work was supported in part by the Basque Country Department of Education, Universities and Research.

6. References

Bagshaw, P., Hiller, S., & Jack, M. (1993). Enhanced pitch tracking and the processing of F0 contours for computer aided intonation teaching. Eurospeech, (págs. 22–25). Berlin.

Baken, P. J., & Orlikoff, R. F. (2000). Clinical Measurement of Speech and Voice. San Diego: Singular Publishing Group.

Burrus, C. S., Goinath, R. A., & Guo, H. (1998). Introduction to wavelets and wavelet transforms, a primer. Upper Saddle River NJ (USA): Prentice Hal.

Chen, J., & Kao, Y. (2001). Pitch marking based on an adaptable filter and a peakvalley estimation method. Computational Linguistics and Chinese Language Processing , 6, 1-112.

Cheveigné, A., & Kawahara, H. (2002). Yin, a fundamental frequency estimator for speech and music, 111 (4) (2002). Journal of the Acoustical Society of America, 11 (4).

Daubechies, I. (1992). Ten lectures on wavelets. Philadelphia: 2nd ed. Philadelphia: SIAM.

Deliyski, D. D. (1993). MDVP Acoustic Model and Evaluation of Pathological Voice Production. Eurospeech. Berlin.

Dorken, E., & Nawab, S. H. (1994). Improved musical pitch tracking using principal decomposition analysis. ICASSP, (págs. 217-220).

Doval, B., & Rodet, X. (1991). Estimation of fundamental frequency of musical sound signals. ICASSP, (págs. 3657–3660).

Doval, B., & Rodet, X. (1993). Fundamental frequency estimation and tracking using maximum likelihood harmonic matching and HMMs. ICASSP, (págs. 221–224).

Flanagan, J. L. (1965). Speech Analysis, Synthesis and Perception. Springer .
García, B., Vicente, J., Ruiz, I., Alonso, A., & Loyo, E. (2005). Esophageal Voices: Glottal Flow Restoration. ICASSP, (págs. 141-144).
García, B., Vicente, J., Ruiz, I., Angulo, J. M., & Aramendi, E. (2002). Esoimprove: Esophageal Voices Characterization and Transformation', :. BIOSIGNAL 2002, (págs. 142-144).
Gibiat, V. (1988). Phase space representations of acoustical musical signals, Journal of Sound and Vibration. Journal of Sound and Vibration , 123 (3), 537–572.
Hagmüller, M., & Kubina, G. (2006). Poincaré pitch marks, 48 (12). Speech Communication , 48 (12), 1650-1665.
Kadambe, S., & Bourdreaux-Bartels, G. F. (1991). A Comparison of a Wavelet Functions for Pitch Detection of Speech Signals. ICASSP , (págs. 449-452).
Kadambe, S., & Bourdreaux-Bartels, G. F. (1992). Application Of The Wavelet Transform For Pitch Detection Of Speech Signals. IEEE Transaction On Information Theory (38), 917-924.
Kedem, B. (1986). Spectral analysis and discrimination by zero-crossings. Proceedings of the IEEE , 74 (11), 1477-1493.
Lió, P. (2003). Wavelets in bioinformatics and computational biology: state of art and perspectives. Bioinformatics Review , 19 (1), 2-9.
Mallat, S. (1989). A theory for multiresolution signal decomposition: the wavelet representation. IEEE Transactions on Pattern Analysis and Machine Intelligence , 11 (7), 674- 693.
Mallat, S. (1999). A Wavelet Tour of Signal Processing. A. Press.
Nadeu, C., Pascual, J., & Herdondo, J. (1991). Pitch Determination Using The Cepstrum Of The One-Sided Autocorrelation Sequence. ICASSP.
Ortolan, R. L., Mori, R. N., Pereira, R. R., Cabral, C. M., Pereira, J. C., & Cliquet, A. (2003). Evaluation of Adaptive/Nonadaptive Filtering and Wavelet Transform Techniques for Noise Reduction in EMG Mobile Acquisition Equipment. IEEE transactions on neural systems and rehabilitation engineering, 11 (1), 60-69.
Piszczalski, M., & Galler, B. A. (1979). Predicting musical pitch from component frequency ratios. Journal of the Acoustical Society of America , 66 (3), 710–720.
Sano, H., & Jenkins, B. K. (1989). A neural network model for pitch perception. Computer Music Journal , 13 (3), 41–48.
Sheng, Y. (1996). Wavelet transform. En The transforms and applications handbook Series (págs. 747-827). Boca Raton, Fl (USA): CRC Press.
Tohidypour, H. R., Seyyedsalehi, S. A., & Behbood, H. (2010). Comparison between Wavelet Packet Transform, Bark Wavelet & MFCC for Robust Speech Recongnition tasks. International Conference on Industrial Mechatronics and Automation (ICIMA). Wuhan.
Wing-kei, Y., Kwong-sak, & Kin-hong, W. (1995). Pitch Detection Of Speech Signal In Noisy Environment By Wavelet. SPIE , 2491, 604-614.

Yumoto, E., & Gould, W. J. (1982). Harmonics-to-noise ratio as an index of the degree of hoarseness. Journal of the Acoustical Society of America , 71 (6), 1544–1550.

9

Poisson Noise Removal in Spherical Multichannel Images: Application to Fermi Data

Jérémy Schmitt[1], Jean-Luc Starck[1], Jalal Fadili[2] and Seth Digel[3]
[1]*Laboratoire AIM, CEA/DSM-CNRS-Universite Paris Diderot, IRFU/SEDI-SAP, CEA Saclay, Orme des Merisiers, Gif-sur-Yvette*
[2]*GREYC CNRS-ENSICAEN-Université de Caen, 6, Bd du Maréchal Juin, 14050 Caen Cedex*
[3]*Kavli Institute for Particle Astrophysics and Cosmology, SLAC National Accelerator Laboratory, Menlo Park, CA*
[1,2]*France*
[3]*USA*

1. Introduction

The Fermi Gamma-ray Space Telescope, which was launched by NASA in June 2008, is a powerful space observatory which studies the high-energy gamma-ray sky Atwood (2009). Fermi's main instrument, the Large Area Telescope (LAT), detects photons in an energy range between 20 MeV to greater than 300 GeV. The LAT is much more sensitive than its predecessor, the EGRET telescope on the Compton Gamma Ray Observatory, and is expected to find several thousand gamma-ray point sources, which is an order of magnitude more than its predecessor EGRET Hartman et al. (1999).

Even with its relatively large acceptance ($\sim$2 m^2 sr), the number of photons detected by the LAT outside the Galactic plane and away from intense sources is relatively low and the sky overall has a diffuse glow from cosmic-ray interactions with interstellar gas and low-energy photons that makes a background against which point sources need to be detected. In addition, the per-photon angular resolution of the LAT is relatively poor and strongly energy dependent, ranging from more than 10° at 20 MeV to $\sim$0.1° above 100 GeV. Consequently, the spherical photon count images obtained by Fermi are degraded by the fluctuations on the number of detected photons. **This kind of noise is strongly signal dependent : on the brightest parts of the image like the galactic plane or the brightest sources, we have a lot of photons per pixel, so the photon noise is low. Outside the galactic plane, the number of photons per pixel is low, which means that the photon noise is high. Such a signal-dependent noise can't be accurately modeled by a Gaussian distribution.** The basic photon-imaging model assumes that the number of detected photons at each pixel location is Poisson distributed.

More specifically, the image is considered as a realization of an inhomogeneous Poisson process. This statistical noise makes the source detection more difficult, consequently it is highly desirable to have an efficient denoising method for spherical Poisson data.

Several techniques have been proposed in the literature to estimate Poisson intensity in 2D. A major class of methods adopt a multiscale bayesian framework specifically tailored for Poisson data Nowak & Kolaczyk (2000), independently initiated by Timmerman & Nowak (1999) and Kolaczyk (1999). Lefkimmiaits et al. (2009) proposed an improved bayesian framework for analyzing Poisson processes, based on a multiscale representation of the Poisson process in which the ratios of the underlying Poisson intensities in adjacent scales are modeled as mixtures of conjugate parametric distributions. Another approach includes preprocessing the count data by a variance stabilizing transform (VST) such as the Anscombe Anscombe (1948) and the Fisz Fisz (1955) transforms, applied respectively in the spatial Donoho (1993) or in the wavelet domain Fryźlewicz & Nason (2004). The transform reforms the data so that the noise approximately becomes Gaussian with a constant variance. Standard techniques for independant identically distributed Gaussian noise are then used for denoising. Zhang et al. (2008) proposed a powerful method called Multi-Scale Variance Stabilizing Tranform (MS-VST). It consists in combining a VST with a multiscale transform (wavelets, ridgelets or curvelets), yielding asymptotically normally distributed coefficients with known variances. **The interest of using a multiscale method is to exploit the sparsity properties of the data : the data is transformed into a domain in which it is sparse, and, as the noise is not sparse in any transform domain, it is easy to separate it from the signal. When the noise is Gaussian of known variance, it is easy to remove it with a high thresholding in the wavelet domain.** The choice of the multiscale transform depends on the morphology of the data. Wavelets represent more efficiently regular structures and isotropic singularities, whereas ridgelets are designed to represent global lines in an image, and curvelets represent efficiently curvilinear contours. Significant coefficients are then detected with binary hypothesis testing, and the final estimate is reconstructed with an iterative scheme. In Starck et al. (2009), it was shown that sources can be detected in 3D LAT data (2D+time or 2D+energy) using a specific 3D extension of the MS-VST.

To denoise Fermi maps, we need a method for Poisson intensity estimation on spherical data. It is possible to decompose the spherical data into several 2D projections, denoise each projection and reconstitute the denoised spherical data, but the projection induces some caveats like visual artifacts on the borders or deformation of the sources.

In the scope of the Fermi mission, two of the main scientific objectives are in a sense complementary:

- Detection of point sources to build the catalog of gamma ray sources,
- Study of the Milky Way diffuse background.

The first objective implies the extraction of the Galactic diffuse background. Consequently, we want a method to suppress Poisson noise while extracting a model of the diffuse background. The second objective implies the suppression of the point sources: we want to apply a binary mask on the data (equal to 0 on point sources, and to 1 everywhere else) and to denoise the data while interpolating the missing part. Both objectives are linked: a better knowledge of the Milky Way diffuse background enables us to improve our background model, which leads to a better source detection, while the detected sources are masked to study the diffuse background.

The aim of this chapter is to present a multi-scale representation for spherical data with Poisson noise called Multi-Scale Variance Stabilizing Transform on the Sphere

(MS-VSTS) J.Schmitt et al. (2010), combining the MS-VST Zhang et al. (2008) with various multi-scale transforms on the sphere (wavelets and curvelets) Abrial et al. (2007); Abrial et al. (2008); Starck et al. (2006). Section 1.2 presents some multi-scale transforms on the sphere. Section 1.3 introduces a new multi-scale representation for data with Poisson noise called MS-VSTS. Section 1.4 applies this representation to Poisson noise removal on Fermi data. Section 1.5 presents applications to missing data interpolation and source extraction. Section 1.6 extends the method to multichannel data.

All experiments were performed on HEALPix maps with $nside = 128$ Górski et al. (2005), which corresponds to a good pixelisation choice for the GLAST/FERMI resolution.

2. Wavelets and curvelets on the sphere

New multi-scale transforms on the sphere were developed by Starck et al. (2006). These transforms can be inverted and are easy to compute with the HEALPix pixellisation, and were used for denoising, deconvolution, morphological component analysis and inpainting applications Abrial et al. (2007). In this chapter, here we use the Isotropic Undecimated Wavelet Transform (IUWT) and the Curvelet Transform.

2.1 The HEALPix pixellisation for spherical data

Different kinds of pixellization scheme exist for data on the sphere. For Fermi data, we use the HEALPix representation (Hierarchical Equal Area isoLatitude Pixellization of a sphere) Górski et al. (2005), a curvilinear hierarchical partition of the sphere into quadrilateral pixels of exactly equal area but with varying shape. The base resolution divides the sphere into 12 quadrilateral faces of equal area placed on three rings around the poles and equator. Each face is subsequently divided into $nside^2$ pixels following a quadrilateral multiscale tree structure. (Fig. 1) The pixel centers are located on iso-latitude rings, and pixels from the same ring are equispaced in azimuth. This is critical for computational speed of all operations involving the evaluation of spherical harmonic transforms, including standard numerical analysis operations such as convolution, power spectrum estimation, etc. HEALPix is a standard pixelization scheme in astronomy.

2.2 Isotropic Undecimated Wavelet Transform on the sphere

The Isotropic Undecimated Wavelet Transform on the sphere (IUWT) is a wavelet transform on the sphere based on the spherical harmonics transform and with a very simple reconstruction algorithm. At scale j, we denote $a_j(\theta, \varphi)$ the scale coefficients, and $d_j(\theta, \varphi)$ the wavelet coefficients, with θ denoting the longitude and φ the latitude. Given a scale coefficient a_j, the smooth coefficient a_{j+1} is obtained by a convolution with a low pass filter h_j : $a_{j+1} = a_j * h_j$. The wavelet coefficients are defined by the difference between two consecutive resolutions : $d_{j+1} = a_j - a_{j+1}$. A straightforward reconstruction is then given by:

$$a_0(\theta, \varphi) = a_J(\theta, \varphi) + \sum_{j=1}^{J} d_j(\theta, \varphi) \tag{1}$$

Since this transform is redundant, the procedure for reconstructing an image from its coefficients is not unique and this can be profitably used to impose additional constraints

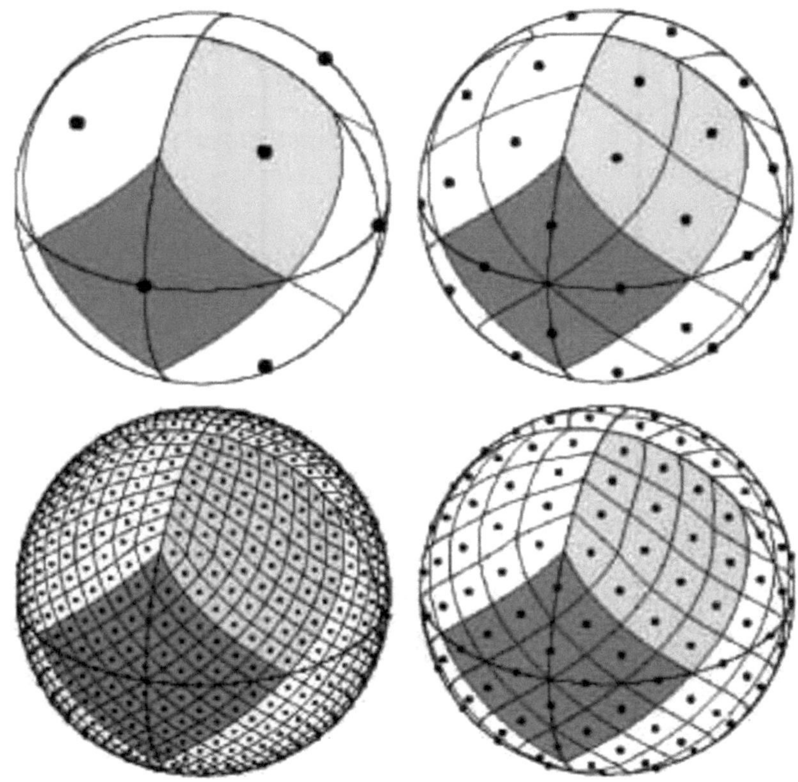

Fig. 1. The HEALPix sampling grid for four different resolutions.

on the synthesis functions (e.g. smoothness, positivity). A reconstruction algorithm based on a variety of filter banks is described in Starck et al. (2006). Figure 2 shows the result of the IUWT on WMAP data (Cosmic Microwave Background).

2.3 Curvelet transform on the sphere

The curvelet transform enables the directional analysis of an image in different scales. The data undergo an Isotropic Undecimated Wavelet Transform on the sphere. Each scale j is then decomposed into smoothly overlapping blocks of side-length B_j in such a way that the overlap between two vertically adjacent blocks is a rectangular array of size $B_j \times B_j/2$, using the HEALPix pixellisation. Finally, the ridgelet transform Candes & Donoho (1999) is applied on each individual block. The method is best for the detection of anisotropic structures and smooth curves and edges of different lengths. The principle of the curvelet transform is schematized on Figure 3. More details can be found in Starck et al. (2006).

2.4 Application to Gaussian denoising on the sphere

Multiscale transforms on the sphere have been used successfully for Gaussian denoising via non-linear filtering or thresholding methods. Hard thresholding, for instance, consists of

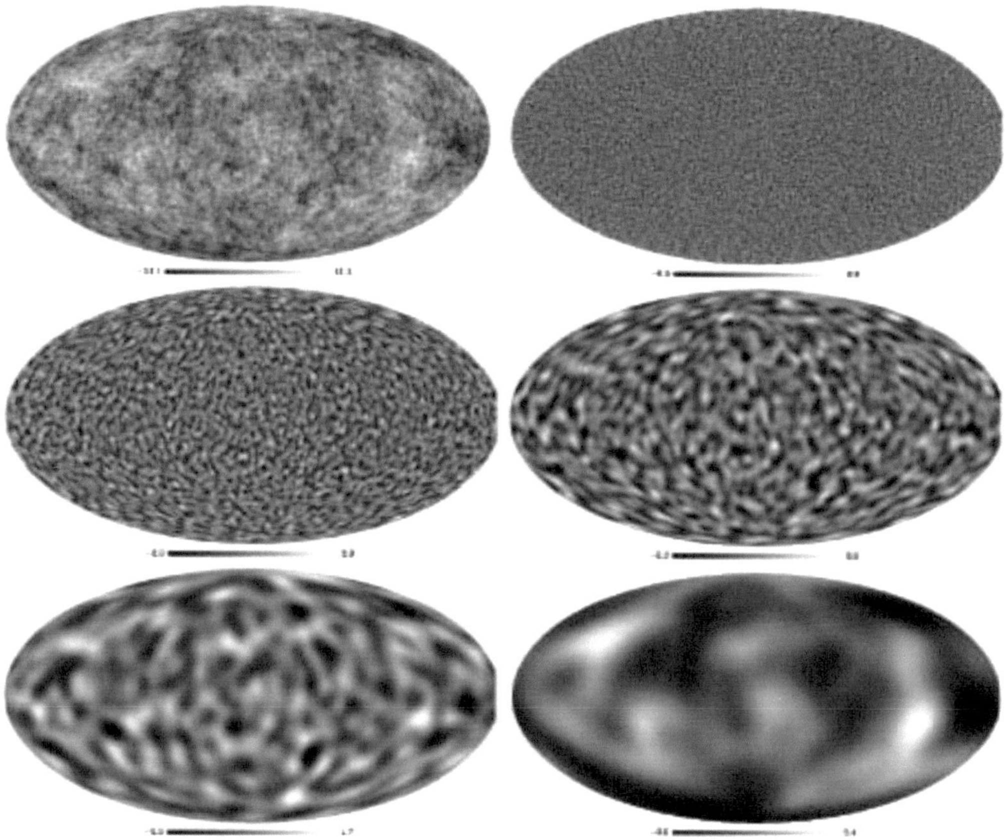

Fig. 2. WMAP data and its wavelet transform on the sphere using five resolution levels (4 wavelet scales and the coarse scale). The sum of these five maps reproduces exactly the original data (top left). *Top*: original data and the first wavelet scale. *Middle*: the second and third wavelet scales. *Bottom*: the fourth wavelet scale and the last smoothed array.

setting all insignificant coefficients (i.e. coefficients with an absolute value below a given threshold) to zero. In practice, we need to estimate the noise standard deviation σ_j in each band j and a coefficient w_j is significant if $|w_j| > \kappa\sigma_j$, where κ is a parameter typically chosen between 3 and 5. Denoting $\mathbf{Y}$ the noisy data and HT_λ the thresholding operator, the filtered data $\mathbf{X}$ are obtained by:

$$\mathbf{X} = \mathbf{\Phi} HT_\lambda(\mathbf{\Phi}^T \mathbf{Y}), \tag{2}$$

where $\mathbf{\Phi}^T$ is the multiscale transform (IUWT or curvelet) and $\mathbf{\Phi}$ is the multiscale reconstruction. λ is a vector which has the size of the number of bands in the used multiscale transform. The thresholding operation thresholds all coefficients in band j with the threshold $\lambda_j = \kappa\sigma_j$.

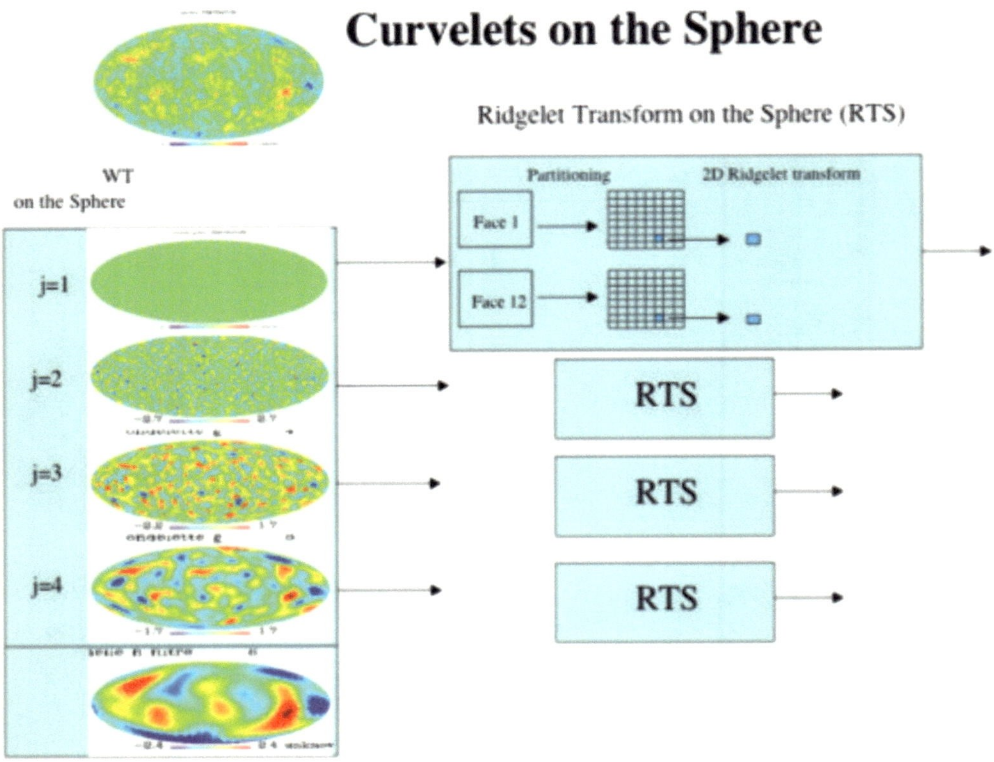

Fig. 3. Principle of curvelets transform on the sphere.

3. Multi-scale transforms on the sphere and Poisson noise

3.1 Principle of the Multi-Scale Variance Stabilizing Transform on the Sphere (MS-VSTS)

In this section, we propose a multi-scale representation designed for data with Poisson noise. The idea is to combine the spherical multi-scale transforms with a variance stabilizing transform (VST), in order to have a multi-scale representation of the data where the noise on multi-scale coefficients behaves like Gaussian noise of known variance. With this representation, it is easy to denoise the data using standard Gaussian denoising methods.

VST of a Poisson process

Given Poisson data $\mathbf{Y} := (Y_i)_i$, each sample $Y_i \sim \mathcal{P}(\lambda_i)$ has a variance $\mathrm{Var}[Y_i] = \lambda_i$. Thus, the variance of $\mathbf{Y}$ is signal-dependant. The aim of a VST $\mathbf{T}$ is to stabilize the data such that each coefficient of $\mathbf{T}(\mathbf{Y})$ has an (asymptotically) constant variance, say 1, irrespective of the value of λ_i. In addition, for the VST used in this study, $T(\mathbf{Y})$ is asymptotically normally distributed. Thus, the VST-transformed data are asymptotically stationary and Gaussian.

The Anscombe Anscombe (1948) transform is a widely used VST which has a simple square-root form

$$\mathbf{T}(Y) := 2\sqrt{Y + 3/8}. \tag{3}$$

We can show that $\mathbf{T}(Y)$ is asymptotically normal as the intensity increases.

$$\mathbf{T}(Y) - 2\sqrt{\lambda} \xrightarrow[\lambda \to +\infty]{\mathcal{D}} \mathcal{N}(0,1) \quad (4)$$

It can be shown that the Anscombe VST requires a high underlying intensity to well stabilize the data (typically for $\lambda \geqslant 10$) Zhang et al. (2008).

VST of a filtered Poisson process

Let $Z_j := \sum_i h[i] Y_{j-i}$ be the filtered process obtained by convolving $(Y_i)_i$ with a discrete filter h. We will use Z to denote any of the Z_j's. Let us define $\tau_k := \sum_i (h[i])^k$ for $k = 1, 2, \cdots$. In addition, we adopt a local homogeneity assumption stating that $\lambda_{j-i} = \lambda$ for all i within the support of h.

We define the square-root transform T as follows:

$$T(Z) := b \cdot \text{sign}(Z + c)|Z + c|^{1/2}, \quad (5)$$

where b is a normalizing factor. It is proven in Zhang et al. (2008) that T is a VST for a filtered Poisson process (with a nonzero-mean filter) in that $T(Y)$ is asymptotically normally distributed with a stabilized variance as λ becomes large.

The Multi-Scale Variance Stabilizing Transform on the Sphere (MS-VSTS) consists in combining the square-root VST with a spherical multi-scale transform (wavelets, curvelets...).

3.2 Wavelets and Poisson noise

This subsection describes the MS-VSTS + IUWT, which is a combination of a square-root VST with the IUWT. The recursive scheme is:

$$\begin{aligned} &\text{IUWT} \begin{cases} a_j = h_{j-1} * a_{j-1} \\ d_j = a_{j-1} - a_j \end{cases} \\ \Longrightarrow\ &\begin{matrix}\text{MS-VSTS} \\ +\ \text{IUWT}\end{matrix} \begin{cases} a_j = h_{j-1} * a_{j-1} \\ d_j = T_{j-1}(a_{j-1}) - T_j(a_j) \end{cases} . \end{aligned} \quad (6)$$

In (6), the filtering on a_{j-1} can be rewritten as a filtering on $a_0 := \mathbf{Y}$, i.e., $a_j = h^{(j)} * a_0$, where $h^{(j)} = h_{j-1} * \cdots * h_1 * h_0$ for $j \geqslant 1$ and $h^{(0)} = \delta$, where δ is the Dirac pulse ($\delta = 1$ on a single pixel and 0 everywhere else). T_j is the VST operator at scale j:

$$T_j(a_j) = b^{(j)} \text{sign}(a_j + c^{(j)}) \sqrt{|a_j + c^{(j)}|}. \quad (7)$$

Let us define $\tau_k^{(j)} := \sum_i (h^{(j)}[i])^k$. In Zhang et al. (2008), it has been shown that, to have an optimal convergence rate for the VST, the constant $c^{(j)}$ associated to $h^{(j)}$ should be set to:

$$c^{(j)} := \frac{7\tau_2^{(j)}}{8\tau_1^{(j)}} - \frac{\tau_3^{(j)}}{2\tau_2^{(j)}}. \quad (8)$$

The MS-VSTS+IUWT procedure is directly invertible as we have:

$$a_0(\theta,\varphi) = T_0^{-1}\left[T_J(a_J) + \sum_{j=1}^{J} d_j\right](\theta,\varphi). \tag{9}$$

Setting $b^{(j)} := \text{sign}(\tau_1^{(j)})/\sqrt{|\tau_1^{(j)}|}$, if λ is constant within the support of the filter. $h^{(j)}$, then we have Zhang et al. (2008):

$$d_j(\theta,\varphi) \xrightarrow[\lambda \to +\infty]{\mathcal{D}} \mathcal{N}\left(0, \frac{\tau_2^{(j-1)}}{4\tau_1^{(j-1)^2}} + \frac{\tau_2^{(j)}}{4\tau_1^{(j)^2}} - \frac{\langle h^{(j-1)}, h^{(j)}\rangle}{2\tau_1^{(j-1)}\tau_1^{(j)}}\right), \tag{10}$$

where $\langle .,. \rangle$ denotes inner product.

This means that the detail coefficients issued from locally homogeneous parts of the signal follow asymptotically a central normal distribution with an intensity-independant variance which relies solely on the filter h and the current scale for a given filter h. Let us define $\sigma_{(j)}^2$ the stabilized variance at scale j for a locally homogeneous part of the signal:

$$\sigma_{(j)}^2 = \frac{\tau_2^{(j-1)}}{4\tau_1^{(j-1)^2}} + \frac{\tau_2^{(j)}}{4\tau_1^{(j)^2}} - \frac{\langle h^{(j-1)}, h^{(j)}\rangle}{2\tau_1^{(j-1)}\tau_1^{(j)}}. \tag{11}$$

To compute the $\sigma_{(j)}$, $b^{(j)}$,$c^{(j)}$,$\tau_k^{(j)}$, we only have to know the filters $h^{(j)}$. We compute these filters thanks to the formula $a_j = h^{(j)} * a_0$, by applying the IUWT to a Dirac pulse $a_0 = \delta$. Then, the $h^{(j)}$ are the scaling coefficients of the IUWT. The $\sigma_{(j)}$ have been precomputed for a 6-scaled IUWT (Table 1).

Wavelet scale j	Value of σ_j
1	0.484704
2	0.0552595
3	0.0236458
4	0.0114056
5	0.00567026

Table 1. Precomputed values of the variances σ_j of the wavelet coefficients.

We have simulated Poisson images of different constant intensities λ, computed the IUWT with MS-VSTS on each image and observed the variation of the normalized value of $\sigma_{(j)}$ ($(\mathbf{œ}_{(\mathbf{j})})_{\text{simulated}}/(\sigma_{(j)})_{\text{theoretical}}$) as a function of λ for each scale j (Fig. 4). We see that the wavelet coefficients are stabilized when $\lambda \gtrsim 0.1$ except for the first wavelet scale, which is largely noise. In Fig. 5, we compare the result of MS-VSTS with Anscombe + wavelet shrinkage, on sources of varying intensities. We see that MS-VSTS works well on sources of very low intensities, whereas Anscombe does not work when the intensity is too low.

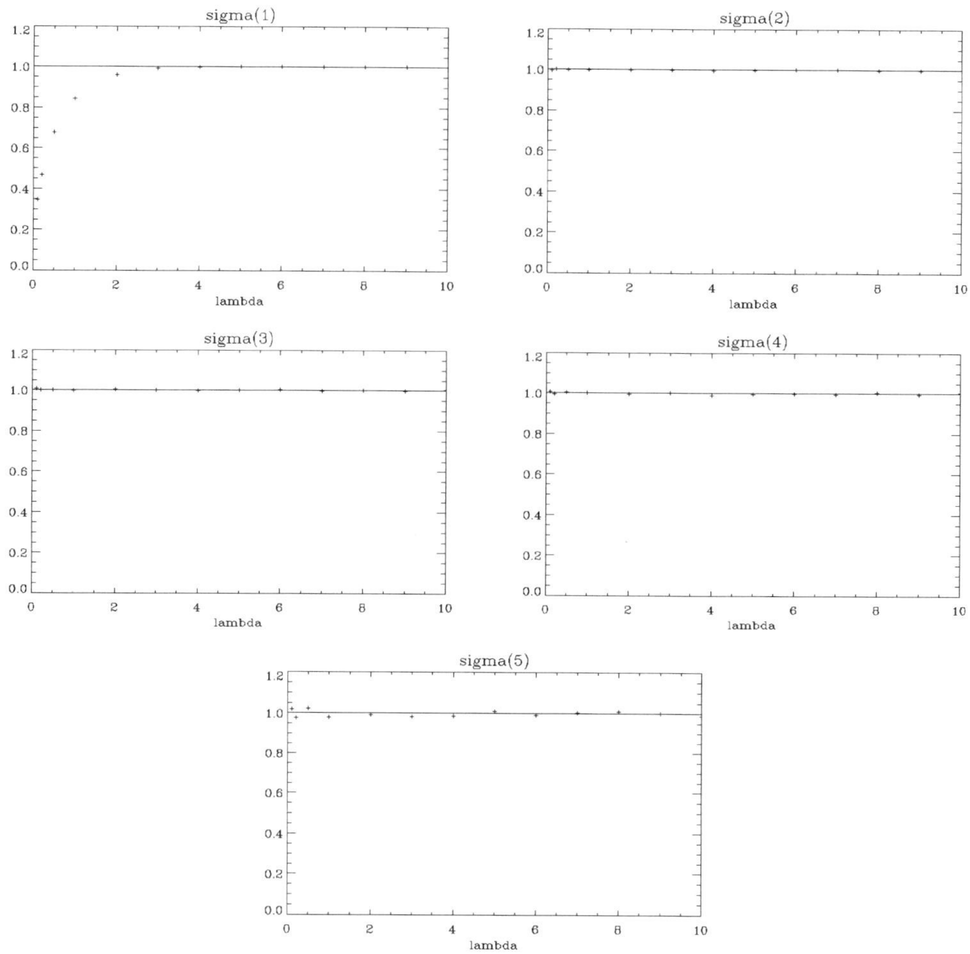

Fig. 4. Normalized value $((\text{œ}_{(j)})_{\text{simulated}}/(\sigma_{(j)})_{\text{theoretical}})$ of the stabilized variances at each scale j as a function of λ.

3.3 Curvelets and Poisson noise

As the first step of the algorithm is an IUWT, we can stabilize each resolution level as in Equation 6. We then apply the local ridgelet transform on each stabilized wavelet band.

It is not as straightforward as with the IUWT to derive the asymptotic noise variance in the stabilized curvelet domain. In our experiments, we derived them using simulated Poisson data of stationary intensity level λ. After having checked that the standard deviation in the curvelet bands becomes stabilized as the intensity level increases (which means that the stabilization is working properly), we stored the standard deviation $\sigma_{j,l}$ for each wavelet scale j and each ridgelet band l (Table 2).

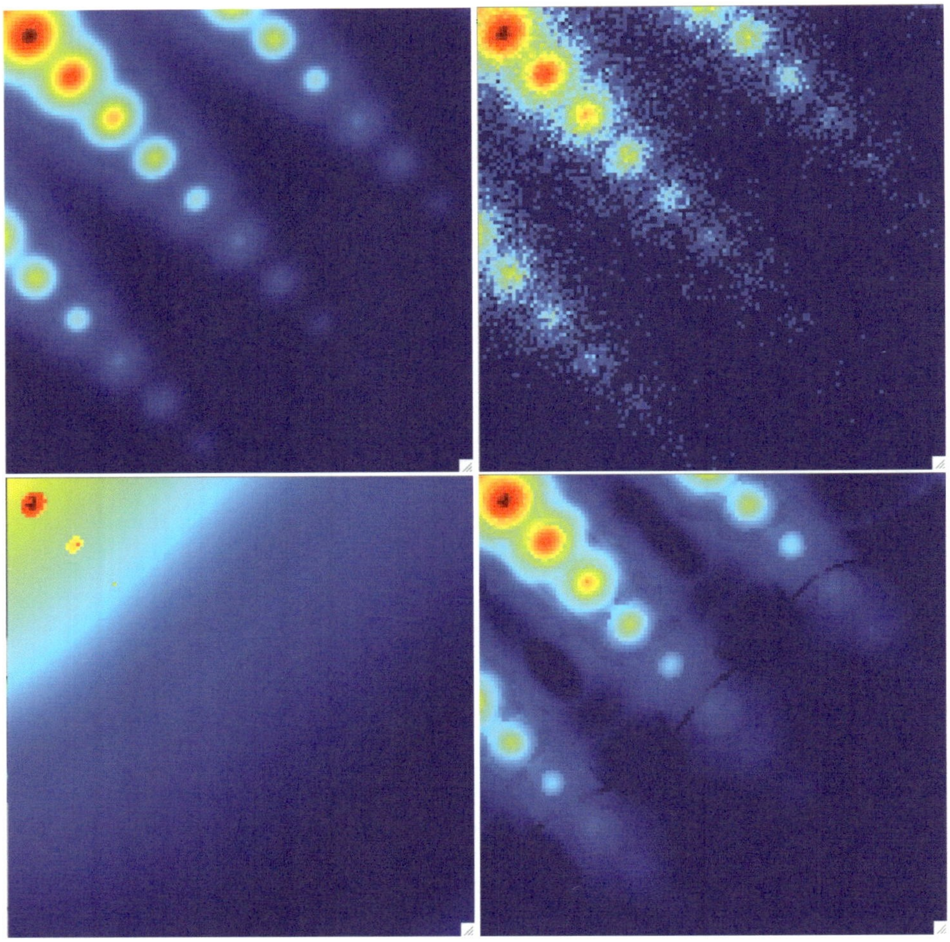

Fig. 5. Comparison of MS-VSTS with Anscombe + wavelet shrinkage on a single face of the first scale of the HEALPix pixelization (angular extent: $\pi/3sr$). *Top Left* : Sources of varying intensity. *Top Right* : Sources of varying intensity with Poisson noise. *Bottom Left* : Poisson sources of varying intensity reconstructed with Anscombe + wavelet shrinkage. *Bottom Right* : Poisson sources of varying intensity reconstructed with MS-VSTS.

j	$l=1$	$l=2$	$l=3$	$l=4$
1	1.74550	0.348175		
2	0.230621	0.248233	0.196981	
3	0.0548140	0.0989918	0.219056	
4	0.0212912	0.0417454	0.0875663	0.20375
5	0.00989616	0.0158273	0.0352021	0.163248

Table 2. Asymptotic values of the variances $\sigma_{j,k}$ of the curvelet coefficients.

4. Application to Poisson denoising on the sphere

4.1 MS-VSTS + IUWT

Under the hypothesis of homogeneous Poisson intensity, the stabilized wavelet coefficients d_j behave like centered Gaussian variables of standard deviation $\sigma_{(j)}$. We can detect significant coefficients with binary hypothesis testing as in Gaussian denoising.

Under the null hypothesis $\mathcal{H}_0$ of homogeneous Poisson intensity, the distribution of the stabilized wavelet coefficient $d_j[k]$ at scale j and location index k can be written as:

$$p(d_j[k]) = \frac{1}{\sqrt{2\pi}\sigma_j} \exp(-d_j[k]^2/2\sigma_j^2). \tag{12}$$

The rejection of the hypothesis $\mathcal{H}_0$ depends on the double-sided p-value:

$$p_j[k] = 2\frac{1}{\sqrt{2\pi}\sigma_j} \int_{|d_j[k]|}^{+\infty} \exp(-x^2/2\sigma_j^2)dx. \tag{13}$$

Consequently, to accept or reject $\mathcal{H}_0$, we compare each $|d_j[k]|$ with a critical threshold $\kappa\sigma_j$, $\kappa = 3, 4$ or 5 corresponding respectively to significance levels. This amounts to deciding that:

- if $|d_j[k]| \geqslant \kappa\sigma_j$, $d_j[k]$ is significant.
- if $|d_j[k]| < \kappa\sigma_j$, $d_j[k]$ is not significant.

Then we have to invert the MS-VSTS scheme to reconstruct the estimate. However, although the direct inversion is possible (Eq. (**??**)), it can not guarantee a positive intensity estimate, while the Poisson intensity is always nonnegative. A positivity projection can be applied, but important structures could be lost in the estimate. To tackle this problem, we reformulate the reconstruction as a convex optimisation problem and solve it iteratively with an algorithm based on Hybrid Steepest Descent (HSD) Yamada (2001).

We define the multiresolution support $\mathcal{M}$, which is determined by the set of detected significant coefficients after hypothesis testing:

$$\mathcal{M} := \{(j,k) | \text{if } d_j[k] \text{ is declared significant}\}. \tag{14}$$

We formulate the reconstruction problem as a convex constrained minimization problem:

$$\begin{gathered} \operatorname*{Arg\,min}_{\mathbf{X}} \|\mathbf{\Phi}^T\mathbf{X}\|_1, \text{s.t.} \\ \begin{cases} \mathbf{X} \geqslant 0, \\ \forall (j,k) \in \mathcal{M}, (\mathbf{\Phi}^T\mathbf{X})_j[k] = (\mathbf{\Phi}^T\mathbf{Y})_j[k], \end{cases} \end{gathered} \tag{15}$$

where $\mathbf{\Phi}$ denotes the IUWT synthesis operator.

This problem is solved with the following iterative scheme: the image is initialised by $\mathbf{X}^{(0)} = 0$, and the iteration scheme is, for $n = 0$ to $N_{\max} - 1$:

$$\tilde{\mathbf{X}} = P_+[\mathbf{X}^{(n)} + \mathbf{\Phi} P_{\mathcal{M}} \mathbf{\Phi}^T(\mathbf{Y} - \mathbf{X}^{(n)})] \tag{16}$$

$$\mathbf{X}^{(n+1)} = \mathbf{\Phi}\mathrm{ST}_{\lambda_n}[\mathbf{\Phi}^T\tilde{\mathbf{X}}] \tag{17}$$

where P_+ denotes the projection on the positive orthant, $P_{\mathcal{M}}$ denotes the projection on the multiresolution support $\mathcal{M}$:

$$P_{\mathcal{M}} d_j[k] = \begin{cases} d_j[k] & \text{if } (j,k) \in \mathcal{M}, \\ 0 & \text{otherwise} \end{cases}. \tag{18}$$

and ST_{λ_n} the soft-thresholding with threshold λ_n:

$$\mathrm{ST}_{\lambda_n}[d] = \begin{cases} \mathrm{sign}(d)(|d| - \lambda_n) & \text{if } |d| \geqslant \lambda_n, \\ 0 & \text{otherwise} \end{cases}. \tag{19}$$

We chose a decreasing threshold $\lambda_n = \frac{N_{\max}-n}{N_{\max}-1}, n = 1, 2, \cdots, N_{\max}$.

The final estimate of the Poisson intensity is: $\hat{\mathbf{\Lambda}} = \mathbf{X}^{(N_{\max})}$. Algorithm 1 summarizes the main steps of the MS-VSTS + IUWT denoising algorithm.

Algorithm 1 MS-VSTS + IUWT Denoising

Require: data $a_0 := \mathbf{Y}$, number of iterations $N_{\max}$, threshold κ

Detection

for $j = 1$ to J **do**
  Compute a_j and d_j using (6).
  Hard threshold $|d_j[k]|$ with threshold $\kappa\sigma_j$ and update $\mathcal{M}$.
end for

Estimation

Initialize $\mathbf{X}^{(0)} = 0$, $\lambda_0 = 1$.
for $n = 0$ to $N_{\max} - 1$ **do**
  $\tilde{\mathbf{X}} = P_+[\mathbf{X}^{(n)} + \mathbf{\Phi} P_{\mathcal{M}} \mathbf{\Phi}^T(\mathbf{Y} - \mathbf{X}^{(n)})]$.
  $\mathbf{X}^{(n+1)} = \mathbf{\Phi}\mathrm{ST}_{\lambda_n}[\mathbf{\Phi}^T\tilde{\mathbf{X}}]$.
  $\lambda_{n+1} = \frac{N_{\max}-(n+1)}{N_{\max}-1}$.
end for
Get the estimate $\hat{\mathbf{\Lambda}} = \mathbf{X}^{(N_{\max})}$.

4.2 Multi-resolution support adaptation

When two sources are too close, the less intense source may not be detected because of the negative wavelet coefficients of the brightest source. To avoid such a drawback, we may update the multi-resolution support at each iteration. The idea is to withdraw the detected sources and to make a detection on the remaining residual, so as to detect the sources which may have been missed at the first detection.

At each iteration n, we compute the MS-VSTS of $\mathbf{X}^{(n)}$. We denote $d_j^{(n)}[k]$ the stabilised coefficients of $\mathbf{X}^{(n)}$. We make a hard thresholding on $(d_j[k] - d_j^{(n)}[k])$ with the same thresholds as in the detection step. Significant coefficients are added to the multiresolution support $\mathcal{M}$.

The main steps of the algorithm are summarized in Algorithm 2. In practice, we use Algorithm 2 instead of Algorithm 1 in our experiments.

Algorithm 2 MS-VSTS + IUWT Denoising + Multiresolution Support Adaptation

Require: data $a_0 := \mathbf{Y}$, number of iterations $N_{\max}$, threshold κ
Detection
for $j = 1$ to J **do**
  Compute a_j and d_j using (6).
  Hard threshold $|d_j[k]|$ with threshold $\kappa\sigma_j$ and update $\mathcal{M}$.
end for
Estimation
Initialize $\mathbf{X}^{(0)} = 0$, $\lambda_0 = 1$.
for $n = 0$ to $N_{\max} - 1$ **do**
  $\tilde{\mathbf{X}} = P_+[\mathbf{X}^{(n)} + \mathbf{\Phi} P_{\mathcal{M}} \mathbf{\Phi}^T(\mathbf{Y} - \mathbf{X}^{(n)})]$.
  $\mathbf{X}^{(n+1)} = \mathbf{\Phi}\mathrm{ST}_{\lambda_n}[\mathbf{\Phi}^T \tilde{\mathbf{X}}]$.
  Compute the MS-VSTS on $\mathbf{X}^{(n)}$ to get the stabilised coeffcients $d_j^{(n)}$.
  Hard threshold $|d_j[k] - d_j^{(n)}[k]|$ and update $\mathcal{M}$.
  $\lambda_{n+1} = \frac{N_{\max} - (n+1)}{N_{\max} - 1}$.
end for
Get the estimate $\hat{\mathbf{\Lambda}} = \mathbf{X}^{(N_{\max})}$.

4.3 MS-VSTS + curvelets

Insignificant coefficients are zeroed by using the same hypothesis testing framework as in the wavelet scale. At each wavelet scale j and ridgelet band k, we make a hard thresholding on curvelet coefficients with threshold $\kappa\sigma_{j,k}$, $\kappa = 3, 4$ or 5. Finally, a direct reconstruction can be performed by first inverting the local ridgelet transforms and then inverting the MS-VST + IUWT (Equation (9)). An iterative reconstruction may also be performed.

Algorithm 3 summarizes the main steps of the MS-VSTS + Curvelets denoising algorithm.

Algorithm 3 MS-VSTS + Curvelets Denoising

Apply the MS-VST + IUWT with J scales to get the stabilized wavelet subbands d_j.
Set $B_1 = B_{\min}$.
for $j = 1$ to J **do**
  Partition the subband d_j with blocks of side-length B_j and apply the digital ridgelet transform to each block to obtain the stabilized curvelet coefficients.
  if j modulo $2 = 1$ **then**
    $B_{j+1} = 2B_j$
  else
    $B_{j+1} = B_j$
  end if
  HTs on the stabilized curvelet coefficients.
end for
Invert the ridgelet transform in each block before inverting the MS-VST + IUWT.

4.4 Experiments

The method was tested on simulated Fermi data. The simulated data are the sum of a Milky Way diffuse background model and 1000 gamma ray point sources. We based our

Galactic diffuse emission model intensity on the model *gll_iem_v02* obtained at the Fermi Science Support Center Myers (2009) . This model results from a fit of the LAT photons with various gas templates as well as inverse Compton in several energy bands. We used a realistic point-spread function for the sources, based on Monte Carlo simulations of the LAT and accelerator tests, that scale with energy approximately as $0.8(E/1GeV)^{-0.8}$ degrees (68% containment angle). The positions of the 205 brightest sources were taken from the Fermi 3-month source list Abdo et al. (2009). The positions of the 795 remaining sources follow the LAT 1-year Point Source Catalog Myers (2010) source distribution: each simulated source was randomly sorted in a box in Galactic coordinates of Δl=5° and Δb=1° around a LAT 1-year catalog source. We simulated each source assuming a power-law dependence with its spectral index given by the 3-month source list and the first year catalog. We used an exposure of $3.10^{10} s.cm^2$ corresponding approximatively to one year of Fermi all-sky survey around 1 GeV. The simulated counts map shown in this section correspond to photons energy from 150 MeV to 20 GeV.

Fig. 6 compares the result of denoising with MS-VST + IUWT (Algorithm 1), MS-VST + curvelets (Algorithm 3) and Anscombe VST + wavelet shrinkage on a simulated Fermi map. Fig. 7 shows the results on one single face of the first scale of the HEALPix pixelization(angular extent: $\pi/3sr$). As expected from theory, the Anscombe method produces poor results to denoise Fermi data, because the underlying intensity is too weak. Both wavelet and curvelet denoising on the sphere perform much better. For this application, wavelets are slightly better than curvelets ($SNR_{wavelets} = 65.8dB$, $SNR_{curvelets} = 37.3dB$, $SNR(dB) = 20\log(\sigma_{signal}/\sigma_{noise})$). As this image contains many point sources, this result is expected. Indeed wavelets are better than curvelets to represent isotropic objects.

5. Application to inpainting and source extraction

5.1 Milky way diffuse background study: denoising and inpainting

In order to extract from the Fermi photon maps the Galactic diffuse emission, we want to remove the point sources from the Fermi image. As our HSD algorithm is very close to the MCA (Morphological Component Analysis) algorithm Starck et al. (2004), an idea is to mask the most intense sources and to modify our algorithm in order to interpolate through the gaps exactly as in the MCA-Inpainting algorithm Abrial et al. (2007). This modified algorithm can be called MS-VSTS-Inpainting algorithm. **What we want to do is to remove the information due to point sources from the maps, in order to keep only the information due to the galactic background. The MS-VSTS-Inpainting algorithm interpolates the missing data to reconstruct a map of the galactic background, which can now be fitted by a theoretical model. The interpolation uses the sparsity of the data in the wavelet domain : the gaps are filled so that the result is the sparsiest possible in the wavelet domain.**

The problem can be reformulated as a convex constrained minimization problem:

$$\operatorname{Arg}\min_{\mathbf{X}} \|\mathbf{\Phi}^T\mathbf{X}\|_1, \text{s.t.} \quad \begin{cases} \mathbf{X} \geqslant 0, \\ \forall (j,k) \in \mathcal{M}, (\mathbf{\Phi}^T\Pi\mathbf{X})_j[k] = (\mathbf{\Phi}^T\mathbf{Y})_j[k], \end{cases} \tag{20}$$

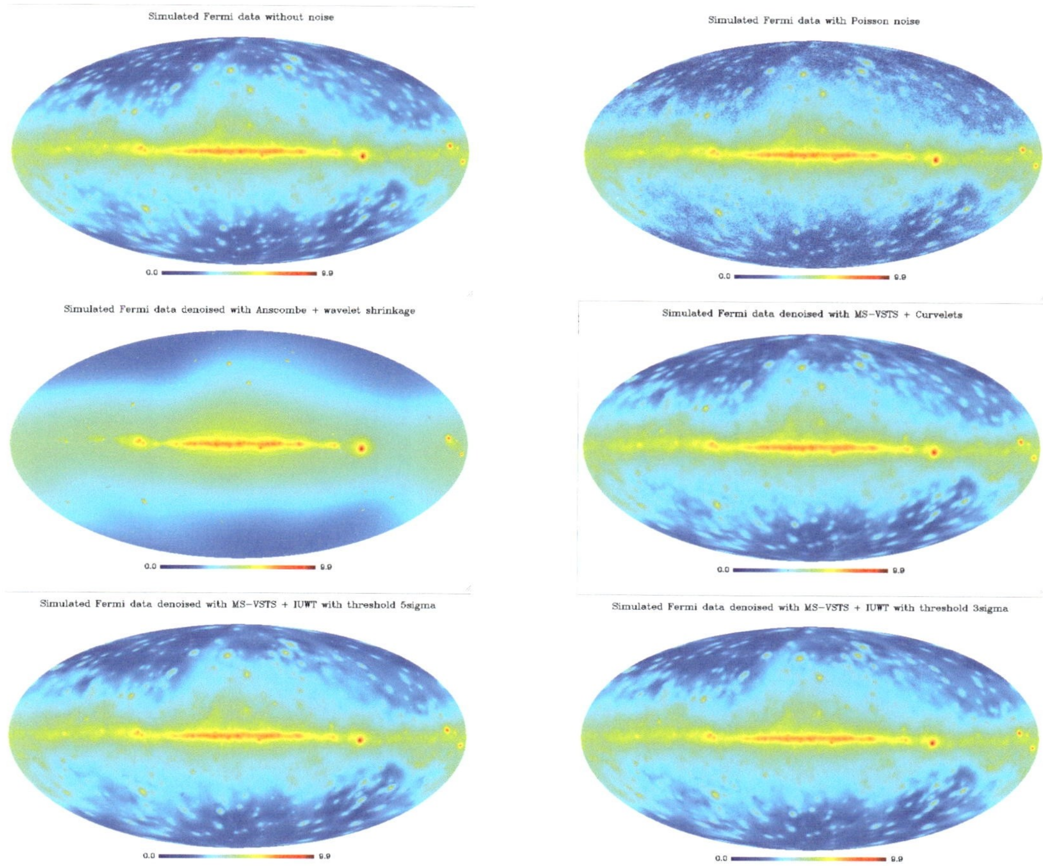

Fig. 6. *Top Left*: Fermi simulated map without noise. *Top Right*: Fermi simulated map with Poisson noise. *Middle Left*: Fermi simulated map denoised with Anscombe VST + wavelet shrinkage. *Middle Right*: Fermi simulated map denoised with MS-VSTS + curvelets (Algorithm 3). *Bottom Left*: Fermi simulated map denoised with MS-VSTS + IUWT (Algorithm 1) with threshold $5\sigma_j$. *Bottom Right*: Fermi simulated map denoised with MS-VSTS + IUWT (Algorithm 1) with threshold $3\sigma_j$. Pictures are in logarithmic scale.

where Π is a binary mask (1 on valid data and 0 on invalid data).

The iterative scheme can be adapted to cope with a binary mask, which gives:

$$\tilde{\mathbf{X}} = P_+[\mathbf{X}^{(n)} + \mathbf{\Phi} P_{\mathcal{M}} \mathbf{\Phi}^T \Pi(\mathbf{Y} - \mathbf{X}^{(n)})], \tag{21}$$

$$\mathbf{X}^{(n+1)} = \mathbf{\Phi} \mathrm{ST}_{\lambda_n}[\mathbf{\Phi} \tilde{\mathbf{X}}]. \tag{22}$$

The thresholding strategy has to be adapted. Indeed, for the inpainting task we need to have a very large initial threshold in order to have a very smooth image in the beginning and to

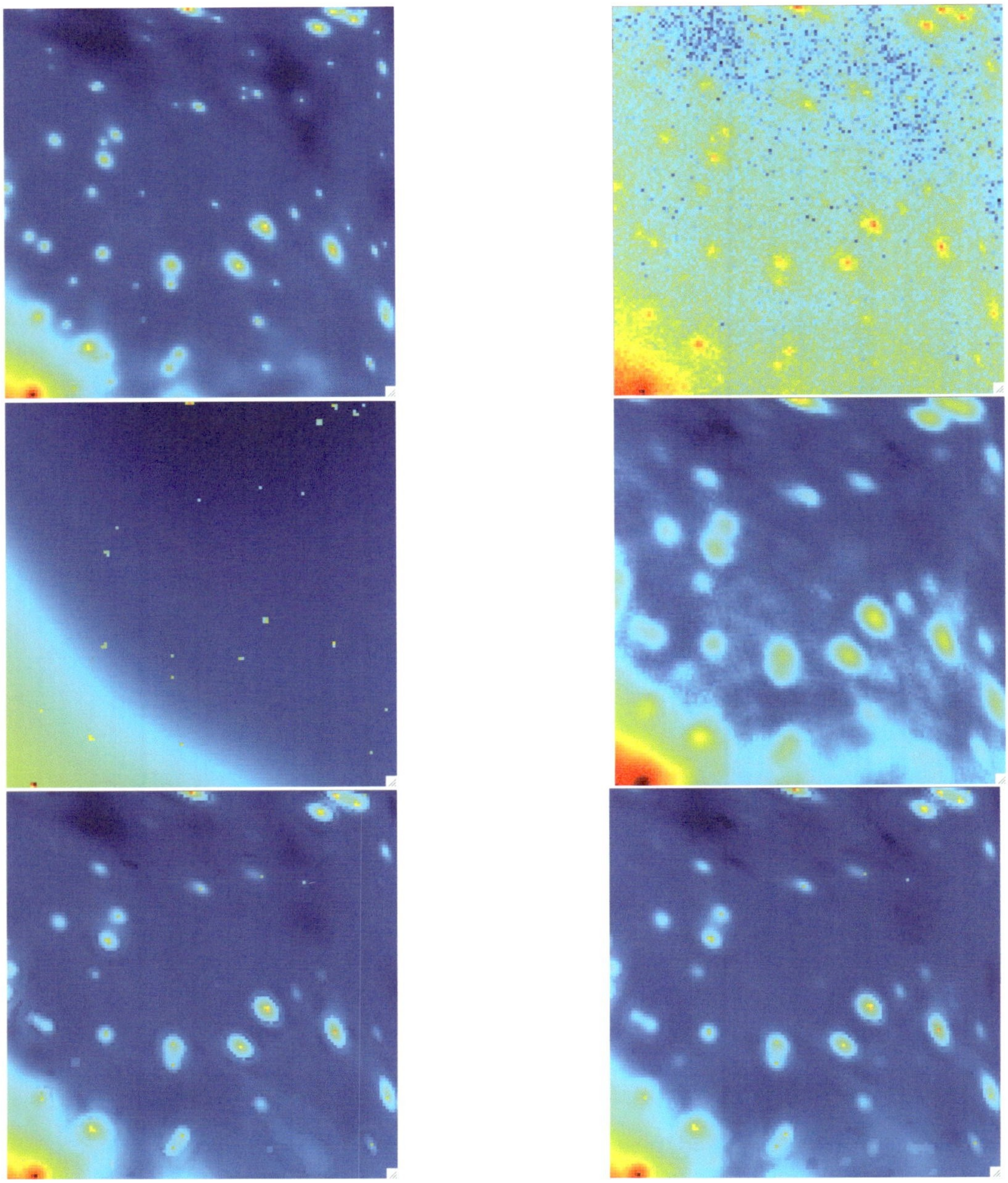

Fig. 7. View of a single HEALPix face (angular extent: $\pi/3sr$) from the results of Figure 6. *Top Left*: Fermi simulated map without noise. *Top Right*: Fermi simulated map with Poisson noise. *Middle Left*: Fermi simulated map denoised with Anscombe VST + wavelet shrinkage. *Middle Right*: Fermi simulated map denoised with MS-VSTS + curvelets (Algorithm 3). *Bottom Left*: Fermi simulated map denoised with MS-VSTS + IUWT (Algorithm 1) with threshold $5\sigma_j$. *Bottom Right*: Fermi simulated map denoised with MS-VSTS + IUWT (Algorithm 1) with threshold $3\sigma_j$. Pictures are in logarithmic scale.

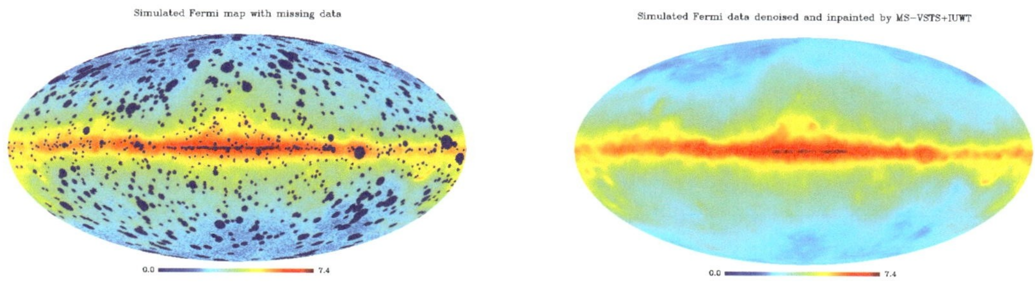

Fig. 8. MS-VSTS - Inpainting. *Left*: Fermi simulated map with Poisson noise and the most luminous sources masked. *Right*: Fermi simulated map denoised and inpainted with wavelets (Algorithm 4). Pictures are in logarithmic scale.

refine the details progressively. We chose an exponentially decreasing threshold:

$$\lambda_n = \lambda_{\max}(2^{(\frac{N_{\max}-n}{N_{\max}-1})} - 1), n = 1, 2, \cdots, N_{\max}, \qquad (23)$$

where $\lambda_{\max} = \max(\mathbf{\Phi}^T \mathbf{X})$.

Algorithm 4 MS-VST + IUWT Denoising + Inpainting

Require: data $a_0 := \mathbf{Y}$, mask Π, number of iterations $N_{\max}$, threshold κ.

<u>***Detection***</u>

for $j = 1$ to J **do**
  Compute a_j and d_j using (6).
  Hard threshold $|d_j[k]|$ with threshold $\kappa\sigma_j$ and update $\mathcal{M}$.
end for

<u>***Estimation***</u>

Initialize $\mathbf{X}^{(0)} = 0$, $\lambda_0 = \lambda_{\max}$.
for $n = 0$ to $N_{\max} - 1$ **do**
  $\tilde{\mathbf{X}} = P_+[\mathbf{X}^{(n)} + \mathbf{\Phi} P_{\mathcal{M}} \mathbf{\Phi}^T \Pi(\mathbf{Y} - \mathbf{X}^{(n)})]$.
  $\mathbf{X}^{(n+1)} = \mathbf{\Phi}^{\mathrm{ST}}_{\lambda_n}[\mathbf{\Phi}^T \tilde{\mathbf{X}}]$.
  $\lambda_{n+1} = \lambda_{\max}(2^{(\frac{N_{\max}-(n+1)}{N_{\max}-1})} - 1)$
end for
Get the estimate $\hat{\mathbf{\Lambda}} = \mathbf{X}^{(N_{\max})}$.

Experiment

We applied this method on simulated Fermi data where we masked the **500** most luminous sources. **(with the highest photon per pixel flux) The other sources are not intense enough to be differencied from the background.**

The results are on Figure 8. The MS-VST + IUWT + Inpainting method (Algorithm 4) interpolates the missing data very well. Indeed, the missing part can not be seen anymore in the inpainted map, which shows that the diffuse emission component has been correctly reconstructed.

5.2 Source detection: denoising and background modeling

5.2.1 Method

In the case of Fermi data, the diffuse gamma-ray emission from the Milky Way, due to interaction between cosmic rays and interstellar gas and radiation, makes a relatively intense background. We have to extract this background in order to detect point sources. This diffuse interstellar emission can be modelled by a linear combination of gas templates and inverse compton map. We can use such a background model and incorporate a background removal in our denoising algorithm.

We denote $\mathbf{Y}$ the data, $\mathbf{B}$ the background we want to remove, and $d_j^{(b)}[k]$ the MS-VSTS coefficients of $\mathbf{B}$ at scale j and position k. We determine the multi-resolution support by comparing $|d_j[k] - d_j^{(b)}[k]|$ with $\kappa\sigma_j$.

We formulate the reconstruction problem as a convex constrained minimization problem:

$$\operatorname*{Arg\,min}_{\mathbf{X}} \|\mathbf{\Phi}^T\mathbf{X}\|_1, \text{s.t.} \quad \begin{cases} \mathbf{X} \geqslant 0, \\ \forall (j,k) \in \mathcal{M}, (\mathbf{\Phi}^T\mathbf{X})_j[k] = (\mathbf{\Phi}^T(\mathbf{Y}-\mathbf{B}))_j[k], \end{cases} \tag{24}$$

Then, the reconstruction algorithm scheme becomes:

$$\tilde{\mathbf{X}} = P_+[\mathbf{X}^{(n)} + \mathbf{\Phi} P_{\mathcal{M}} \mathbf{\Phi}^T(\mathbf{Y} - \mathbf{B} - \mathbf{X}^{(n)})], \tag{25}$$

$$\mathbf{X}^{(n+1)} = \mathbf{\Phi}\mathrm{ST}_{\lambda_n}[\mathbf{\Phi}^T\tilde{\mathbf{X}}]. \tag{26}$$

The algorithm is illustrated by the theoretical study in Figure 9. We denoise Poisson data while separating a single source, which is a Gaussian of standard deviation equal to 0.01, from a background, which is a sum of two Gaussians of standard deviation equal to 0.1 and 0.01 respectively.

Algorithm 5 MS-VSTS + IUWT Denoising + Background extraction

Require: data $a_0 := \mathbf{Y}$, background B, number of iterations $N_{\max}$, threshold κ.

Detection

for $j = 1$ to J **do**
  Compute a_j and d_j using (6).
  Hard threshold $(d_j[k] - d_j^{(b)}[k])$ with threshold $\kappa\sigma_j$ and update $\mathcal{M}$.
end for

Estimation

Initialize $\mathbf{X}^{(0)} = 0$, $\lambda_0 = 1$.
for $n = 0$ to $N_{\max} - 1$ **do**
  $\tilde{\mathbf{X}} = P_+[\mathbf{X}^{(n)} + \mathbf{\Phi} P_{\mathcal{M}} \mathbf{\Phi}^T(\mathbf{Y} - \mathbf{B} - \mathbf{X}^{(n)})]$.
  $\mathbf{X}^{(n+1)} = \mathbf{\Phi}\mathrm{ST}_{\lambda_n}[\mathbf{\Phi}^T\tilde{\mathbf{X}}]$.
  $\lambda_{n+1} = \frac{N_{\max} - (n+1)}{N_{\max} - 1}$.
end for
Get the estimate $\hat{\mathbf{\Lambda}} = \mathbf{X}^{(N_{\max})}$.

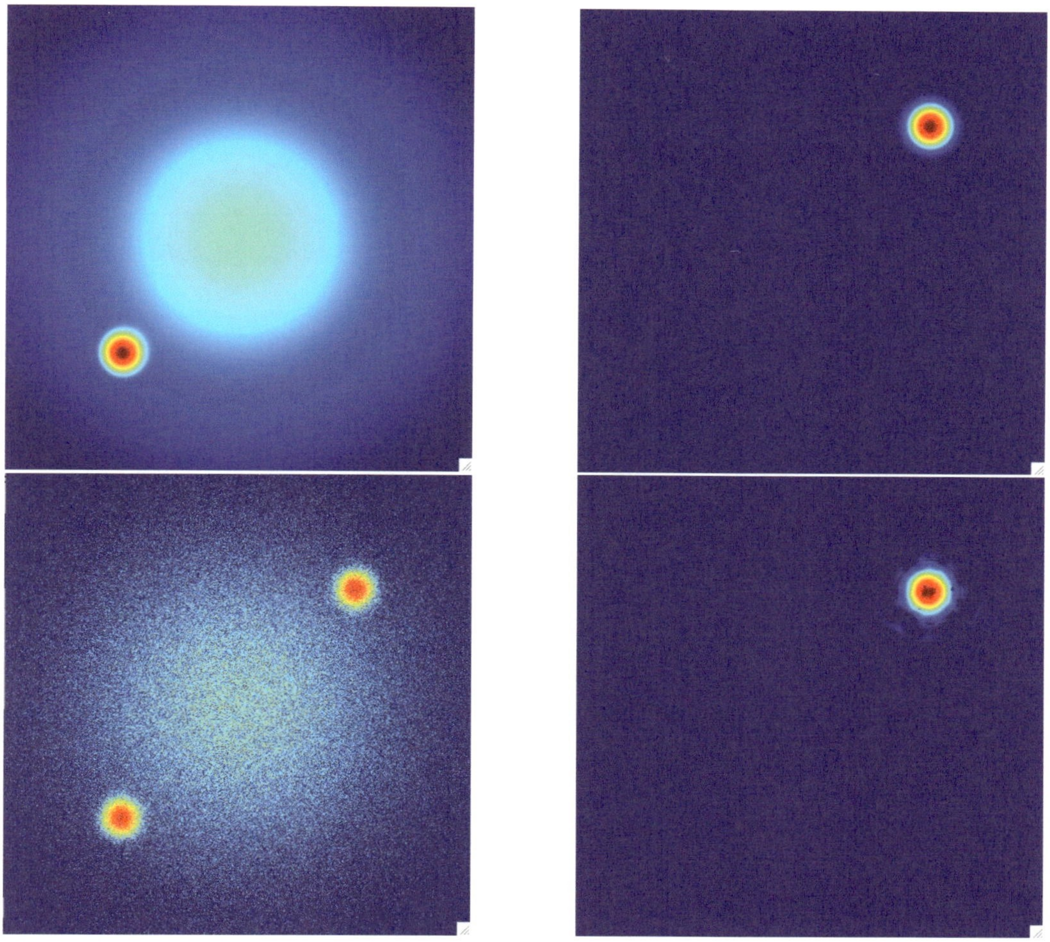

Fig. 9. Theoretical testing for MS-VSTS + IUWT denoising + background removal algorithm (Algorithm 5). View on a single HEALPix face. *Top Left*: Simulated background : sum of two Gaussians of standard deviation equal to 0.1 and 0.01 respectively. *Top Right*: Simulated source: Gaussian of standard deviation equal to 0.01. *Bottom Left*: Simulated poisson data. *Bottom Right*: Image denoised with MS-VSTS + IUWT and background removal.

Like Algorithm 1, Algorithm 5 can be adapted to make multiresolution support adaptation.

5.2.2 Experiment

We applied Algorithms 5 on simulated Fermi data. To test the efficiency of our method, we detect the sources with the SExtractor routine Bertin & Arnouts (1996), and compare the detected sources with the input source list to get the number of true and false detections. Results are shown on Figures 10 and 11. The SExtractor method was applied on the first wavelet scale of the reconstructed map, with a detection threshold equal to 1. It has been chosen to optimise the number of true detections. SExtractor makes 593 true detections and

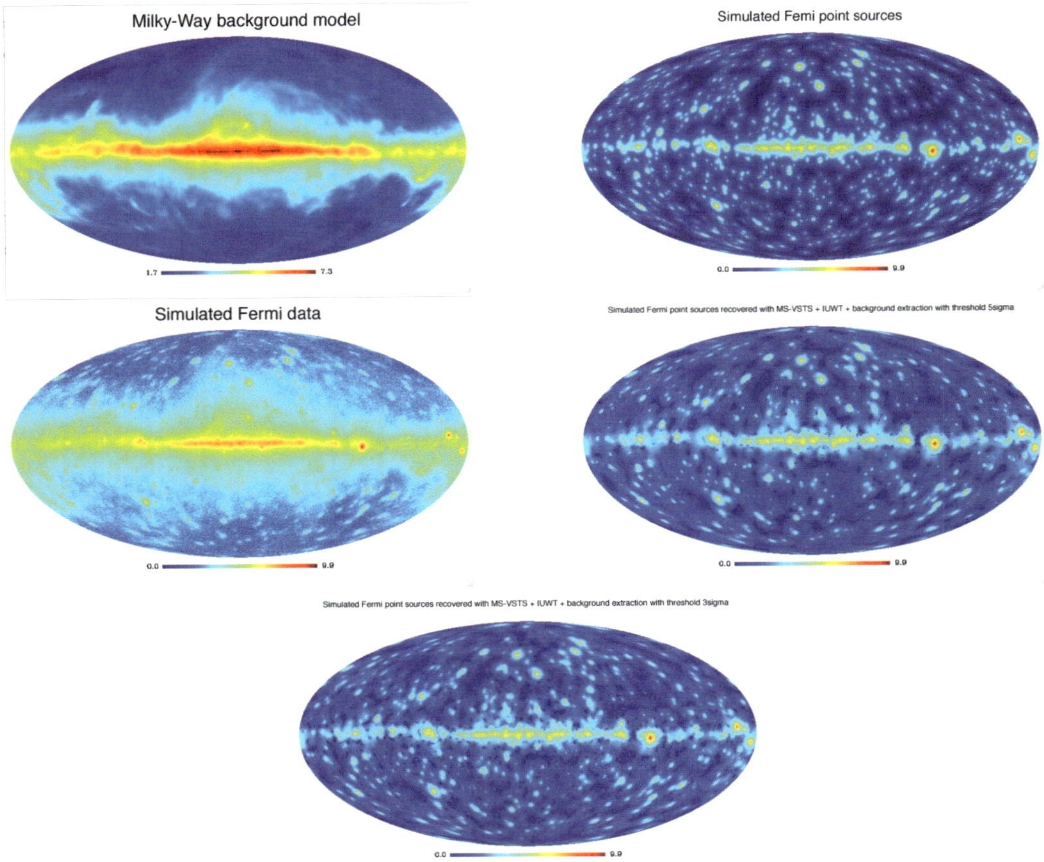

Fig. 10. *Top Left*: Simulated background model. *Top Right*: Simulated Gamma Ray sources. *Middle Left*: Simulated Fermi data with Poisson noise. *Middle Right*: Reconstructed Gamma Ray Sources with MS-VSTS + IUWT + background removal (Algorithm 5) with threshold $5\sigma_j$. *Bottom*: Reconstructed Gamma Ray Sources with MS-VSTS + IUWT + background removal (Algorithm 5) with threshold $3\sigma_j$. Pictures are in logarithmic scale.

71 false detections on the Fermi simulated map restored with Algorithm 2 among the 1000 sources of the simulation. On noisy data, many fluctuations due to Poisson noise are detected as sources by SExtractor, which leads to a big number of false detections (more than 2000 in the case of Fermi data).

Sensitivity to model errors

As it is difficult to model the background precisely, it is important to study the sensitivity of the method to model errors. We add a stationary Gaussian noise to the background model, we compute the MS-VSTS + IUWT with threshold $3\sigma_j$ on the simulated Fermi Poisson data with extraction of the noisy background, and we study the percent of true and false detections with respect to the total number of sources of the simulation and the signal-noise ratio ($\mathrm{SNR}(dB) = 20\log(\sigma_{signal}/\sigma_{noise})$) versus the standard deviation of the Gaussian perturbation. Table 3

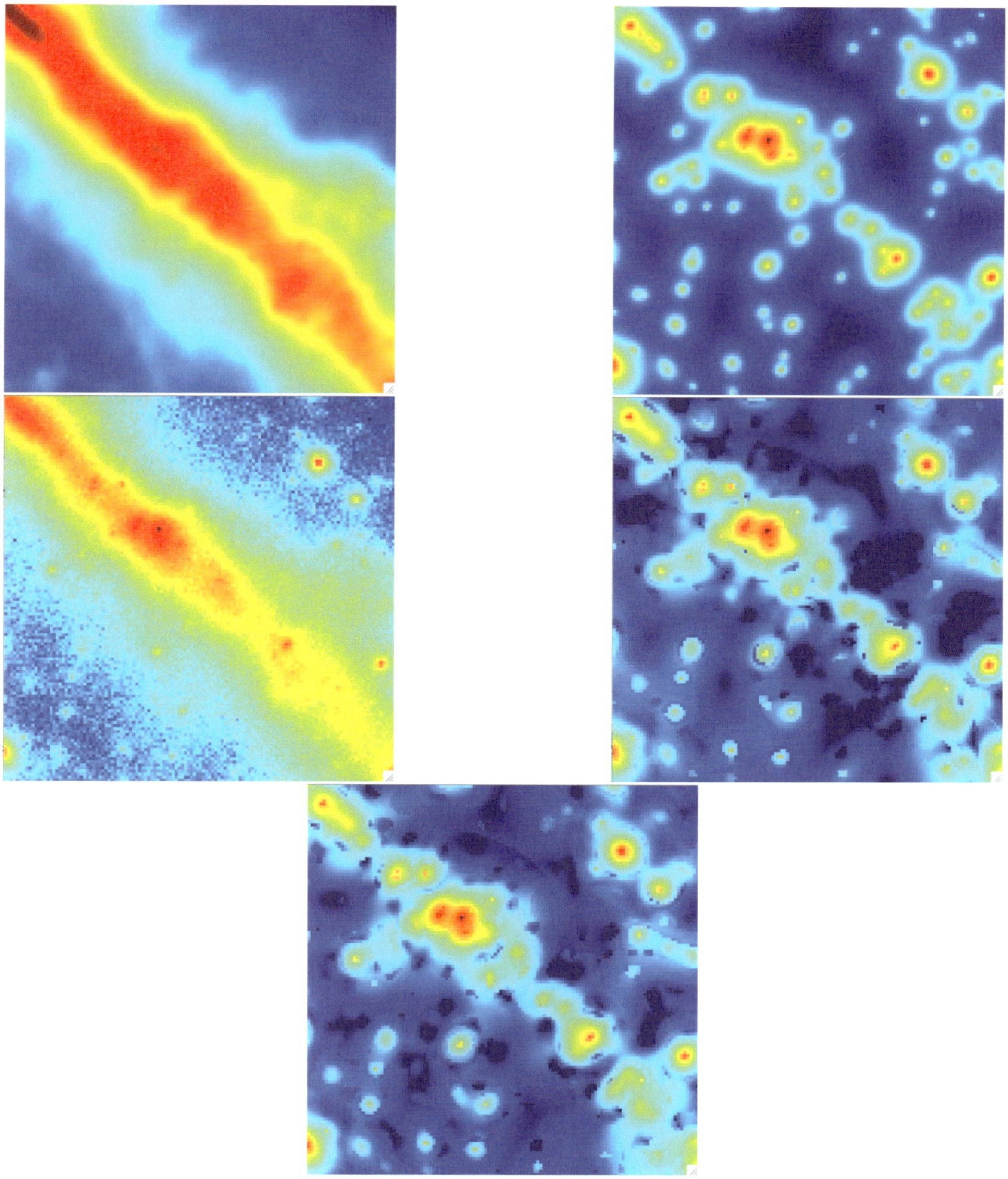

Fig. 11. View of a single HEALPix face (angular extent: $\pi/3sr$) from the results of Figure 10.*Top Left*: Simulated background model. *Top Right*: Simulated Gamma Ray sources. *Middle Left*: Simulated Fermi data with Poisson noise. *Middle Right*: Reconstructed Gamma Ray Sources with MS-VSTS + IUWT + background removal (Algorithm 5) with threshold $5\sigma_j$. *Bottom*: Reconstructed Gamma Ray Sources with MS-VSTS + IUWT + background removal (Algorithm 5) with threshold $3\sigma_j$. Pictures are in logarithmic scale.

Model error std dev	% of true detect	% of false detect	SNR (dB)
0	59.3%	7.1%	23.8
10	57.0%	11.0%	23.2
20	53.2%	18.9%	22.6
30	49.1%	43.5%	21.7
40	42.3%	44.3%	21.0
50	34.9%	39.0%	20.3
60	30.3%	37.5%	19.5
70	25.0%	34.6%	18.9
80	23.0%	28.5%	18.7
90	23.6%	27.1%	18.3

Table 3. Percent of true and false detection and signal-noise ratio versus the standard deviation of the Gaussian noise on the background model.

shows that, when the standard deviation of the noise on the background model becomes of the same range as the mean of the Poisson intensity distribution ($\lambda_{\text{mean}} = 68.764$), the number of false detections increases, the number of true detections decreases and the signal noise ratio decreases. While the perturbation is not too strong (standard deviation < 10), the effect of the model error remains low.

6. Extension to multichannel data

6.1 Gaussian noise

6.1.1 2D-1D Wavelet Transform on the sphere

We propose a denoising method for 2D - 1D data on the sphere, where the two first dimensions are spatial (longitude and latitude) and the third dimension is either the time or the energy. We need to analyze the data with a non-isotropic wavelet, where the time or energy scale is not connected to the spatial scale. An ideal wavelet function would be defined by:

$$\psi(\theta, \varphi, t) = \psi^{(\theta,\varphi)}(\theta, \varphi)\psi^{(t)}(t) \tag{27}$$

where $\psi^{(\theta,\varphi)}$ is the spatial wavelet and $\psi^{(t)}$ is the temporal (or energy) wavelet. In the following, we will consider only isotropic and dyadic spatial scales, and we denote j_1 the spatial resolution index (i.e. scale 2^{j_1}), j_2 the time (or energy) resolution index. We thus define the scaled spatial and temporal (or energy) wavelets $\psi_{j_1}^{(\theta,\varphi)}(\theta, \varphi) = \frac{1}{2^{j_1}}\psi^{(\theta,\varphi)}(\frac{\theta}{2^{j_1}}, \frac{\varphi}{2^{j_1}})$ and $\psi_{j_1}^{(t)} = \frac{1}{2^{j_2}}\psi^{(t)}(\frac{t}{2^{j_1}})$.

Hence, we derive the wavelet coefficients $w_{j_1,j_2}[k_\theta, k_\varphi, k_t]$ from a given data set D (k_θ and k_φ are spatial index and k_z a time (or energy) index. In continuous coordinates, this amounts to the formula

$$\begin{aligned} w_{j_1,j_2}[k_\theta, k_\varphi, k_t] = \frac{1}{2^{j_1}}\frac{1}{\sqrt{2^{j_2}}}\int\int\int_{-\infty}^{+\infty} D(\theta, \varphi, t) \\ \times\psi^{(\theta,\varphi)}(\frac{\theta - k_\theta}{2^{j_1}}, \frac{\varphi - k_\varphi}{2^{j_1}}\psi^{(t)}(\frac{t - k_t}{2^{j_2}})dxdydz = D * \bar{\psi}_{j_1}^{(\theta,\varphi)} * \bar{\psi}_{j_2}^{(t)}(\theta, \varphi, t) \end{aligned} \tag{28}$$

where $*$ is the convolution and $\bar{\psi}(t) = \psi(-t)$.

6.1.2 Fast undecimated 2D-1D decomposition/reconstruction

In order to have a fast algorithm for discrete data, we use wavelet functions associated to filter banks. Hence, our wavelet decomposition consists in applying first a IUWT on the sphere for each frame k_z. Using the spherical IUWT, we have the reconstruction formula:

$$D[k_\theta, k_\varphi, k_t] = a_{J_1}[k_\theta, k_\varphi] + \sum_{j_1=1}^{J1} w_{j_1}[k_\theta, k_\varphi, k_t], \forall k_t \tag{29}$$

where J_1 is the number of spatial scales. To have simpler notations, we replace the two spatial indexes by a single index k_r which corresponds to the pixel index:

$$D[k_r, k_t] = a_{J_1}[k_r] + \sum_{j_1=1}^{J1} w_{j_1}[k_r, k_t], \forall k_t \tag{30}$$

Then, for each spatial location k_r and for each 2D wavelet scale j_1, we apply a 1D wavelet transform along t on the spatial wavelet coefficients $w_{j_1}[k_r, k_t]$ such that

$$w_{j_1}[k_r, k_t] = w_{j_1,J_2}[k_r, k_t] + \sum_{j_2=1}^{J_2} w_{j_1,j_2}[k_r, k_t], \forall (k_r, k_t) \tag{31}$$

where j_2 is the number of scales along t.The same processing is also applied on the coarse spatial scale $a_{J_1}[k_r, k_t]$ and we have

$$a_{J_1}[k_r, k_t] = a_{J_1,J_2}[k_r, k_t] + \sum_{j_2=1}^{J_2} w_{J_1,j_2}[k_r, k_t], \forall (k_r, k_t) \tag{32}$$

Hence, we have a 2D-1D spherical undecimated wavelet representation of the input data D:

$$D[k_r, k_t] = a_{J_1,J_2}[k_r, k_t] + \sum_{j_1=1}^{J_1} w_{j_1,J_2}[k_r, k_t] + \sum_{j_2=1}^{J_2} w_{J_1,j_2}[k_r, k_t] + \sum_{j_1=1}^{J_1} \sum_{j_2=1}^{J_2} w_{j_1,j_2}[k_r, k_t] \tag{33}$$

From this expression, we distinguish four kinds of coefficients:

- Detail-Detail coefficients ($j_1 \leqslant J_1$ and $j_2 \leqslant J_2$):

$$w_{j_1,j_2}[k_r, k_t] = (\delta - \bar{h}_{1D}) \star (\bar{h}_{1D}^{(j_2-1)} \star a_{j_1-1}[k_r, \cdot] - h_{1D}^{(j_2-1)} \star a_{j_1}[k_r, \cdot]) \tag{34}$$

- Approximation-Detail coefficients ($j_1 = J_1$ and $j_2 \leqslant J_2$):

$$w_{J_1,j_2}[k_r, k_t] = h_{1D}^{(j_2-1)} \star a_{J_1}[k_r, \cdot] - h_{1D}^{(j_2)} \star a_{J_1}[k_r, \cdot] \tag{35}$$

- Detail-Approximation coefficients ($j_1 \leqslant J_1$ and $j_2 = J_2$):

$$w_{j_1,J_2}[k_r,k_t] = h_{1D}^{(J_2)} \star a_{j_1-1}[k_r,\cdot] - h_{1D}^{(J_2)} \star a_{j_1}[k_r,\cdot] \quad (36)$$

- Approximation-Approximation coefficients ($j_1 = J_1$ and $j_2 = J_2$):

$$a_{J_1,J_2}[k_r,k_t] = h_{1D}^{(J_2)} \star a_{J_1}[k_r,\cdot] \quad (37)$$

6.1.3 Multichannel Gaussian denoising

As the spherical 2D-1D undecimated wavelet transform just described is fully linear, a Gaussian noise remains Gaussian after transformation. Therefore, all thresholding strategies which have been developed for wavelet Gaussian denoising are still valid with the spherical 2D-1D wavelet transform. Denoting TH the thresholding operator, the denoised cube in the case of additive white Gaussian noise is obtained by:

$$\begin{aligned} \tilde{D}[k_r,k_t] = \; & a_{J_1,J_2}[k_r,k_t] + \sum_{j_1=1}^{J_1} \mathrm{TH}(w_{j_1,J_2}[k_r,k_t]) \\ & + \sum_{j_2=1}^{J_2} \mathrm{TH}(w_{J_1,j_2}[k_r,k_t]) + \sum_{j_1=1}^{J_1}\sum_{j_2=1}^{J_2} \mathrm{TH}(w_{j_1,j_2}[k_r,k_t]) \end{aligned} \quad (38)$$

A typical choice of TH is the hard thresholding operator, i.e.

$$\mathrm{TH}(x) = \begin{cases} 0 \; if|x| < \tau \\ x \; if|x| \geqslant \tau \end{cases} \quad (39)$$

The threshold τ is generally chosen between 3 and 5 times the noise standard deviation.

6.2 Poisson Noise

6.2.1 Multi-scale variance stabilzing transform

To perform a Poisson denoising, we have to plug the MS-VST into the spherical 2D-1D undecimated wavelet transform. Again, we distinguish four kinds of coefficients that take the following forms:

- Detail-Detail coefficients ($j_1 \leqslant J_1$ and $j_2 \leqslant J_2$):

$$w_{j_1,j_2}[k_r,k_t] = (\delta - \bar{h}_{1D}) \star (T_{j_1-1,j_2-1}[\bar{h}_{1D}^{(j_2-1)} \star a_{j_1-1}[k_r,\cdot]] - T_{j_1,j_2-1}[h_{1D}^{(j_2-1)} \star a_{j_1}[k_r,\cdot]]) \quad (40)$$

- Approximation-Detail coefficients ($j_1 = J_1$ and $j_2 \leqslant J_2$):

$$w_{J_1,j_2}[k_r,k_t] = T_{J_1,j_2-1}[h_{1D}^{(j_2-1)} \star a_{J_1}[k_r,\cdot]] - T_{J_1,j_2}[h_{1D}^{(j_2)} \star a_{J_1}[k_r,\cdot]] \quad (41)$$

- Detail-Approximation coefficients ($j_1 \leqslant J_1$ and $j_2 = J_2$):

$$w_{j_1,J_2}[k_r,k_t] = T_{j_1-1,J_2}[h_{1D}^{(J_2)} \star a_{j_1-1}[k_r,\cdot]] - T_{j_1,J_2}[h_{1D}^{(J_2)} \star a_{j_1}[k_r,\cdot]] \quad (42)$$

- Approximation-Approximation coefficients ($j_1 = J_1$ and $j_2 = J_2$):

$$a_{J_1,J_2}[k_r, k_t] = h_{1D}^{(J_2)} \star a_{J_1}[k_r, \cdot] \tag{43}$$

Hence, all 2D-1D wavelet coefficients w_{j1,j_2} are now stabilized, and the noise on all these wavelet coefficients is Gaussian with known scale-dependent variance that depends solely on h. Denoising is however not straightforward because there is no explicit reconstruction formula available because of the form of the stabilization equations above. Formally, the stabilizing operators T_{j_1,j_2} and the convolution operators along the spatial and temporal dimensions do not commute, even though the filter bank satisfies the exact reconstruction formula. To circumvent this difficulty, we propose to solve this reconstruction problem by using an iterative reconstruction scheme.

6.2.2 Detection-reconstruction

As the noise on the stabilized coefficients is Gaussian, and without loss of generality, we let its standard deviation equal to 1, we consider that a wavelet coefficient $w_{j_1,j_2}[k_r, k_t]$ is significant, i.e., not due to noise, if its absolute value is larger than a critical threshold τ, where τ is typically between 3 and 5.

The multiresolution support will be obtained by detecting at each scale the significant coefficients. The multiresolution support for $j_1 \leqslant J_1$ and $j_2 \leqslant J_2$ is defined as:

$$\mathcal{M}_{j_1,j_2}[k_r, k_t] = \begin{cases} 1 \text{ if } w_{j_1,j_2}[k_r, k_t] \text{is significant} \\ 0 \qquad\qquad \text{otherwise} \end{cases} \tag{44}$$

We denote $\mathcal{W}$ the spherical 2D-1D undecimated wavelet transform described above, and $\mathcal{R}$ the inverse wavelet transform. We want our solution X to preserve the significant structures of the original data by reproducing exactly the same coefficients as the wavelet coefficients of the input data Y, but only at scales and positions where significant signal has been detected. At other scales and positions, we want the smoothest solution with the lowest budget in terms of wavelet coefficients.Furthermore, as Poisson intensity functions are positive by nature, a positivity constraint is imposed on the solution. It is clear that there are many solutions satisfying the positivity and multiresolution support consistency requirements, e.g. Y itself. Thus, our reconstruction problem based solely on these constraints is an ill-posed inverse problem that must be regularized. Typically, the solution in which we are interested must be sparse by involving the lowest budget of wavelet coefficients. Therefore our reconstruction is formulated as a constrained sparsity-promoting minimization problem that can be written as follows

$$\min_{\mathbf{X}} \|\mathcal{W}\mathbf{X}\|_1 \text{ subject to } \begin{cases} \mathcal{M}\mathcal{W}\mathbf{X} = \mathcal{M}\mathcal{W}\mathbf{Y} \\ \mathbf{X} \geqslant 0 \end{cases} \tag{45}$$

where $\| \cdot \|$ is the L_1-norm playing the role of regularization and is well known to promote sparsityDonoho (2004). This problem can be solved efficently using the hybrid steepest descent algorithm Yamada (2001)Zhang et al. (2008), and requires about 10 iterations in practice. Transposed into our context, its main steps can be summarized as follows:

Require: Input noisy data $\mathbf{Y}$, a low-pass filter h, multiresolution support $\mathcal{M}$ from the detection step, number of iterations $N_{\max}$
1: Initialize $\mathbf{X}^{(0)} = \mathcal{M}\mathcal{W}\mathbf{Y} = \mathcal{M}w_Y$,
2: **for** $n = 1$ to $N_{\max}$ **do**
3: $\tilde{d} = \mathcal{M}w_Y + (1 - \mathcal{M})\mathcal{W}X^{(n-1)}$,
4: $\mathbf{X}^{(n)} = P_+(\mathcal{R}\mathrm{ST}_{\beta_n}[\tilde{d}])$,
5: Update the step $\beta_n = (N_{\max} - n)/(N_{\max} - 1)$
6: **end for**

where P_+ is the projector onto the positive orthant, i.e. $P_+(x) = max(x, 0)$, ST_{β_n} is the soft-thresholding operator with threshold β_n, i.e. $\mathrm{ST}_{\beta_n}[x] = x - \beta_n \mathrm{sign}(x)$ if $|x| \geqslant \beta_n$, and 0 otherwise.

The final spherical MSVST 2D-1D wavelet denoising algorithm is the following:

Require: Input noisy data $\mathbf{Y}$, a low-pass filter h, threshold level τ
1: *Spherical 2D-1D MSVST:* Apply the spherical 2D-1D-MSVST to the data using (40)-(43).
2: *Detection:* Detect the significant wavelet coefficients that are above τ, and compute the multiresolution support $\mathcal{M}$.
3: *Reconstruction:* Reconstruct the denoised data using the algorithm above.

7. Conclusion

This chapter presented new methods for restoration of spherical data with noise following a Poisson distribution. A denoising method was proposed, which used a variance stabilization method and multiscale transforms on the sphere. Experiments have shown it is very efficient for Fermi data denoising. Two spherical multiscale transforms, the wavelet and the curvelets, were used. Then, we have proposed an extension of the denoising method in order to take into account missing data, and we have shown that this inpainting method could be a useful tool to estimate the diffuse emission. Then, we have introduced a new denoising method on the sphere which takes into account a background model. The simulated data have shown that it is relatively robust to errors in the model, and can therefore be used for Fermi diffuse background modeling and source detection. Finally, we introduced an extension for multichannel data.

8. Acknowledgement

This work was partially supported by the European Research Council grant ERC-228261.

9. References

Abdo, A. A., Ackermann, M. & al (2009). Fermi/large area telescope bright gamma-ray source list, *ApJS* 183: 46–66.

Abrial, P., Moudden, Y., Starck, J., Bobin, J., Fadili, M., Afeyan, B. & Nguyen, M. (2007). Morphological component analysis and inpainting on the sphere: Application in physics and astrophysics, *Journal of Fourier Analysis and Applications* 13(6). 729–748.

Abrial, P., Moudden, Y., Starck, J., Fadili, M. J., Delabrouille, J. & Nguyen, M. (2008). CMB data analysis and sparsity, *Statistical Methodology* 5(4). 289-298.

Anscombe, F. (1948). The transformation of poisson, binomial and negative-binomial data, *Biometrika* vol. 35, no. 3: pp. 246–254.

Atwood, W. B. e. a. (2009). The Large Area Telescope on the Fermi Gamma-Ray Space Telescope Mission, *Astrophysical Journal* 697: 1071–1102.

Bertin, E. & Arnouts, S. (1996). Sextractor: Software for source extraction, *Astronomy and Astrophysics Supplement* 117: 393–404.

Candes, E. & Donoho, D. (1999). Ridgelets: the key to high dimensional intermittency?, *Philosophical Transactions of the Royal Society of London* A 357: 2495.

Donoho, D. (1993). Nonlinear wavelet methodes for recovery of signals, densitites, and spectra from indirect and noisy data, *Proc. Symposia in Applied Mathematics* vol. 47: pp. 173–205.

Donoho, D. (2004). For most large underdetermined systems of linear equations, the minimal ℓ^1-norm near-solution approximates the sparsest near-solution, *Technical report*, Department of Statistics of Stanford University.

Fisz, M. (1955). The limiting distribution of a function of two independant random variables and its statistical application, *Colloquium Mathematicum* vol. 3: pp. 138–146.

Fryźlewicz, P. & Nason, G. (2004). A haar-fisz algorithm for poisson intensity estimation, *J. Comp. Graph. Stat.* vol. 13: pp. 621–638.

Górski, K. M., Hivon, E., Banday, A. J., Wandelt, B. D., Hansen, F. K., Reinecke, M. & Bartelmann, M. (2005). HEALPix: A framework for high-resolution discretization and fast analysis of data distributed on the sphere, *Astrophysical Journal* 622. 759–771.

Hartman, R. C., Bertsch, D. L., Bloom, S. D., Chen, A. W., Deines-Jones, P., Esposito, J. A., Fichtel, C. E., Friedlander, D. P., Hunter, S. D., McDonald, L. M., Sreekumar, P., Thompson, D. J., Jones, B. B., Lin, Y. C., Michelson, P. F., Nolan, P. L., Tompkins, W. F., Kanbach, G., Mayer-Hasselwander, H. A., Mucke, A., Pohl, M., Reimer, O., Kniffen, D. A., Schneid, E. J., von Montigny, C., Mukherjee, R. & Dingus, B. L. (1999). Third EGRET catalog (3EG) (Hartman+, 1999), *VizieR Online Data Catalog* 212: 30079–+.

J.Schmitt, Starck, J., Casandjian, J., Fadili, J. & Grenier, I. (2010). Poisson denoising on the sphere: Application to the fermi gamma ray space telescope, *Astronomy and Astrophysics* 517.

Kolaczyk, E. (1999). Bayesian multiscale models for poisson processes, *J. Amer. Stat. Assoc.* vol. 94, no. 447: pp. 920–933.

Lefkimmiaits, S., Maragos, P. & Papandreou, G. (2009). Bayesian inference on multiscale models for poisson intensity estimation: Applications to photon-limited image denoising, *IEEE Transactions on Image Processing* vol. 20, no. 20.

Myers, J. (2009). Lat background models. `http://fermi.gsfc.nasa.gov/ssc/data/access/lat/BackgroundModels.html`.

Myers, J. (2010). Lat 1-year point source catalog. `http://fermi.gsfc.nasa.gov/ssc/data/access/lat/1yr_catalog/`.

Nowak, R. & Kolaczyk, E. (2000). A statistical multiscale framework for poisson inverse problems, *IEEE Trans. Inf. Theory* vol. 45, no. 5: pp. 1811–1825.

Starck, J., Fadili, J. M., Digel, S., Zhang, B. & Chiang, J. (2009). Source detection using a 3D sparse representation: application to the Fermi gamma-ray space telescope, *Astronomy and Astrophysics* 504: 641–652.

Starck, J.-L., Elad, M. & Donoho, D. (2004). Redundant multiscale transforms and their application for morphological component analysis, *Advances in Imaging and Electron Physics* 132.

Starck, J.-L., Moudden, Y., Abrial, P. & Nguyen, M. (2006). Wavelets, ridgelets and curvelets on the sphere, *Astronomy and Astrophysics* 446: 1191–1204.

Timmerman, K. & Nowak, R. (1999). Multiscale modeling and estimation of poisson processes with application to photon-limited imaging, *IEEE Trans. Inf. Theory* vol. 45, no. 3: pp. 846–862.

Yamada, I. (2001). The hybrid steepest descent method for the variational inequality problem over the intersection of fixed point sets of nonexpansive mappings, *Inherently Parallel Algorithm in Feasibility and Optimization and their Applications*, Elsevier, pp. pp. 473–504.

Zhang, B., Fadili, J. & Starck, J.-L. (2008). Wavelets, ridgelets and curvelets for poisson noise removal, *IEEE Transactions on Image Processing* vol. 11, no. 6: pp. 1093–1108.

10

A Wavelet Multiscale De-Noising Algorithm Based on Radon Transform

Xueling Zhu[1,*], Xiaofeng Yang[2,*], Qinwu Zhou[3], Liya Wang[4], Fulai Yuan[5] and Zhengzhong Bian[3]
[1]*The School of Humanities and Social Sciences, National University of Defense Technology, Changsha,*
[2]*Department of Radiation Oncology, Emory University, Atlanta, GA*
[3]*Department of Biomedical Engineering, School of Life Science and Technology, Xi'an Jiaotong University, Xi'an, Shaanxi,*
[4]*Department of Radiology and Imaging Sciences, Emory University, Atlanta, GA,*
[5]*Department of Stomatology, Xiangya Hospital, Central South University, Changsha,*
[1,3,5]*China*
[2,4]*USA*

1. Introduction

The rapid development of medical imaging technology and the introduction of new imaging modalities, call for new image processing methods which include specialized noise filtering, enhancement, classification and segmentation techniques. Denoising is a particularly delicate and difficult task in medical images. A tradeoff between noise reduction and the preservation of actual image features has to be made in a way that enhances the diagnostically relevant image content [1].

Single scale filtering methods, like linear (Wiener) or nonlinear (median) are commonly used to eliminate noise. They do not take into account different noise levels and count distributions. Linear filters can reduce the noise variance increasing the signal-to-noise ratio (SNR). However, they smooth images to the point of degrading the image contrast and detail. Nonlinear filters also reduce the noise variance, while preserving the edges to some extent, but image contrast and detail degradation still occurs. In order to avoid such problem many edge preserving filters, like the Anisotropic Diffusion (AD) Filter [2, 3] have been used. Such filters respect edges by averaging pixels in the orthogonal direction of the local gradient. More recently, the trilateral filter [4], an evolution of the bilateral filter [5], has been proposed to take into account local structure in addition to intensity and geometric features.

On the other hand, multiscale filtering methods have also been proposed for image filtering. Unlike single scale filtering methods, multiscale expansions offer the possibility of separating features of interest and noise components into distinct sub-band coefficients.

---

[*] Yang X. and Zhu X. contributed equally as co-first authors.

Specifically, multiscale methods, based on wavelet transforms, have widely been applied for noise reduction in medical images. Among the wavelet-based noise reduction techniques, nonlinear thresholding is simple yet very effective. Healy et al. [6] were the first to apply wavelet techniques, based on soft thresholding, for filtering MR images. Nowak [7] squared the magnitude MR image and used a wavelet-based Wiener-filter-like filtering method. Donoho [8] showed a universal threshold of asymptotically optimal in the minimax sense, but it is well-known that the universal threshold over-smoothes images. Pan et al. [9] presented a hard threshold with a nonorthogonal wavelet expansion. At the same time the correlation between wavelet coefficients in several scales has also been employed to filter. Paul et al. [10] presented a multiscale thresholding scheme that incorporated the merits of interscale dependencies into the thresholding technique for filtering, and then applied thresholding to the multiscale products instead of the wavelet coefficients. The interscale correlation information is exploited by Pizurica et al. [11] to classify the wavelet coefficients. Xu et al. [12] developed a spatially selective filtering technique by iteratively selecting edge pixels in the multiscale products. Zhang et al. [13] used Radon transform and an adaptive median filter based on Walsh list in the Laplacian pyramid domain to denoise medical image. These methods are effective for noise suppression. However, there are some significant problems such as blurring, losing of detail texture, the changing of edge points, and generating artificial smear.

In this paper we present a medical image filtering method based on the Radon and wavelet transforms. We perform Radon transform for input images to get sinograms. Then we apply 1D non-orthogonal wavelets transform along s in sinograms, and use threshold-based methods to filter it. Dissimilar to the traditional threshold schemes that apply the same threshold to the wavelet coefficients at every scale, the proposed method can get robust and adaptive noise threshold at every scale. We can take advantage of the interscale dependency information between wavelet scales to evaluate original noise variance. Finally we apply the inverse Radon transform algorithm to reconstruct the original images. Our method has been validated using MR images. The detailed steps of our method and its evaluative results are reported in the following sections.

2. Materials and methods

2.1 Radon transform

The Radon transform of a 2D function is defined as:

$$Rf(\alpha,s)=\int_{-\infty}^{\infty}\int_{-\infty}^{\infty}f(x,y)\delta(s-x\cos\alpha-y\sin\alpha)dxdy \quad \alpha\in[0,\pi] \quad s\in R \tag{1}$$

where s is the perpendicular distance of a line from the origin and α is the angle formed by the distance vector. Fourier slice theorem states that for a 2D function $f(x,y)$, the 1D Fourier transforms of the Radon transform along s, are the 1D radial samples of the 2D Fourier transform of $f(x,y)$ at the corresponding angles [14].

2.1.1 Noise robustness

Since the Radon transform is a line integral of the image, the Radon transform of noise is constant for all of the points and directions and is equal to the mean value of the noise.

Therefore, this means zero-mean noise has no effect on the Radon transform of the image. Here we suppose the image is a square with side length $m=\sqrt{2}a$ ($a>0$), where a is the half diagonal length of the square area in terms of pixels. We add up intensity values of the pixels to calculate the Radon transform as shown in Fig. 1. Suppose $f(x,y)$ is a 2D discrete image whose intensity values are random variables with mean μ and variance σ^2. For each point of the projection p_s, we add up n_s pixels of $f(x,y)$. Therefore, $var(p_s)=n_s\cdot\sigma^2$, $mean(p_s)=n_s\cdot\mu$, and $n_s=2(a-s)$, and the integer s is the projection index, which varies from $-a$ to a.

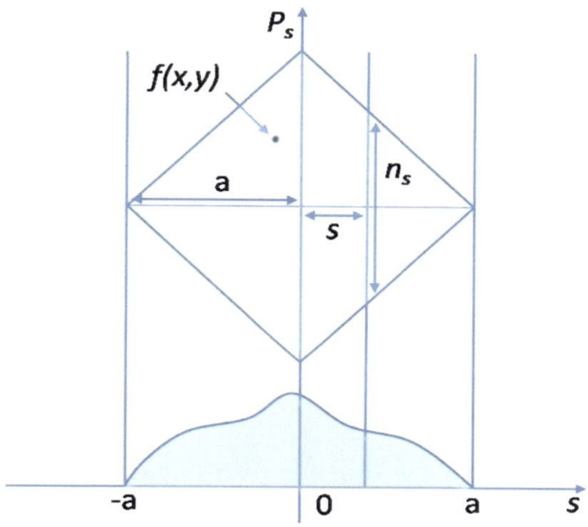

Fig. 1. Diagram for calculations in the Radon domain

The average of p_s^2 is

$$A(p_s^2)=\frac{1}{2a}\sum_{s=-a}^{a}p_s^2 \tag{2}$$

Its expected value is defined as $E_p=E\{A(p_s^2)\}$. Then,

$$E_p=\frac{1}{2a}\sum_{s=-a}^{a}E\{p_s^2\}=\frac{1}{2a}\sum_{s=-a}^{a}\left[\mathrm{var}(p_s)+(E(p_s))^2\right]=\frac{1}{2a}\sum_{s=-a}^{a}(n_s\sigma^2+n_s^2\mu^2) \tag{3}$$

For large a we can write equation (3) as

$$E_p=\frac{1}{2a}\int_{-a}^{a}(n_s\sigma^2+n_s^2\mu^2)ds \tag{4}$$

So

$$E_p=\frac{1}{a}\int_{0}^{a}(2(a-s)\sigma^2+4(a-s)^2\mu^2)ds \tag{5}$$

$$E_p = \frac{2}{a}\int_0^a (2u^2s^2 - (\sigma^2 + 4a\mu^2)s + a\sigma^2 + 2\mu^2a^2)ds \tag{6}$$

$$E_p = (\frac{4u^2s^3}{3a})|_0^a - (\sigma^2 + 4a\mu^2)\frac{s^2}{a}|_0^a + (a\sigma^2 + 2\mu^2a^2)\frac{2s}{a}|_0^a$$

$$E_p = a\sigma^2 + \frac{4}{3}a^2\mu^2 \tag{7}$$

From equation (7) we can get $E_s = a\sigma_s^2 + \frac{4}{3}a^2\mu_s^2$ for the signal, and get $E_n = a\sigma_n^2$ for white noise with zero mean and variance σ_n^2. Then,

$$SNR_{Radon} = \frac{E_s}{E_n} = \frac{\sigma_s^2}{\sigma_n^2} + \frac{4a}{3}\cdot\frac{\mu_s^2}{\sigma_n^2} \tag{8}$$

In original image we define $SNR_{image} = (\sigma_s^2 + \mu_s^2)/\sigma_n^2$. Therefore,

$$SNR_{Radon} = \frac{\sigma_s^2 + \mu_s^2}{\sigma_n^2} + (\frac{4a}{3} - 1)\cdot\frac{\mu_s^2}{\sigma_n^2} = SNR_{image} + (\frac{4a}{3} - 1)\cdot\frac{\mu_s^2}{\sigma_n^2} \tag{9}$$

For medical images like MRI or CT the signal is distributed over the whole intensity range we can always get $\mu_s^2 \geq \sigma_s^2$, and we may write $SNR_{image} \leq 2\mu_s^2/\sigma_n^2$. Therefore,

$$SNR_{Radon} \geq (\frac{4a+3}{6})SNR_{image} = \frac{2\sqrt{2}m+3}{6}SNR_{image} \tag{10}$$

This shows that SNR_{Radon} has been increased by a factor of at least $\frac{2\sqrt{2}m+3}{6}$, which is in practice much too large a quantity. If the image size is 256×256, and then $SNR_{Radon} = \frac{2\sqrt{2}\times256+3}{6} \cong 121.2$ or $(10\log_{10}121.2 = 20.8dB)$.

2.1.2 The results for noise reducing

From above, we can see that the Radon transform is very robust in reducing the effect of additive noise. Here we add 3% and 10% Gaussian noise with a zero mean to the original phantom and at the same time we perform the Radon transform to get corresponding sinograms as shown in Fig. 2. Fig. 3(a) shows the profiles through the original image and the noised image (the profile position is shown in the top row of Fig. 2), Fig. 3(b) shows the profiles through their corresponding sinograms (the profile position is shown in the bottom row of Fig. 2). In Fig. 3(a) it is easy to see that the image with 10% noise has already been contaminated severely, and we cannot see any image details in the noised image. In Fig. 3(b) we can clearly see the profiles through the corresponding sinograms are very close, and it shows the transform has greatly decreased the noise in the original image. Since the Radon transform is a linear transform and is invertible, it does not lose any texture information. And rotation of the input image corresponds to the translation of the Radon transform along α , which is more tractable.

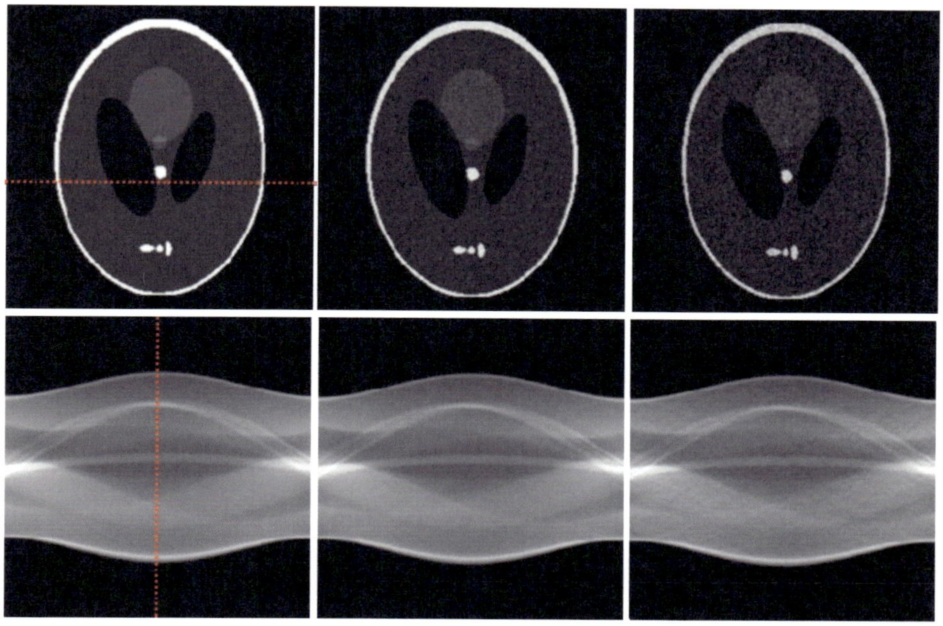

Fig. 2. Phantoms and corresponding sinograms. The top row consists of phantoms without noise and with 3% and 10% noise respectively. The bottom row is the corresponding sinograms of the top row phantoms.

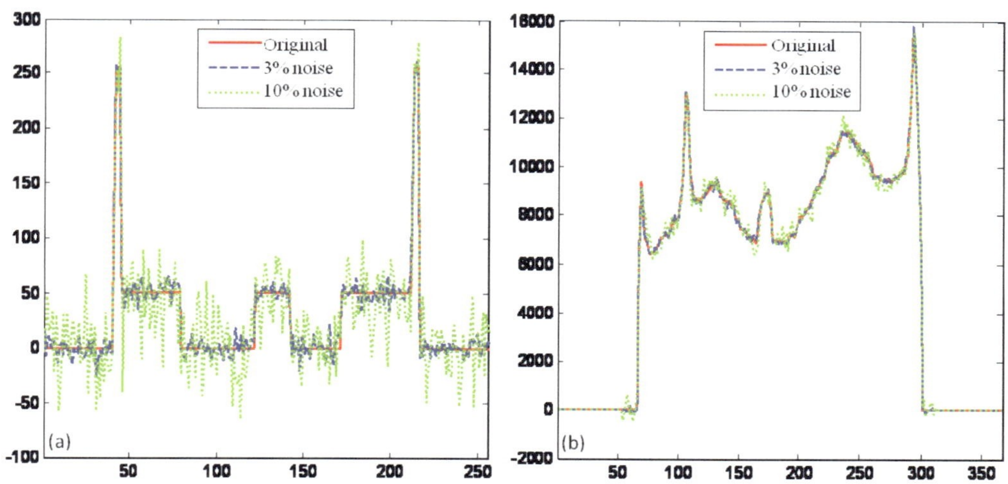

Fig. 3. Quantitative comparison of profiles through images. (a) The profiles between the original and the noised images. (b) The profiles of the corresponding sinograms.

2.2 Wavelet transform

2.2.1 Wavelet theory

Wavelets are mathematical functions that decompose data into different frequency components. Then each component is performed with a resolution matched to its scale [15]. The wavelet analysis decomposes a signal into a hierarchy of scales ranging from the coarsest scale to the finest one. Hence, wavelet transforms which provide representation of an image at various resolutions are a better tool for feature extraction from images. Wavelet coefficients of a signal are the projections of the signal onto the multiresolution subspaces. Wavelets are functions generated from basis function, called the mother wavelet, by dilations and translations in time (frequency) domain. If the mother wavelet is denoted by $\psi(t)$ other wavelets $\psi_{A,B}(t)$ can be represented as

$$\psi_{A,B}(t) = 1 / \sqrt{|A|} * \psi((t-B)/A) \tag{11}$$

where A and B are two arbitrary real numbers. The variables A and B represent the parameters for dilations and translations respectively in the time axis.

2.2.2 Dyadic wavelet transform

The discrete wavelet transform is an implementation of the wavelet transform using a discrete set of the wavelet scales and translation which obey some defined rules. For practical computations, it is necessary to discretize the wavelet transform. The scale parameter A is discretized on a logarithmic grid. The translation parameter B is then discretized with respect to the scale parameter, i.e. sampling is done on the dyadic sampling grid. A dyadic wavelet transform is a semi-discrete wavelet transform, makes scale factor binary discrete, but the displacement factor to change continuously. The discretized scale and translation parameters are given by $A = 2^{-j}$ and $B = k2^{-j}$, where $j,k \in Z$, is the set of all integers. Thus, the family of wavelet functions is represented as

$$\psi_{j,k}(t) = 2^{j/2}\psi(2^j t - k) \tag{12}$$

When analyzing wavelet transforms from a multiresolution point of view, the wavelet decomposition of a discrete time signal $X[n]$ is given by

$$X[n] = \sum_k c_{j_0,k}\varphi_{j_0,k}(n) + \sum_{j=j_0}^{\infty}\sum_k d_{j,k}\psi_{j,k}(n) \tag{13}$$

where $\varphi_{j_0,k}$, $\psi_{j,k}$ are the scaling functions and wavelet functions respectively.

As shown in Fig. 4, for any coarse scale 2^j, a discrete signal sequence $\{S_j f, (W_j f)_{1\le j\le J}\}$ is named as the discrete wavelet transform of the original signal f. In Fig. 4, $H_j(G_j)$ is taken for the $2^j th$ scale expansion of $H_1(G_1)$, that is, inserted $2^j - 1$ zeros among the filter coefficients.

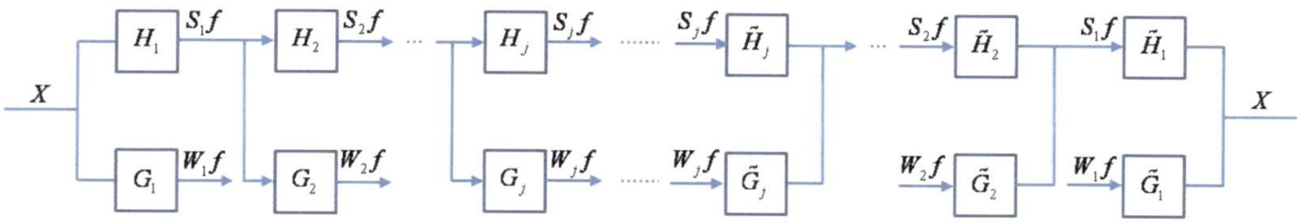

Fig. 4. Dyadic wavelet transforms

Reconstruction is the reverse process of decomposition, so we only need to exchange the decomposition filters for the reconstruction filters. The decomposition and reconstruction of $f(t)$ is completed by the Mallat tower algorithm [16]. Fast wavelet reconstruction (16) and inverse fast wavelet reconstruction (17) are given below

$$\begin{cases} c_{j,k} = \sum_m h(m-2k)c_{j+1,m} \\ d_{j,k} = \sum_m g(m-2k)c_{j+1,m} \end{cases} \tag{14}$$

$$c_{j+1,m} = \sum_k c_{j,k}h(m-2k) + \sum_k d_{j,k}g(m-2k) \tag{15}$$

Where c is the scale factor, d is the wavelet coefficients, and h and g are coefficient of the filter H and G respectively in Fig. 4.

2.2.3 Wavelet base choice

Orthogonal wavelet transforms have fewer coefficients at coarse scales, and therefore the transformed signal is uncorrelated across the scales. Lack of a translation invariant will make filtering by orthogonal wavelet transforms exhibit visual artifacts. The non-orthogonal wavelet offers much better edge detection because the signal is correlated across the scales. Non-orthogonal wavelet is better than orthogonal wavelet for SNR [9]. In this paper we use non-orthogonal wavelets first introduced by Mallat et al. [17]. This base wavelet is shown in Fig. 5.

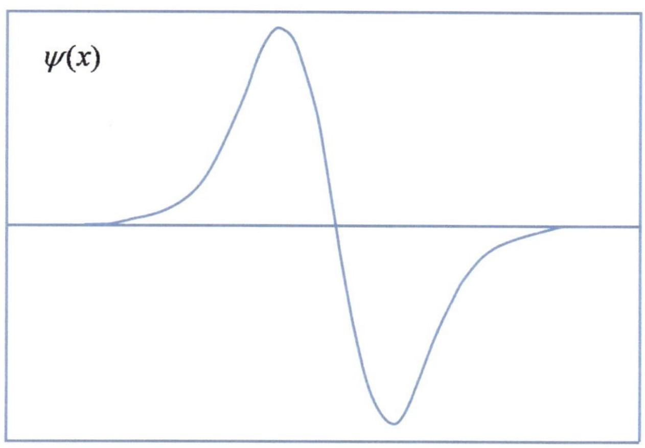

Fig. 5. Plot of a typical quadratic spline wavelet

It is very close to the derivative of a Gaussian function. From the Radon transform, we know the translation along α in the Radon domain corresponds to the rotation of the input image. The shift along s in the Radon domain corresponds to the translation of the input image. So we apply a 1D wavelet transform along s.

2.2.4 Threshold-based filtering

Threshold-based filtering is very simple and gives a satisfactory performance. It can be divided into three steps: (1) transform the noisy signal into wavelet coefficient w, (2) employ a hard or soft threshold t_j at each scale j, (3) transform back to the original domain, and get the estimated signal. Although Donoho [8] proved the optimality of a soft threshold in theory, a hard threshold has shown better results for certain applications.

$$\hat{w}(j,n) = \begin{cases} w(j,n) & w(j,n) \geq t_j \\ 0 & w(j,n) < t_j \end{cases} \tag{16}$$

As can be seen from filtering theory, during the wavelet filtering process a small threshold value will leave behind all the noisy coefficients, and subsequently, the resultant filtered image may still be noisy. On the other hand, a large threshold value will make more of the coefficients zero, and the resultant image may be blurred and have artifacts. So the threshold is key to the signal filtering effect, how the threshold is determined is critical. We choose $t_j = c\sigma_j$, where c is a constant. It is well known that for *i.i.d.* Gaussian noise $X \sim N(0,\sigma^2)$, a threshold $t = \sigma, 2\sigma, 3\sigma, \cdots$ will suppress 68.26%, 95.44%, and 99.74% of its values. Therefore, imposing c between 3~4, will give good results.

From the above discussion we can see the choice of threshold depends on the noise variance σ_j in different scales. So the key is how to get the optimum noise variance σ_j, which is adaptive to different scale characteristics [18].

2.2.5 Noise variance σ_j in different scales

In order to get the noise variance in different scales we suppose white noise is ε, its variance is σ^2, the power spectral density (PSD) is $S_\varepsilon(\omega) = \sigma^2$. After applying filter G_1 (see Fig. 4) the variance of the wavelet coefficient $W_1\varepsilon$ is ${\sigma_1}^2$. Suppose $G_1(\omega) = \sum_{-\infty}^{\infty} g_1(n) e^{-jn\omega}$, where $g_1(n)$ is the pulse response sequence, and $\omega \in [-\pi, \pi]$, and then $S_{W_1}(\omega) = |G_1(\omega)|^2 S_\varepsilon(\omega)$. Therefore, the autocorrelation function of $W_1\varepsilon$ is $R_{W_1}(\tau) = \frac{1}{2\pi}\int_{-\infty}^{\infty} S_{W_1}(\omega) e^{j\omega\tau} d\omega$. After filtering the mean of white noise is zero.

$$\begin{aligned}\sigma_1{}^2 &= R_{W_1}(0) = \frac{1}{2\pi}\int_{-\infty}^{\infty}|G_1(\omega)|^2 S_\varepsilon(\omega)d\omega \\ &= \frac{\sigma^2}{2\pi}\int_{-\pi}^{\pi}\left[\sum_{n=-\infty}^{\infty} g_1(n)e^{-jn\omega}\right]\left[\sum_{n=-\infty}^{\infty} g_1(n)e^{jn\omega}\right]d\omega \\ &= \sigma^2\sum_{n=-\infty}^{\infty}(g_1(n))^2 = \sigma^2\|g_1\|^2\end{aligned} \tag{17}$$

In (19) $\| \;\|$ denotes the norm of a vector, so $\sigma_2{}^2 = \sigma^2\|h_1 * g_2\|^2$, and then

$$\sigma_j{}^2 = \sigma^2\left\|h_1 * h_2 * \cdots * h_{j\text{-}1} * g_j\right\|^2 \tag{18}$$

where h_j and g_j are the unit pulse response of H_j and G_j respectively. h and g depend on the scaling function $\phi(t)$ and wavelet function $\psi(t)$. They are not determined by specific scale.

2.2.6 Noise variance σ estimation

Our algorithms require the calculation of the underlying noise variance σ. In a real world application, the variance is usually unknown a priori, so it must be estimated from the data. Some papers used the dark (signal free) regions at the boundaries of each image to estimate the noise power at each wavelet scale. But this method requires manually choosing such regions and sometimes the noise in medical images is not uniformly distributed in both signal and free regions [19].

Based on [12] we are using the spatial correlation $Corr_2(j,n)$ between the first two scales of wavelet transform to compute the power of $P_{corr}(j)$ and $P_W(j)$

$$P_{corr}(j) = \sum_n Corr_2(j,n)^2 \tag{19}$$

$$P_W(j) = \sum_n W(j,n)^2 \tag{20}$$

$$Corr_2(j,n) = W(j,n) * W(j+1,n) \qquad n = 1,2,\cdots N \tag{21}$$

where $W(j,n)$ denotes the wavelet transform data. The power of $Corr_2(j,n)$ is then rescaled to that of $W_2(j,n)$

$$NewCorr_2(j,n) = Corr_2(j,n) \times \sqrt{P_W(j) / P_{corr}(j)} \tag{22}$$

If $|NewCorr_2(1,n)| > |W_2(1,n)|$, then the corresponding data in $W(1,n)$ is reset to 0. Refer to the remainder of $W(1,n)$ as $\tilde{W}(1,n)$. Suppose k points are killed totally; then $\tilde{W}(1,n)$ can be roughly considered to be produced by noise. It is well known that $\sqrt{P_x / M}$ is an asymptotically unbiased estimation of σ for a sequence $X \sim N(0,\sigma^2)$, where $P_x = \sum_n x(n)^2$ and M is the length of X. From $\sigma = \sigma_1 / \|g_1\|$ and the asymptotically unbiased estimation of σ_1^2, we can get

$$\sigma = \sqrt{\tilde{P}_W(1)/(N-k-1)} \Big/ \|g_1\| \tag{23}$$

From the Radon transform and Fig. 1, the noise variance $\sigma_{s,j}$ of the j th scale in the Radon domain is determined as

$$\sigma_{s,j} \approx \|h_1 * h_2 * \cdots * h_{j-1} * g_j\| n_s \sqrt{\sigma^2 / (JI)} \tag{24}$$

where n_s are the pixels through the image as shown in Fig. 1, and J and I are the size of the image.

Based on all of the above deductions the adaptive threshold in the Radon domain is given as

$$\begin{aligned} t_{s,j} &= c \cdot \sigma_{s,j} = cn_s \|h_1 * h_2 * \cdots * h_{j-1} * g_j\| \sqrt{\sigma^2 / (JI)} \\ &= cn_s \|h_1 * h_2 * \cdots * h_{j-1} * g_j\| \sqrt{\tilde{P}_W(1)/[JI(N-k-1)]} \Big/ \|g_1\| \end{aligned} \tag{25}$$

This threshold can vary adaptively, because it depends on the decomposed scales and the pixels through image in the Radon transform adaptively.

2.3 Inverse radon transform

After obtaining the filtered sinogram, we perform an inverse radon transform in order to get the original image. It is defined as

$$f(x,y) = \int_0^{\pi} R(\alpha, x\cos\alpha + y\sin\alpha) d\alpha \tag{26}$$

where R is the filtered projections. Generally, three different inverse Radon transform methods are direct inverse Radon transform (DIRT), filtered back-projection (FBP) and convolution filtered back-projection (CFBP) [20]. DIRT is computationally efficient, but it introduces some artifact. FBP based on linear filtering model often exhibits degradation in recovering from noisy data [21]. Spline-convolution filtered back-projection (SCFBP) offers better approximation performance than the conventional lower-degree formulation (e.g. piecewise constant or piecewise linear models) [22]. For SCFBP the denoised sinogram in the Radon domain is approximated in the B-spline space, while the resulting image in image domain is in the dual-spline space. We used SCFBP to propagate the deniosed sinogram back into the image space along the projection paths.

3. Experiments and results

Our filtering method has been evaluated by using simulation brain data. We also applied the method to filter real brain MR images. Single scale optimum linear filter - Wiener filter, traditional multiscale wavelet filter [8], anisotropic diffusion (AD) filter [2, 3] and bilateral filter [5, 25] were applied to these dataset in order to compare the performance of these filtering methods. Here we define an average SNR as quality metric. It is given by

$$SNR = 10\log_{10}\frac{\sum_{m,n}\left|X[m,n]\right|^2}{\sum_{m,n}\left|X[m,n]-\hat{X}[m,n]\right|^2} \quad (27)$$

with the results averaged over all images and reported as mean decibels (dB), where $X[m,n]$ is the original image and $\hat{X}[m,n]$ is the filtered image.

We transformed the intensity of all testing images to uint8 (0~255) before all images were filtered with the five methods. Neighborhoods of size 3×3 were used to estimate the local image mean and standard deviation of Wiener filter. A Db3 wavelet was used in the traditional wavelet filter, and the images were decompounded to four levels. The constant was imposed $c = 3.78$ in our method, and the images were also decompounded to four levels. For simulated and real brain data the AD filter was used when diffusion constant was 140. And the bilateral filter was performed when spatial function variance $\sigma_s = 3$, range function variance $\sigma_r = 20$.

3.1 Simulation brain MR data

We obtained the brain MR images from the McGill phantom brain database for comprehensive validation of the different filtering methods. The MR volume was constructed by subsampling and averaging a high-resolution (1-mm isotropic voxels) dataset consisting of 27 aligned scans from one individual in the stererotaxic space. The volume contains 181 × 217 × 181 voxels and covers the entire brain. Based on the realistic phantom, an MR simulator is provided to generate specified MR images [23].

We obtained T1-weighted, T2-weighted and Pd-weighted MR volumes with an isotropic voxel size of 1mm at different noise levels. The noise effect was simulated as zero mean Gaussian noise adding to the MR volume with its standard deviation equal to the noise percentage multiplied by the brightest tissue intensity [24, 26]. Fig. 6 illustrates the visual assessment of filtered results on simulated T1-weighted, T2-weighted and Pd-weighted MR images with 9% noise and the comparison of the five different methods. We can easily see that bilateral filter and our method reduce more noise than other filtering methods. But the results of our method are much closer to original images without noise than those of bilateral filter.

Fig. 7 shows output SNR comparison of different filters for different image types (T1-weighted, T2-weighted and Pd-weighted MR image) and noise levels (1%, 3%, 5%, 7%, 9%, 11%, 13%, 15% and 17%). The output SNRs of the five methods for three types of MR images always decrease as the noise in MR image increases, and even some output SNRs of AD filter is less than input SNRs. From the three figures it can be seen that the output SNR of our method is always higher than other methods for almost all noised MR images except Pd-weighted images with 3% and 5% noise. Moreover, output SNRs of all denoising methods are always less than input SNRs of original MR images when original images have been added 1% Gaussian noise. We also see that the more noise (over 11%) exists in MR images, the better our method performs than others.

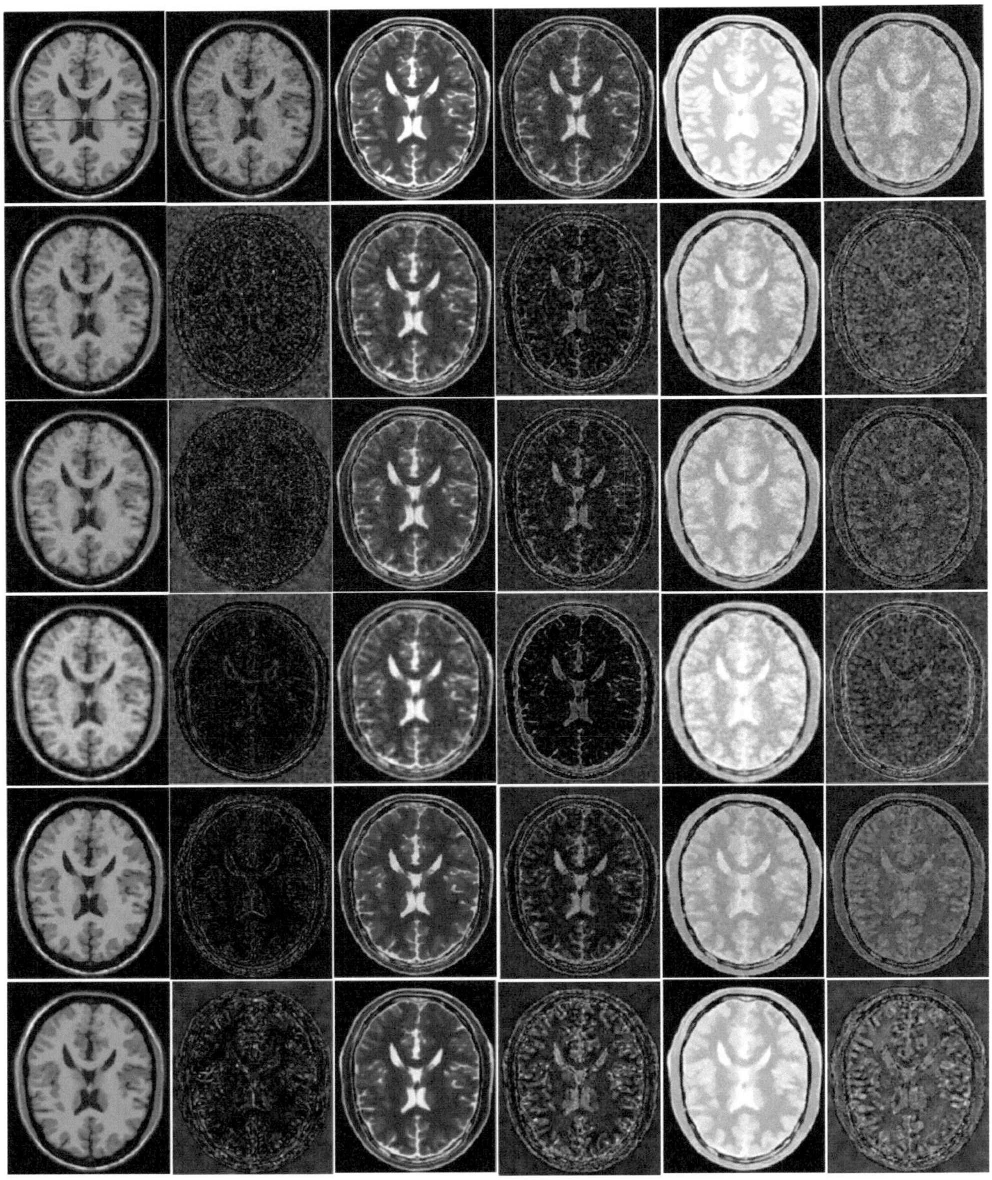

Fig. 6. Filtered results of the simulated brain data. The first row shows original T1-weighted MR image, noised T1-weighted MR image, original T2-weighted MR image, noised T2-weighted MR image, original Pd-weighted MR image and noised Pd-weighted MR image from left to right respectively. The first column shows the T1-weighted filtered results using Wiener filter, wavelet filter, AD filter, bilateral filter and our method from the second row to the bottom respectively. The second column shows the corresponding residuals between the original image and the filtered image. The third and fourth columns are T2-weighted filtered results after using the different methods and the corresponding residuals. And the fifth and sixth columns are the corresponding Pd-weighted results and residuals.

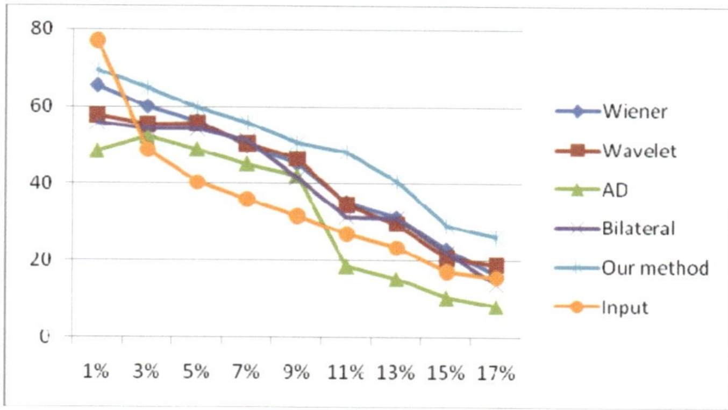

(a) T1-weighted MR image

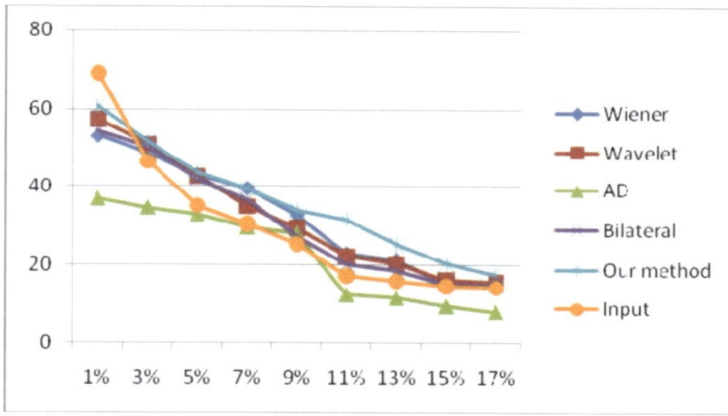

(b) T2-weighted MR image

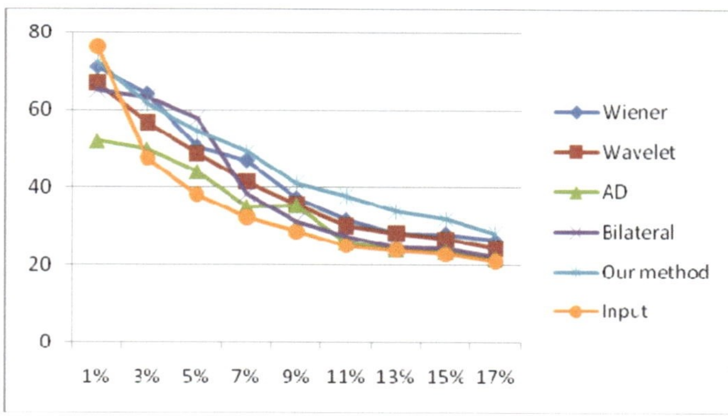

(c) Pd-weighted MR image

Fig. 7. Output SNR comparison of the different filters for different image types and noise levels. In (a), (b) and (c) the horizontal axis is different noise levels. The vertical axis is the output SNRs (dB) of the different filters and the original input SNRs (dB) of different noise levels.

Fig. 8 indicates the quantitative comparison of horizontal profiles between original T1-weighted image and filtered images after using the different methods. The profile position is shown in the original T1-weighted MR image of Fig.6. From Fig.8 we can see that the filtered result of our method is closest to original standard image without noise in all denoised results. It demonstrates our method is effective in reducing noise and preserving details in MR image.

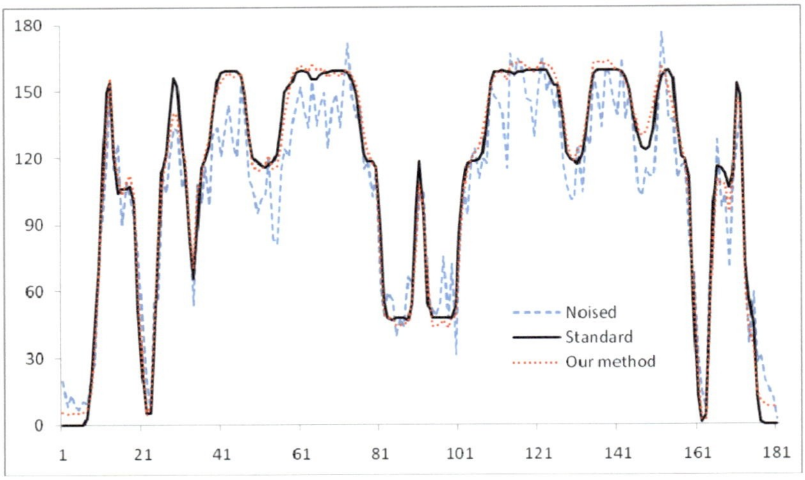

(a) Comparison of horizontal profiles between original standard, noised image and the filtered image after using our method.

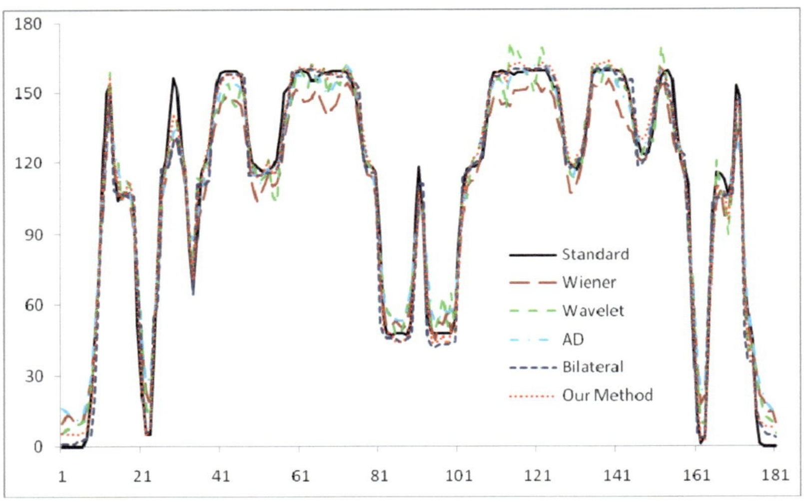

(b) Comparison of horizontal profiles between original standard, the filtered images using Wiener filter, wavelet filter, AD filter, bilateral filter and our method respectively.

Fig. 8. Comparison of horizontal profiles between the original T1-weighted image and the filtered images after using the different methods.

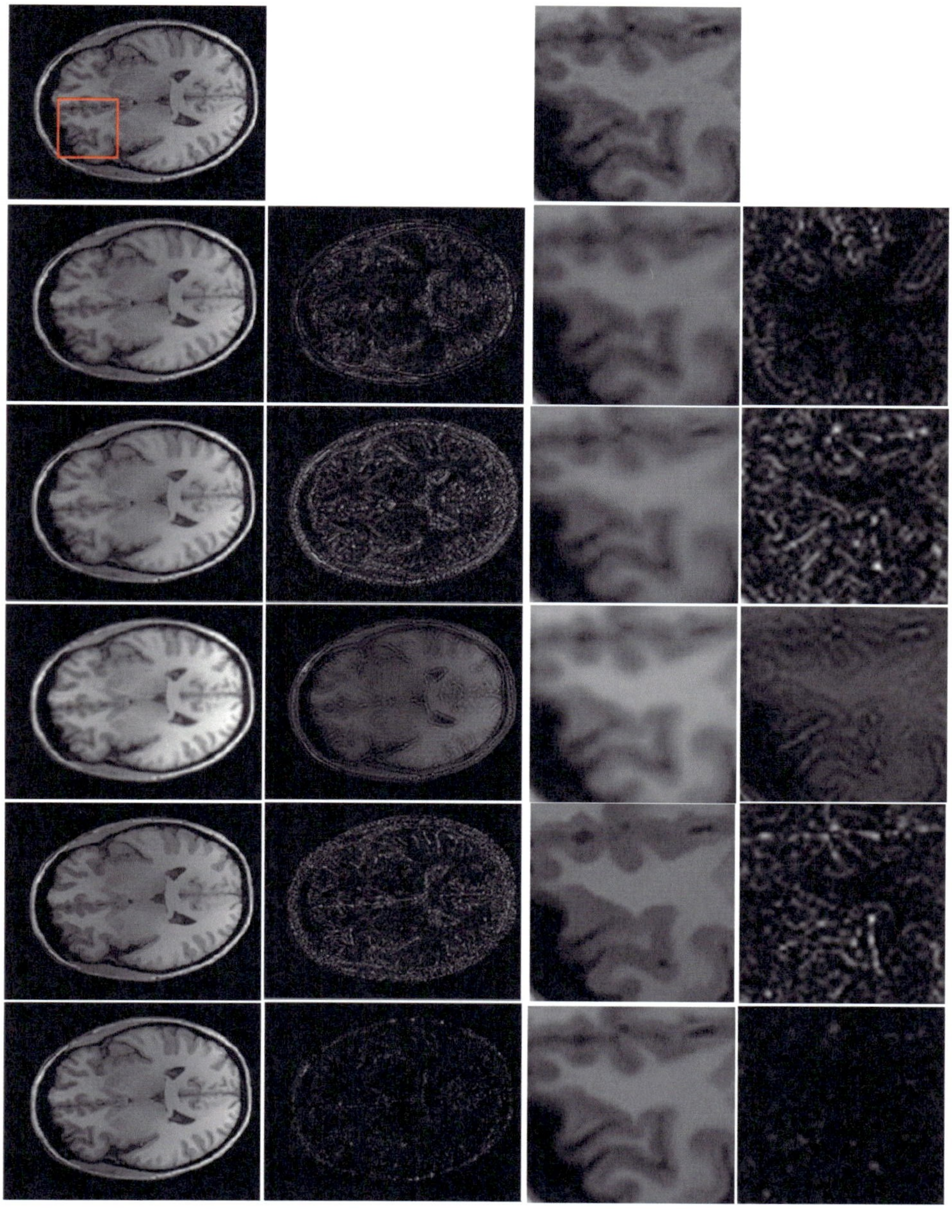

Fig. 9. Qualitative comparison of filtering results after using the different methods for a slice. The first column is the real brain image and the filtered results after using Wiener filter, wavelet filter, AD filter, bilateral filter and our method from the top to the bottom respectively. The second column is the corresponding residuals between the original image and the filtered image. Qualitative comparisons of ROIs are shown in the third and fourth columns. The third column shows the ROIs of real brain image, and the filtered results using Wiener filter, wavelet filter, AD filter, bilateral filter and our method from top to bottom respectively. The fourth column shows the corresponding residuals in ROIs between the original image and the filtered image.

4.2 Real MR brain data

Our filtering method also was applied to real T1 weighted MR images of the human brain. Fig. 9 illustrates the visual assessment of the filtered results of real T1-weighted MR brain images using the five different methods. By comparing the filtered results and corresponding residuals, it can easily be seen that Wiener filter makes image blurred and their residual in the whole image is almost the same except edges. Wavelet filter seems to introduce artifacts in the denoised image. AD filter erases small features and transforms image statistics due to its edge enhancement effect resulting in an unnatural image. And it can also be seen that bilateral filter and our method are excellent in reducing the noise, and enhancing the image contrast. But bilateral filter loses many small edges and features while our method preserves more of the details. The enlarged view of the region of interest (ROI) is also shown in Fig.9. It is evident that our method suppresses noise effectively while keeping more of the image details by comparing the two ROIs (50×50 voxels).

5. Conclusion

A wavelet domain filtering method based on the Radon transform for noise reduction in medical images was presented. We performed the Radon transform for input images, and based on our image model we validated that the Radon transform can improve the SNR and zero mean white noise has no effect on it. Wavelet analysis decomposes a signal into a hierarchy of scales ranging from the coarsest scale to the finest one. Orthogonal wavelet transforms have fewer coefficients at course scales, which prevents the transformed signal from correlated across the scales. Lack of a translation invariant makes filtering by orthogonal wavelet transforms exhibit visual artifacts. Therefore, we apply a 1D non-orthogonal wavelet transform along s, and use a threshold-based method to filter. We use a spatial correlation function to enhance significant structures and dilute noise, and then estimate the original noise variance from the first two scales. Dissimilar to traditional threshold methods that apply the same threshold to the wavelet coefficients in all scales, the proposed method provides an adaptive and robust threshold for every scale. The images in the Radon domain do not have more high-frequency parts or more edges than the corresponding images in the time domain, so our method for noise variance estimation works well. On the other side, inverse radon transform will introduce some artifacts [27]. Currently, our method produced promising results on a small segment of the image in additive Gaussian noise. Further work needs to be performed to shown if this technique is useful in multiplicative noise, on a larger wake image, and the wake detection problem.

Simulated brain MR image and real brain MR image images were used to validate our method. Wiener filter, wavelet filter, AD filter and bilateral filter were compared with our method. Our method performed better than the other methods in improving SNR and in preserving the key image details and features. The experimental results showed the superiority of the proposed method as it outperformed the traditional denoising methods. Our method can also be used in other medical imaging.

6. References

[1] Xiaofeng Y, Baowei F. A wavelet multiscale denoising algorithm for magnetic resonance (MR) images. Measurement Science and Technology, 2011, 22 (2): 025803-025815.

[2] Perona P, Malik J. Scale-space and edge detections using anistropic diffusion. IEEE Transactions on Pattern Analysis and Machine Intelligence. 1990; 12: 629-39.
[3] Samsonov A, Johnson C. Noise-adaptive nonlinear diffusion filtering of MR images with spatially varying noise levels. Magnetic Resonance in Medicine. 2004; 52: 798–806.
[4] Wong W, Chung A. Trilateral filtering: a non-linear noise reduction technique for MRI. International Society for Magnetic Resonance in Medicine. 2004; 2218.
[5] Tomasi C, Manduchi R. Bilateral filtering for gray and color images. Sixth International Conference on Computer Vision. 1998; 839-46.
[6] Healy D M, Weaver J B. Two applications of wavelet transforms in magnetic resonance imaging. IEEE Transactions on Information Theory 1992; 38: 840-60.
[7] Nowak R D. Wavelet-based Rician noise removal for magnetic resonance imaging. IEEE Transactions on Image Processing 1999; 8: 1408-19
[8] Donoho D L, Johnstone I M. Ideal spatial adaptation by wavelet shrinkage. Biometrika Trust 1994; 81425-455.
[9] Quan P, Lei Z, Guanzhong D, Hongai Z. Two denoising methods by wavelet transform. IEEE Transactions on Signal Processing 1999; 47: 3401-06.
[10] Paul B, Lei Z. Noise reduction for magnetic resonance images via adaptive multiscale products thresholding. IEEE Transactions on Medical Imaging 2003; 22: 1089-99.
[11] Pizurica A, Philips W, Lemahieu I, Acheroy M. A versatile wavelet domain noise filtration technique for medical imaging. IEEE Transactions on Medical Imaging 2003; 22: 323-31.
[12] Yansun X, Weaver J B, Healy D M, Jian L. Wavelet transform domain filters: a spatially selective noise filtration technique. IEEE Transactions on Image Processing 1994; 3:747-58.
[13] Deng Z, Toshi H N. Medical Image Noise Reduction Using Radon Transform and Walsh List in Laplacian Pyramid Domain. The 13th IEEE International Symposium on Consumer Electronics (ISCE2009). 2009; 756-760.
[14] Ratnaparkhe V R , Manthalkar R R, Joshi Y V. Texture Characterization of CT Images Based on Ridgelet Transform. ICGST-GVIP 2009; 8: 43-50.
[15] Anand C S, Sahambi J S. MRI denoising using bilateral filter in redundant wavelet domain. TENCON 2008. IEEE Region 10 Conference 2008; 1-6.
[16] Mallat S G. A Theory for Multiresolution Signal Decomposition - the Wavelet Representation. IEEE Transactions on Pattern Analysis and Machine Intelligence 1989; 11: 674-93.
[17] Mallat S, Zhong S. Complete signal representation with multiscale edges. New York University, Institute of Fine Arts Library. 1989.
[18] Xiaofeng Y, Peng L, Xin Z, Zhengzhong B, Bo W. De-Noising of the Doppler Fetal Heart Rate Signal with Wavelet Threshold Filtering Based on Spatial Correlation. Bioinformatics and Biomedical Engineering, 2007. ICBBE 2007. The 1st International Conference on, 928-931, 6-8 July 2007
[19] Delakis I, Hammad O, Kitney R I. Wavelet-based de-noising algorithm for images acquired with parallel magnetic resonance imaging (MRI). Physics in Medicine and Biology 2007; 52: 3741-51.
[20] Yazgan B, Paker S, Kartal M. Image reconstruction with diffraction tomography using different inverse Radon transform algorithms. Proceedings of the 1992 International Biomedical Engineering Days 1992; 170-173

[21] Nam-Yong L, Lucier B J. Wavelet methods for inverting the Radon transform with noisy data IEEE Transactions on Image Processing. 2001; 10: 79-94

[22] Horbelt S, Liebling M, Unser M. Discretization of the radon transform and of its inverse by spline convolutions IEEE Transactions on Medical Imaging. 2002; 21: 363-376.

[23] Kwan R K S, Evans A C, Pike G B. MRI simulation-based evaluation of image-processing and classification methods. IEEE Transactions on Medical Imaging. 1999; 18: 1085-97.

[24] Xiaofeng Y, Baowei F. A multiscale and multiblock fuzzy c-means classification method for brain MR images. Medical Physics, 2011; 38 (6): 2879-2893.

[25] Xiaofeng Y, Sechopoulos I, Baowei F. Automatic tissue classification for high-resolution breast CT images based on bilateral filtering. Proc. SPIE 7962, 79623H (2011); doi:10.1117/ 12.877881

[26] Xiaofeng Y, Baowei F. A MR Brain Classification Method Based on Multiscale and Multiblock Fuzzy C-Means. Bioinformatics and Biomedical Engineering, (iCBBE) 2011 5th International Conference on, 1-4, May 2011

[27] Xiaofeng Y, and Baowei F. A skull segmentation method for brain MR images based on multiscale bilateral filtering scheme. Proc. SPIE 7623, 76233K (2010); doi:10.1117/12.844677

11

Application of Wavelet Transform Method for Textile Material Feature Extraction

Lijing Wang[1], Zhongmin Deng[2,3] and Xungai Wang[2,3]
[1]*School of Fashion and Textiles, RMIT University*
[2]*School of Textile Science and Engineering, Wuhan Textile University*
[3]*Institute for Frontier Materials, Deakin University*
[1,3]*Australia*
[2]*China*

1. Introduction

During textile manufacturing processing, it is vital to inspect, detect, assess and rectify defects as soon as they emerge. This is particularly important for fabric engineering and finishing, as the fabric quality directly affects the quality of their final products. In textile industry history, visual inspection and classification were common practice. The nature of traditional manual inspection was very repetitive and slow, and defects could be missed due to inspector fatigue, misjudgement and neglect, not mentioning the costs of skilled labour for the task. Automated visual inspection of texture content in digital images plays an important role in detecting textile defects for quality control. A computer based inspection system can be designed to perform 100% defect inspection objectively and consistently. It eliminates inspection error due to human frailty, and saves the costs of skilled inspectors.

Ngan, Pang and Yung (2011) reviewed automated fabric defect detection methods developed in recent years. They summarised different approaches for texture defect automatic detection, where computer vision and image processing have been the key for success. Current technology of image acquisition has resulted in inexpensive, but high quality digital images, which can be conveniently streamed to a computer for information extraction. A general procedure for an automated inspection system, as shown in Figure 1, is to process the images to be inspected and extract the features of defects. Image processing should highlight texture defects, such as the location or extension of the defects using various properties of the textures. A classifier, trained and validated by history data and defect definitions, analyses the features to objectively classify the defects. Neural networks are commonly used to combine with texture features obtained from image processing for defect classification.

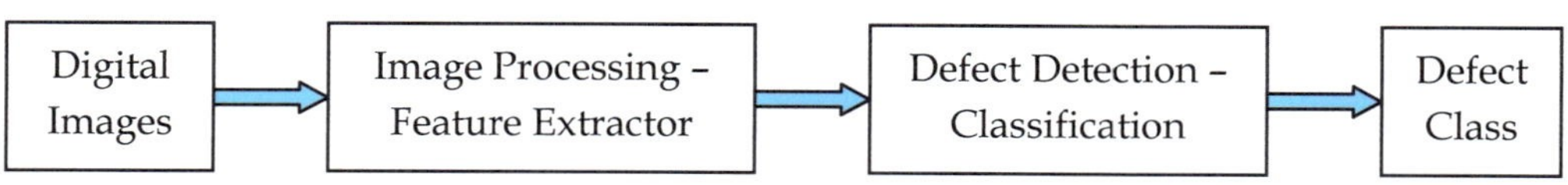

Fig. 1. A textile defect detection process

An industrial automated inspection system must operate in real-time, and produce a low false alarm rate. As computer technology develops, rapid image processing and pattern recognition can be performed quickly and inexpensively. It therefore becomes increasingly popular for computer aided feature extraction, defect detection, quality classification and decision making (Deng et al., 2011; Han & Shi, 2007; Liu et al., 2011a; Liu et al., 2011b; Mak & Peng, 2008; Mak et al., 2009; Wong et al., 2009; Zhang et al., 2010b) in the textile field. Thus, it is the trend that automated image-based inspection replaces human inspection, and improved feature extraction methods will accelerate the process. The region with a defect has a texture different from the background, which has a relatively consistent texture. Texture analysis has played an important role in the automatic visual inspection of textile surfaces. Among many methods, the wavelet transform tool has been studied for analysing images and extracting important information for improved pattern recognition (Deng et al., 2011; Han & Shi, 2007; Liu et al., 2011a; Ngan et al., 2005; Wong et al., 2009; Zhang et al., 2007, 2010b).

Human vision researchers have found that the visual cortex can be modelled as a set of independent channels, each with a particular orientation and spatial frequency tuning (Beck et al., 1987). This forms the basis for texture classification using the wavelet transform. The wavelet transform provides a solid and unified mathematical framework for the analysis and characterization of an image at different scales. It provides both time and frequency information, and has been successfully applied for textile image classification, including the application examples presented in this chapter. The wavelet transform approach appears to improve textile defect detection accuracy and real-time factory implementations.

This chapter presents some examples using wavelet transform for textile material feature extraction and classification. It first outlines how wavelet transforms together with other techniques were used in different textile defect feature extraction and classification. In particular, this chapter briefly summarises some applications examples in literature. It then presents a detailed application example to demonstrate wavelet transform feature extraction from knitted, woven, and nonwoven pilled fabric images for assessing the level of pilling. This approach is based on the multi-scale two-dimensional dual-tree complex wavelet transform to decompose the pilled fabric image with six orientations at different scales and reconstruct fabric pilling sub-images. For pilling objective rating, feature parameters are extracted from the pilling sub-images to describe pill (a small ball of fibres) properties. By using a classifier, the pilling grade can be accurately classified. Another detailed example is the use of the wavelet transform to extract animal fibre surface texture features for fibre identification. The complex wavelet transform decomposition and reconstruction are an effective way to extract cashmere and superfine merino wool fibre features, including cuticular scale height, scale shape and scale interval, which can be used for characterizing different animal fibres and subsequently classifying them.

2. Wavelet transform in textile applications

The hierarchical wavelet transform uses a family of wavelet functions and its associated scaling functions to decompose the original image into different sub-bands, providing both frequency and spatial locality. The wavelet transform method has been investigated in many situations where visual inspection is necessary in the textile industry; in particular, much research has been conducted on fabric defect automatic detection and classification.

A fabric can be seen as a manufactured planar structure made of fibres and/or yarns assembled by various means such as weaving, knitting, tufting, felting, braiding, or bonding of webs. Good quality fabric should have its designed appearance, and defects exist if the appearance was disrupted. Sari-Sarraf and Goddard (1999) described a vision-based on-loom fabric inspection system. The system processed images with a segmentation algorithm based on the concepts of 2-D discrete wavelet transform, image fusion, and the correlation dimension to attenuate the background texture and accentuate the defects. Fabric defects can be determined from the disruption of global homogeneity of the background texture of output images.

Kim and Kang (2007) used the wavelet packet frame decomposition to extract texture features followed by a Gaussian-mixture-based classifier for texture segmentation and classification. The wavelet packet frame decomposition produced texture information in all the frequency sub-channels, and the important texture was concentrated in the intermediate frequency sub-channels, which formed the feature vectors for classification. As a defect comprises a relatively small part of the fabric image, and has a texture different from the background, the advantage of this method is that homogeneous subregions can be regarded as non-defective. Each image was therefore divided into multiple non-overlapping regions for fabric texture/defect segmentation. Kim and Kang (2005) also evaluated fabric pilling based on the wavelet reconstruction scheme to attenuate the periodic background and enhance the pills using un-decimated discrete wavelet transform. Kim and Park (2006) conducted a further study on the quantification and evaluation of fabric pilling using two-dimensional and three-dimensional hybrid imaging methods. Each method was found to have its own merits.

Mak and Peng (2008), and Mak, Peng and Yiu (2009) presented a woven fabrics defect detection system, in which important fabric texture features were extracted optimally from a non-defective fabric image using a pre-trained Gabor wavelet network. The wavelet transforms were used to facilitate the construction of structuring elements in subsequent morphological processing to remove the fabric background and isolate the defects. This method does not need fabric defect information and hence is suitable for supervised defect detection.

Latif-Amet, Ertüzün, and Erçil (2000) reported a method for texture defect detection in fabric images by combining concepts from wavelet theory and co-occurrence matrices, which consider the relative occurrence frequency of pixels separated by a certain distance in a given direction. Detection of defects within the inspected texture was performed first by decomposing fabric images into sub-bands, then by partitioning the textured image into non-overlapping sub-windows and extracting the co-occurrence features, such as coarseness, contrast, homogeneity and texture complexity. They found that a particular band obtained by wavelet transformation has high discriminatory power to improve the detection performance. The wavelet transformation method extracted the particular band and discarding the others, which carried information with low discriminatory power. In general, the defect detection method that applied to the sub-band images is superior to the same method applied to the raw images.

Yang, Pang and Yung (2002) designed an adaptive wavelet-based feature extractor method to characterise fabric images with multi-scale wavelet features by using undecimated discrete wavelet transforms. To minimise the error rate in fabric defect classification, Yang, Pang and Yung (2004) studied six wavelet transform-based feature extractors and classification methods. Compared to those commonly used wavelets, such as Haar wavelets and Daubechies wavelets, the adaptive wavelet designed with the discriminative feature extraction method largely enhanced the discriminant power of the wavelet features, which resulted in much better classification performance for the detection of fabric defects than the other discriminant training methods.

Using the wavelet transform method, Han and Shi (2007) decomposed fabric images with high frequency texture background at various levels to detect defects on images. By selecting an appropriate level at which the approximation sub-image was reconstructed, textures on the background were effectively removed. Together with an adaptive level-selecting scheme for analysing the co-occurrence matrices of the approximation sub-images, non-texture techniques can be used to resolve the texture defect detection problem. Compared with traditional frequency domain low and high pass filters, common wavelet transform methods appear to be much more effective in detecting defects due to the multi-scale analysis ability of wavelet transform.

Tsai and Hsiao (2001), and Tsai and Chiang (2003) proposed a wavelet transform method for the inspection of local defects embedded in homogeneous textured surfaces. Each wavelet decomposition level provided unique information about texture characteristics, and the energy value of the smooth sub-image was relatively low. By properly selecting the smooth sub-image or the combination of detail sub-images in different decomposition levels, backward wavelet transform constructed an efficient image with regular and repetitive texture patterns removed, and local anomalies enhanced for discriminating between defective regions and homogeneous regions. In addition, since this proposed method did not use an advance classifier such as a neural network to identify defects from textural features, it was simple and used less computational power. Tsai and Hsiao (2001) also pointed out that orthogonal wavelet bases captured local deviations in homogeneous textured surfaces better than biorthogonal bases for the application of defect detection in textured surfaces. Based on the energy distributions of wavelet coefficients, Tsai and Chiang (2003) further proposed an automatic band selection procedure to reconstruct images in the wavelet transform domain for automatically determining local defects. The band selection procedure determines the best decomposed sub-images and the number of multi-resolution levels to remove the global repetitive texture pattern, and reinstate image with local anomalies using the wavelet transforms for defect inspection in homogeneously textured images, including fabrics.

Liu *et al* (2010) combined wavelet texture analysis and learning vector quantization neural networks to classify nonwoven uniformity. Nonwoven images were decomposed to generate textural features from wavelet detail (sub-band) coefficients or sub-images at each scale with wavelet texture analysis. As the approximation sub-band coefficients usually represent the lighting or illumination variation, the high frequency sub-bands were used as the input features of the neural network for training and testing the classifier. The image classing system categorised five nonwoven uniformity grades with an overall identification accuracy of more than 87%. Liu *et al* (2011a) further studied combining wavelet energy signatures and robust Bayesian neural network for nonwoven uniformity classification. They found that when 18 features of the nonwoven images decomposed at level 3 with wavelet transform were employed to describe the texture of nonwoven, the average accuracy was over 98% for all visual quality ratings. When 24 features of the nonwoven images decomposed at level 4 were used, the average recognition accuracy was over 99%.

Ngan *et al* (2005) applied the direct thresholding method based on wavelet transform detailed sub-images as an automated visual inspection method for defect detection on patterned fabric. They reported that the wavelet pre-processed golden image subtraction method, which can segment out the defective regions on patterned fabric effectively, provided an overall detection success rate of 96.7% from 30 defect-free images and 30 defective patterned images for one common class of patterned jacquard fabric. In this report wavelet based methods were used to extract detailed and approximation sub-images from a histogram equalized defective image. The sub-images were further processed for defect detection.

Wong *et al* (2009) presented a hybrid approach of stitching defect detection and classification in a fabric image using the wavelet transform and the back propagation neural network. The pyramid wavelet transform was employed to generate an approximation (smooth) image and detailed sub-images at a certain resolution level. The smooth sub-image has high energy concentration (more than 95%). It was further processed for image segmentation, attenuating the background texture and accentuating the anomalies. Through a neural network classifier, the combined method can identify five classes of stitching defects effectively.

Shin, Kim and Kim (2010) used a multi-level wavelet transform to extract pattern features from textile images. The pattern, colour and texture information were used as cues to predict the emotional semantics associated with the image. Although not relevant to defect detection, this work could be applied for efficient indexing annotating and searching/retrieving for diverse textile images on the Web to reduce the semantic gap between low-level features and the high-level perception of users.

Liang *et al* (2012) employed wavelet transform analysis and statistical measurement for objective and automatic evaluation of yarn surface appearance, which is related to yarn quality and traditionally was subjectively performed by visual inspection of the yarn board (parallel yarns on a black background board) in the textile industry. The wavelet transform algorithm (Discrete Wavelet Transform) was used to remove the yarn hairiness (the fibres on the periphery layer of yarn) as noise and separate yarn lines from the background during pre-processing of the yarn board images. The yarn hairiness image, on the other hand, can be reconstructed by subtracting the detected yarn lines from the original yarn image. The extraction of hairiness characteristics from a yarn surface appearance image was carried out by calculating wavelet energies under a certain decomposition level using four types of mother wavelets. Furthermore, statistical measurements extracted other important yarn quality statistics features, such as yarn diameter variation and distribution, yarn faults (neps, thin/thick places) and hairiness from the texture images. Depending on the classification tool, their experimental validation results showed their yarn grading system achieved over 90% classification accuracy for the individual yarn category and 87% for global yarn database.

Qu and Ding (2010) used 2D wavelet transform to extract the edge features of foreign fibres in lint cotton within a complex background. The wavelet transformed the original cotton luminance image to the morphology component and three detail components. By removing the background and noise, the contaminant edge features were obtained from the detail components. Further morphological analysis differentiated the gray and colour features between foreign fibres and cotton.

3. Fabric pilling feature extraction and rating

3.1 Background

Fabric pilling refers to the formation of fibrous balls on the surface of a fabric. The pills, balled or matted particles of fibre that remain attached to the surface of the fabric, spoil the original appearance of the fabric and can cause premature wear. Normally, resistance to pilling is tested in the laboratory by standard methods that simulate accelerated wear, followed by a manual assessment of the degree of pilling based on a visual comparison of the sample with a set of standard pilling images. The degree of pilling is determined on a level ranging from 1 (very severe pilling) to 5 (no pilling). This subjective evaluation can be inconsistent and inaccurate, and pill rating may vary from one laboratory to another. Reliable and accurate objective evaluation methods are desirable for the textile industry.

Current pilling image processing methods cannot effectively eliminate the influence of fabric texture for accurate pilling prediction (Deng et al., 2011). A pilled fabric image consists of brightness variations caused by high frequency noise, randomly distributed fibres, fuzz and pills, fabric surface unevenness, and background illumination variance. Interference from fabric background texture affects the accuracy of pilling rating by directly computer-aided image classification. A pilled fabric often has distinct pills as well as ambiguous fuzz and small pills that are difficult to classify. In addition, the fabric surface ruggedness may add further difficulty to fabric pilling feature extraction and assessment. Many researchers have tried to separate the pills from the background by image analysis techniques, such as pixel-based brightness threshold (Kim, S. & Park, 2006; Konda *et al.*, 1988; Xin *et al.*, 2002) and region-based template matching (Xin et al., 2002). However, these methods can not eliminate the influence of fabric texture.

The two-dimensional discrete Fourier transform (DFT) has been used (Palmer & Wang, 2004; Tsai & Chiang, 2003; Zhang et al., 2007) to separate periodic structures in fabric image from non-periodic structures in the image (the pills) in the frequency domain. The DFT can provide only gross summary of spatial frequency information about the entire image, not location information since localized pills in nature cannot be easily identified directly by Fourier transform (Xu, 1997). The multi-scale transform can effectively analyse the images at different scales of decomposition. For example, the wavelet transform measures the image brightness variations at different scales (Kim & Kang, 2005). It has been applied to objective pilling grading in recent studies. Palmer and Wang (Palmer & Wang, 2003, 2004) suggested that the pilling intensity can be classified by the standard deviation of the horizontal detail coefficients of a two-dimensional discrete wavelet transform at one given scale. When the analysis scale closely matched the fabric texture frequency, the discrimination was the largest. More pills and fuzz on the fabric surface gave higher standard deviation of the horizontal detail coefficients. However, the energy method for pilling analysis can only give frequency information, not the location information.

Otsu (1979) proposed a simple threshold to separate pills from the background in the reconstructed smooth (approximation) sub-image at an appropriate decomposition level.

However, the approximation sub-image comprises not only pilling information but also surface unevenness and illuminative variation, which will influence the determination of the threshold.

Using the two-dimensional dual-tree complex wavelet transform, the pilled nonwoven fabric image can be decomposed to sub-images of different frequency components, and the fabric texture and the pilling information are presented in different frequency bands. The energies of the six direction detail sub-images, which capture brightness variation caused by fuzz and pills of different sizes, quantitatively characterised the pilling volume distribution at different directions and scales. By extracting the pilling sub-images, a pilled image can be reconstructed for objective pilling evaluation by the combination of pilling identification, characterization method and an appropriate neural network supervised classifier (Zhang et al., 2010b).

3.2 The complex wavelet transform

The complex wavelet transform (CWT) is an enhancement to the two-dimensional discrete wavelet transform (DWT), which yields nearly perfect reconstruction, an approximately analytic wavelet basis and directional selectiveness in two dimensions. The most important property of CWT is that it can separate more directions than the real wavelet transform. The CWT can provide six sub-images in two adjacent spectral quadrants at each level, which are oriented at angles of ±15°, ±45°, ±75°. The orientation selectivity is clearly shown in Figure 2. The strong orientation occurs because the complex filters are asymmetric responses. They can separate positive frequencies from negative ones vertically and horizontally without aliasing positive and negative frequencies.

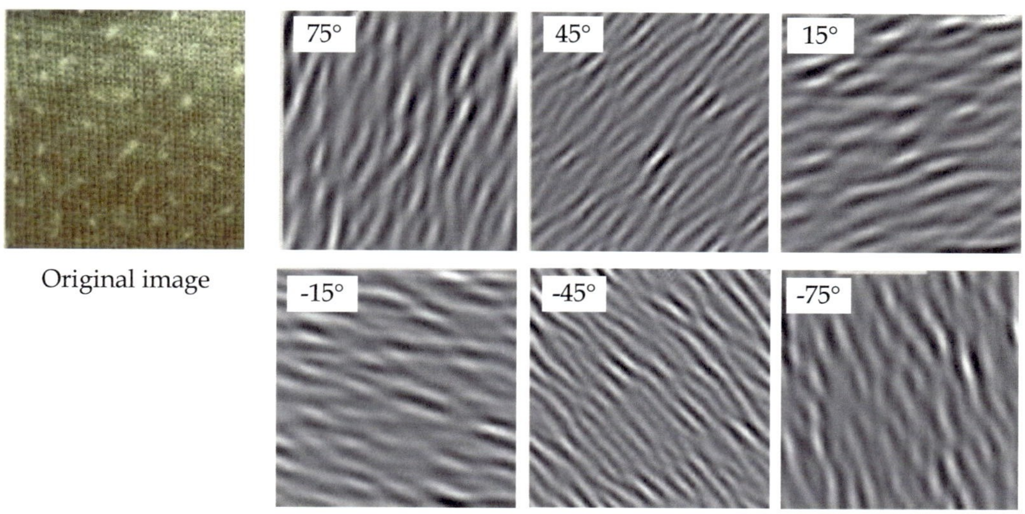

Fig. 2. 2-D impulse responses of the complex wavelets at scale 4 (6 bands at angles from +75° to -75°)

Figure 3 illustrates the difference between DWT and CWT. The orientation selectivity of CWT method is much clearer under each scale in comparison with the classical wavelet

DWT. There is an obvious mosaic phenomenon for Scales 5-6 sub-images from the DWT method. As a result, the reconstructed pilling image is not as clear as that from the CWT method.

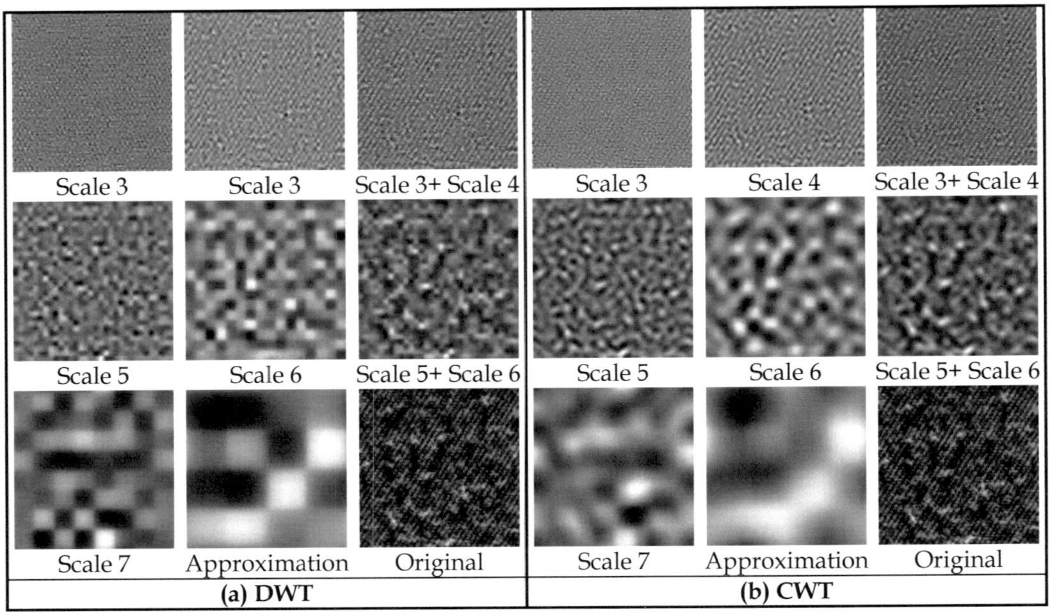

Fig. 3. Decomposition and reconstruction effects from DWT and CWT

3.3 Separate pilling from fabric background texture and wavelet reconstruction

Pills and fuzz are the vital features in the pilling rating procedure. Extracting them from the fabric image is the most important task for objective pilling assessment. To achieve this, the pilled fabric image has to be decomposed and reconstructed at different scales (see Figure 3 and Figure 4). After a fabric image is processed by CWT, the frequency band in each sub-image decreases with the decomposition scale. The attenuation of fabric patterns by CWT reconstruction is simple and effective since it can be attained by simply setting the relative detail coefficients to zero and reconstructing the image (Figure 4) without the delicate selection of the threshold and filter design. The sub-images contain information of high frequency signals, the fabric textures and the low frequency signals such as the background illuminative variation and the fabric surface unevenness, which are normally irrelevant to fabric pilling. The sub-images also contain information of fuzz and different size pills, which are the features for pilling classification. The key to pilling image reconstruction is to determine the decomposition level of original image and select appropriate Scales (sub-images) for invert complex wavelet transform (Figure 4).

The pilling feature can be determined by examining the relationship between energy and the decomposition scales. The energies of the wavelet coefficients distributed in different frequency channels at various decomposition levels provide unique information about texture characteristics. The choice of a proper decomposition scale is based on the energy relative gradient of detail sub-images in two consecutive scales. The decomposition scale for

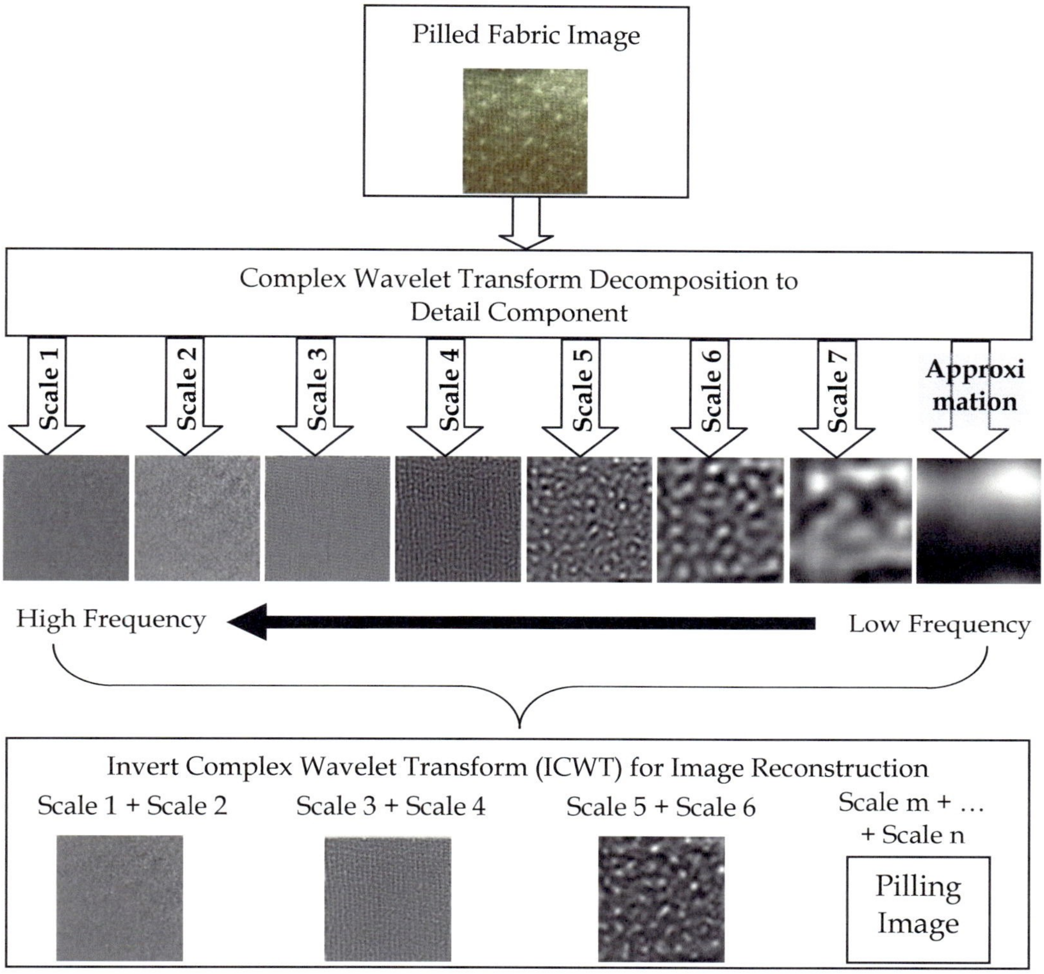

Fig. 4. Block scheme of extracting pilling image from pilled fabric image with CWT

obtaining an optimum pilling image is different for different frequency backgrounds. In general, fabrics with a rough background have more low frequency components and a greater decomposition scale than those with a fine background. By the optimum scales, the pilling information can be separated from the high-frequency noise, fabric texture, surface unevenness, and illuminative variation of the pilled fabric images (Deng et al., 2011). The examples in Figure 5 indicate that a pilled fabric can be successfully decomposed into pilling image and background image after filtering out disturbance information. The pilling image can be further converted to a binary image for feature extraction and classification.

3.4 Extraction of pilling features

Pilling features can be extracted in the spatial domain, in the frequency domain, or a combination of both frequency and spatial domains. In this example, the combination method was used.

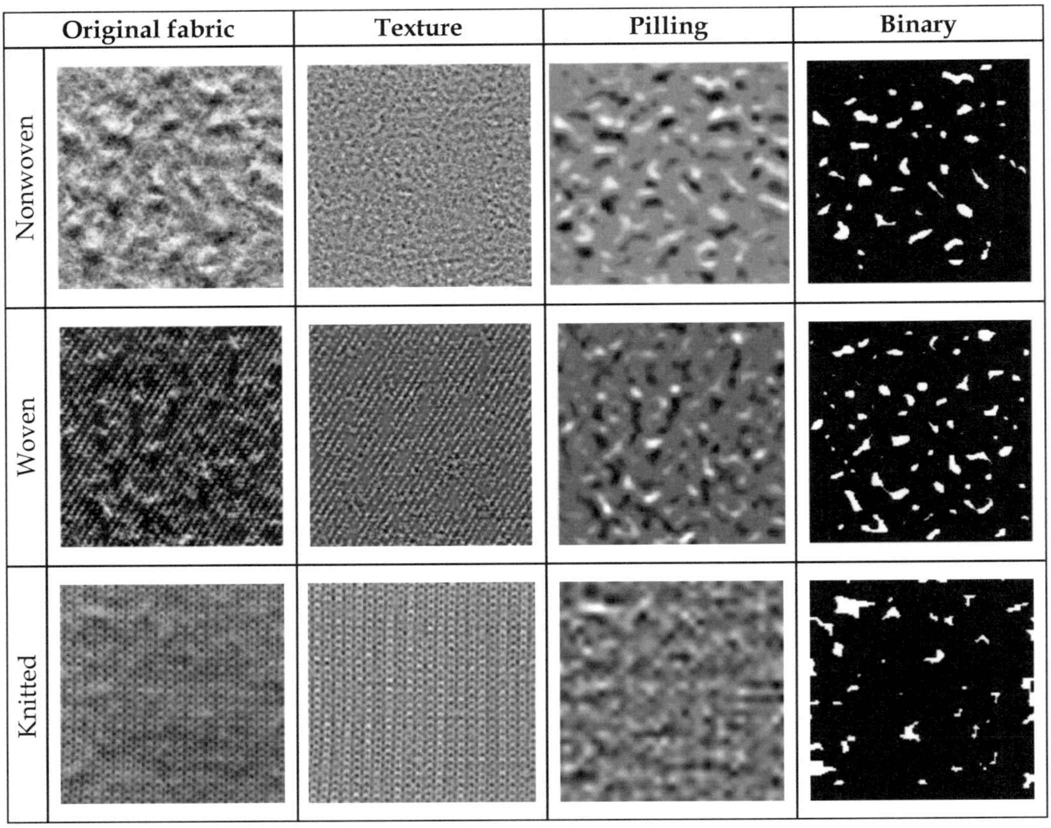

Fig. 5. Original, texture, pilling and binary images

3.4.1 Energy feature of pilling

Under lateral illumination specified in the testing standard (determination of the resistance to pilling and change of appearance of fabrics), pills can be easily differentiated from the bright (pills) to dark (background) gray value variation. This local contrast between a pill and its immediate surrounding region highlights the size and height of the pill as shown in the reconstructed images of Figure 3 and Figure 5. Larger pills lead to higher energy (the sum of the gray value squared), which must be used as one of elements of the pilling feature vector to characterize the pilling intensity and differentiate samples of different grades.

3.4.2 Shape feature of pilling

Pill density, size, and height are the main pill properties that observers use to rate the pilling grade of a tested fabric. They have a decreasing trend when the pilling grade increases, and linear and non-linear relationships have been observed in woven and knitted fabrics respectively (Kang et al., 2004; Konda et al., 1988). Therefore, pill shape feature (Figure 6a) should be used as a pilling feature vector to characterize the pilling degree. The total volume, height of pills, and standard deviation of height of pills, which can be calculated directly from the extracted gray pilling images shown in Figure 6, indicate the gray value magnitude and deviation of sample images and thus reflect the 3D pilling information of the pilled fabric surface. The greater the value of magnitude and deviation, the more severe the fabric pilling will be. The pill number and pill area show the pilling 2D information and they can be calculated from the binary images shown in Figure 5 and Figure 6b. These feature indexes increase when fabric pilling becomes more severe, and also should be used for objective pilling rating.

3.4.3 Pill binary image

In Figure 6c, the semi-binary images are obtained by replacing those pixel gray values that are lower than a threshold with 0, but keeping the other original gray values. The binary images, however, are developed by substituting the gray value higher than the threshold with 1 and those lower than the threshold with 0. Every small area of the white region in the binary image in Figure 6b represents a pill. The pilling area parameters, including total area, standard deviation of each pill area, and the pill location deviation coefficient of pills can be used for indexing the pill shape features.

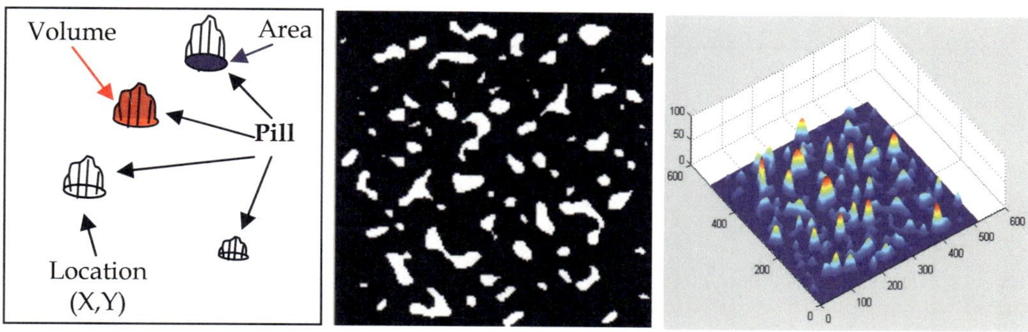

(a) Shape & location of pills (b) Pill binary image (c) Semi binary image

Fig. 6. Pill features and binary image

3.5 Neural network pilling objective classification

To objectively rate fabric pilling, an artificial neural network was trained to establish and generalize the relationship between pilling grade and objective pilling assessment parameters obtained through image processing. There are six input neurons corresponding to six feature indexes and five output nodes representing five pilling grades. Pilled fabric samples including woven, nonwoven and knitted fabrics were rated for fabric pilling rating according to their respective standards. All pilled fabric samples were digitised with a digital camera in the same way. Each fabric image was cropped to a 512×512 pixel 256 colour image or 512×512 pixel 8 bit gray scale image for feature extraction with wavelet transform. The features obtained included the pilling intensity and location, the energy ratio of pilling sub-image to background texture sub-image, the ratios of total pill area, the height standard deviation and volume to the image size, standard deviation of area and volume, and location deviation of pills. Information from forty different kinds of pilled fabric images was used to train the neural network and information from additional twenty pilled fabric images was used to test the trained model. The results indicated that, once the classification rules have been established, they can be used as a tool for objective classifying pilling grades (Deng et al., 2011).

In another attempt at objective pilling rating (Zhang et al., 2010a), a large set of 203 commercially rated pilled fabric samples were imaged using a digital camera. The two-dimensional dual-tree complex wavelet transform was used to decompose and reconstruct the sample images into their single-scale detail and approximation images. From each of the 203 fabric images, a texture feature vector consisting of 12 energy features was developed. Principal component analysis revealed that 87% of the variation in the texture feature vector accounted for the first principal component, and only minor proportions of the variation distributed amongst the remaining components. Based on this result, the single transformed first principal component consisting 12 pilling texture features was used as the basis for classification. Two thirds of the fabric image sample sets were used to train the neural network. Following training, the remaining 68 image samples were presented to the neural network as test samples for automatic pilling classification. Figure 7 shows the test sample rating results from the neural network classifier and paired with the original human expert rating for the same fabric sample, ranked according to the expert pilling ratings from 1 to 5.

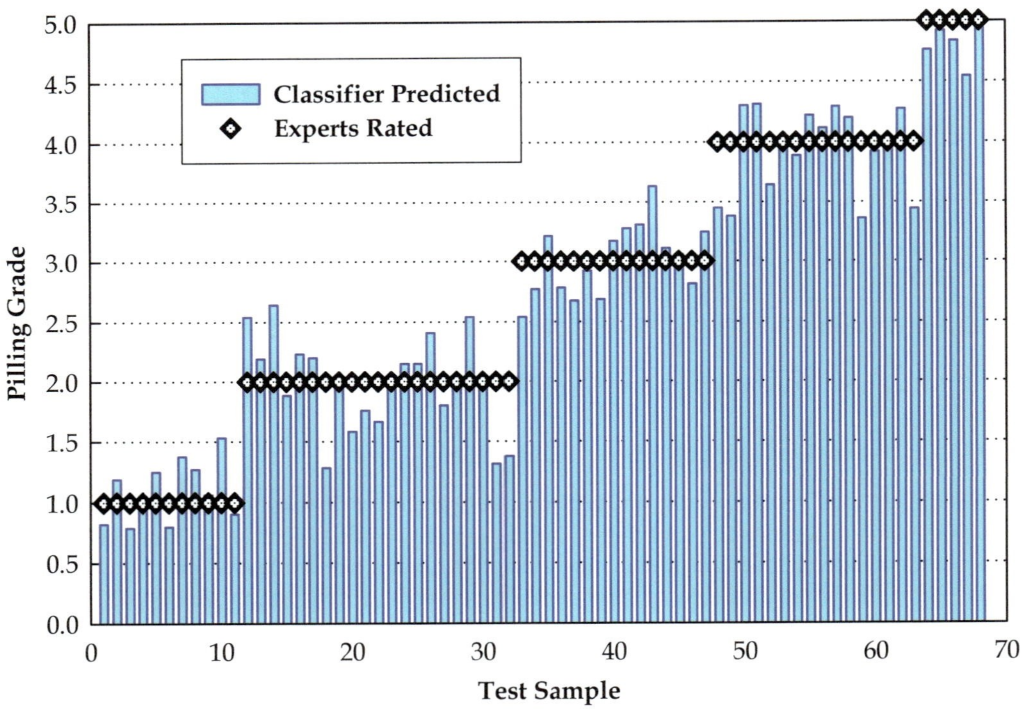

Fig. 7. Artificial neural network pilling classification results (not to a pilling grade) VS expert ratings (Data source: (Zhang et al., 2010a))

It can be seen from Figure 7 that the difference between the classifier test results and the expert measured grades for the test subset samples ranges ±0.7 pilling grades. When the classifier test results are converted to integer values and compared to the expert ratings, only a handful of test samples were misclassified. Re-examining the misclassified samples revealed that it was difficult to visually discern the difference between the pilling ratings of the fabric samples in question. This perhaps raises as many questions about the human expert rating ability as it does about the accuracy of the automatic classification. Many human experts of pilling rating claim the ability to interpolate half-interval pilling intensity ratings based on comparisons of fabric samples to a standard pilling image set. The ability of the neural network classifier to produce a floating point output rating can match this purported precision rating precision.

4. Identification of animal fibres with wavelet texture analysis

Cashmere is fine, downy wool growing beneath the outer hair of the Cashmere goat. Cashmere products are soft, luxurious and expensive. Cashmere and fine sheep's wool blends produces a lower cost product while exploiting the positive market perceptions associated with the luxury cashmere content. Correct labelling cashmere composition in such blends is required by law in most countries. Though cashmere and wool fibres have similar scaly surface morphology as shown in Figure 8, they may be classified through the following main cuticular scale features (Wildman, 1954):

- the form of the scale margins, e.g., smooth, crenate (scalloped) or rippled;
- the distance apart of the external margins of the scales, e.g., close, distant or near;
- the type of overall pattern, e.g., regular, irregular mosaic, waved or chevron; and scale height.

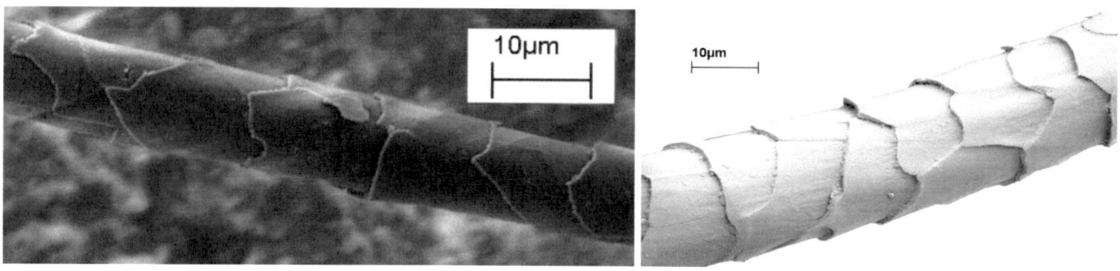

Fig. 8. SEM images of an Australian cashmere fibre (Left) and Merino wool fibre (Right)

To visually identify cashmere and sheep's wool in their blends, International Wool Textile Organisation test method (IWTO-58-00) uses scanning electron microscopic (SEM) analysis, while both American Association of Textile Chemists and Colorists Test method (20A-2000) and American Society for Testing and Materials method (D629-88) use light microscopy. However, the test accuracy that can be achieved depends largely on the operator's expertise with the microscopic appearances of different fibres. The operator-based assessment is subjective, tedious and costly. An automatic method is desirable to objectively identify animal fibres.

A computer based classification method for animal fibre identification may use combinations of microscopy and image analysis together with statistical and feature classification techniques (Robson, 1997, 2000; Robson et al., 1989; She et al., 2002). However, it is believed that wavelet texture analysis provides a reliable fibre classification system for the discrimination between cashmere fibre and the superfine merino fibre (Zhang et al., 2010a).

To demonstrate this, 13 cashmere fibre images and 15 superfine merino wool fibre images were scanned from the reference collection Cashmere Fibre Distinction Atlas (Zhang, 2005). By using the two-dimensional dual-tree complex wavelet transform, a scanned fibre image was decomposed and reconstructed into single-scale only detail and approximation images as shown in Figure 9. The lowest frequency approximation image represents the brightness variation, the lighting or illumination variation, so it is not used to generate a textural feature. The Scales 1 to 4 detail images measure the brightness variations of the cuticular scale edges at different scales/frequencies. The cuticular scale's height, shape and interval are directly related to the brightness variation at scale edges. Therefore, the texture features extracted from these detail images are intended to be a comprehensive measurement of the scale height, scale shape and scale interval.

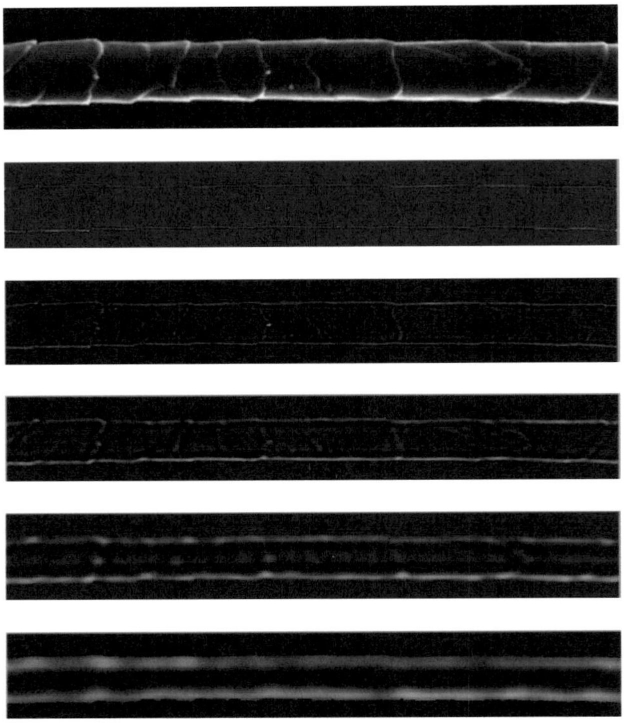

Cashmere fibre

Reconstructed Scale 1 detail image

Reconstructed Scale 2 detail image

Reconstructed Scale 3 detail image

Reconstructed Scale 4 detail image

Reconstructed Approximation image

Fig. 9. Reconstructed Scale 1 to 4 detail images and approximation image

From each of the 28 fibre images, textural features were generated from the six directional detail sub-images at scales 1 to 4. A texture feature vector consisting of 24 (6 orientations × 4 scales) energy features was developed. Principal component analysis (Krzanowski, 1988) was used to reduce the dimension of the texture feature vector, and generate a new set of variables, called principal components. Each principal component was a linear combination of the original variables. All the principal components were orthogonal to each other, so there was no redundant information. The principal components, as a whole, formed an orthogonal basis for the space of the data. Eight principal components represented more than 99.8% the actual dimensionality of the 28×24 texture feature vector data, which were used as the input of a classifier. When using 24 or 26 fibre samples as training data, the rest as testing sets, 27 samples out of the 28 samples were correctly classified, suggesting that wavelet transform based features contained vital information for fibre identification. Visual comparison revealed that the misclassified cashmere fibre has the same range diameter of a wool fibre, and their scale characteristics such as scale frequency and scale length are difficult to discern. If a large sample size was used for classifier training, it might be possible for the classifier to pick up small details of fibre surface feature; hence the accuracy of fibre classification could improve. With the enhanced classifier, this method could be further developed to a completely automated and objective system for animal fibre identification.

5. Summary

This chapter presents selected examples of textile related applications using the wavelet transform method, which extracts surface texture features for objective defect identification and quality classification. Wavelet transform is often used together with other methods to create a process that can best identify the required features. The general procedure for an automated inspection/classification system is to first process the digital images to be inspected and extract the features of significant for highlighting defects in the original image or key identification parameters. Then, a classifier, trained and validated by history experimental data and vector definitions, analyses the features to objectively classify the defects. Neural networks are commonly used to combine with texture features obtained from image processing for defect classification.

The multi-scale two-dimensional dual-tree complex wavelet transform method can effectively decompose a textile image with six orientations at different scales and reconstruct the textile background texture and defect sub-images. An energy analysis method provides important information of sub-images, which can be used to dynamically search for an optimum image decomposition scale and dynamically discriminate different features from the textile image. It has been proved that this method is an effective way to extract pilling features from pilled fabric images, which consist of noise, fabric texture, surface unevenness, and background illuminative variation. Normally, the fabric texture information is obtained from high frequency detail images and the pilling information from low frequency images. Using the complex wavelet transform, pills of different sizes can be identified by the reconstructed detail sub-images with six orientations at different scales. With the energy analysis method, the optimum decomposition scale can be obtained for the distinction between pilling and fabric texture. The energy of the sub-image is coherent with the given size and pill height that describe the pilling intensity. By using the pilling parameters and an artificial neuron network method, classification rules can be established to comprehensively evaluate knitted, woven and nonwoven pilling test images for fabric pilling, and successfully classify them into the five pilling ratings.

This chapter also demonstrates the feasibility of using wavelet texture analysis in classifying cashmere and superfine merino wool fibres. The two-dimensional dual-tree complex wavelet provides an effective way to extract features that represent cuticular scale height, scale shape and scale interval. This may provide the essential foundations to develop an objective automated system for animal fibre distinction.

It is evident that the wavelet transform method is a robust and adaptable tool that is able to produce enhanced results when integrated with other feature extraction processes. Already, wavelet transform has been used for a number of practical purposes in the textiles industry for defect identification and surface characteristic classification, but there is still much potential for further development of more accurate detection methods and new areas of application.

6. References

Beck, J., Sutter, A., & Ivry, R. (1987). Spatial Frequency Channels and Perceptual Grouping in Texture Segregation. *Comput. Vision Graph. Image Process., 37*(2), 299-325.

Deng, Z., Wang, L., & Wang, X. (2011). An Integrated Method of Feature Extraction and Objective Evaluation of Fabric Pilling. *Journal of the Textile Institute, 102*(1), 1-13.

Han, Y., & Shi, P. (2007). An Adaptive Level-Selecting Wavelet Transform for Texture Defect Detection. *Image and Vision Computing, 25*(8), 1239-1248.

Kang, T. J., Cho, D. H., & Kim, S. M. (2004). Objective Evaluation of Fabric Pilling Using Stereovision. *Textile Research Journal, 74*(11), 1013-1017.

Kim, S., & Park, C. K. (2006). Evaluation of Fabric Pilling Using Hybrid Imaging Methods. *Fibers and Polymers, 7*(1), 57-61.

Kim, S. C., & Kang, T. J. (2005). Image Analysis of Standard Pilling Photographs Using Wavelet Reconstruction. *Textile Research Journal, 75*(12), 801-811.

Kim, S. C., & Kang, T. J. (2007). Texture Classification and Segmentation Using Wavelet Packet Frame and Gaussian Mixture Model. *Pattern Recognition, 40*(4), 1207-1221.

Konda, A., Xin, L. C., Takadera, M., Okoshi, Y., & Toriumi, K. (1988). Evaluation of Pilling by Computer Image Analysis. *Journal of the Textile Machinery Society of Japan, 36*(3), 96-107.

Krzanowski, W. J. (1988). *Principles of Multivariate Analysis*. New York: Oxford University Press.

Latif-Amet, A., Ertüzün, A., & Erçil, A. (2000). An Efficient Method for Texture Defect Detection: Sub-Band Domain Co-Occurrence Matrices *Image and Vision Computing, 18* (6-7), 543-553.

Liang, Z., Xu, B., Chi, Z., & Feng, D. (2012). Intelligent Characterization and Evaluation of Yarn Surface Appearance Using Saliency Map Analysis, Wavelet Transform and Fuzzy ARTMAP Neural Network. *Expert Systems with Applications, 39*(4), 4201–4212.

Liu, J., Zuo, B., Zeng, X., Vroman, P., & Rabenasolo, B. (2010). Nonwoven Uniformity Identification Using Wavelet Texture Analysis and LVQ Neural Network. *Expert Systems with Applications, 37*(3), 2241-2246.

Liu, J., Zuo, B., Zeng, X., Vroman, P., & Rabenasolo, B. (2011a). Wavelet Energy Signatures and Robust Bayesian Neural Network for Visual Quality Recognition of Nonwovens. *Expert Systems with Applications, 38*(7), 8497-8508.

Liu, J., Zuo, B., Zeng, X., Vroman, P., Rabenasolo, B., & Zhang, G. (2011b). A Comparison of Robust Bayesian and LVQ Neural Network for Visual Uniformity Recognition of Nonwovens. *Textile Research Journal, 81*(8), 763-777.

Mak, K. L., & Peng, P. (2008). An Automated Inspection System for Textile Fabrics Based on Gabor Filters. *Robotics and Computer-Integrated Manufacturing, 24*(3), 359-369.

Mak, K. L., Peng, P., & Yiu, K. F. C. (2009). Fabric Defect Detection Using Morphological Filters. *Image and Vision Computing, 27*(10), 1585-1592.

Ngan, H. Y. T., Pang, G. K. H., & Yung, N. H. C. (2011). Automated Fabric Defect Detection--A Review. *Image and Vision Computing, 29*(7), 442-458.

Ngan, H. Y. T., Pang, G. K. H., Yung, S. P., & Ng, M. K. (2005). Wavelet Based Methods on Patterned Fabric Defect Detection. *Pattern Recognition, 38*(4), 559-576.

Otsu, N. (1979). A Threshold Selection Method From Gray-Level Histograms. *IEEE Transactions on Systems, Man and Cybernetics, 9*(1), 62-66.

Palmer, S., & Wang, X. (2003). Objective Classification of Fabric Pilling Based on the Two-Dimensional Discrete Wavelet Transform. *Textile Research Journal, 73*(8), 713-720.

Palmer, S., & Wang, X. (2004). Evaluating the Robustness of Objective Pilling Classification with the Two-Dimensional Discrete Wavelet Transform. *Textile Research Journal, 74*(2), 140-145.

Qu, X., & Ding, T.-H. (2010). A Fast Feature Extraction Algorithm for Detection of Foreign Fiber in Lint Cotton within a Complex Background. *Acta Automatica Sinica, 36*(6), 785-790.

Robson, D. (1997). Animal Fiber Analysis Using Imaging Techniques: Part I: Scale Pattern Data. *Textile Research Journal, 67*(10), 747-752.

Robson, D. (2000). Animal Fiber Analysis Using Imaging Techniques: Part II: Addition of Scale Height Data. *Textile Research Journal, 70*(2), 116-120.

Robson, D., Weedall, P. J., & Harwood, R. J. (1989). Cuticular Scale Measurements Using Image Analysis Techniques. *Textile Research Journal, 59*(12), 713-717.

Sari-Sarraf, H., & Goddard, J. S., Jr. (1999). Vision System for On-Loom Fabric Inspection. *IEEE Transactions On Industry Applications, 35*(6), 1252 - 1259.

She, F. H., Kong, L. X., Nahavandi, S., & Kouzani, A. Z. (2002). Intelligent Animal Fiber Classification with Artificial Neural Networks. *Textile Research Journal, 72*(7), 594-600.

Shin, Y., Kim, Y., & Kim, E. Y. (2010). Automatic Textile Image Annotation By Predicting Emotional Concepts From Visual Features. *Image and Vision Computing, 28*(3), 526-537.

Tsai, D.-M., & Chiang, C.-H. (2003). Automatic Band Selection for Wavelet Reconstruction in the Application of Defect Detection. *Image and Vision Computing, 21*(5), 413-431.

Tsai, D.-M., & Hsiao, B. (2001). Automatic Surface Inspection Using Wavelet Reconstruction. *Pattern Recognition, 34*(6), 1285-1305.

Wildman, A. B. (1954). *The Microscopy of Animal Textile Fibres*. Leeds, England: Wool Industry Research Association.

Wong, W. K., Yuen, C. W. M., Fan, D. D., Chan, L. K., & Fung, E. H. K. (2009). Stitching Defect Detection and Classification Using Wavelet Transform and BP Neural Network. *Expert Systems with Applications, 36*(2, Part 2), 3845-3856.

Xin, B., Hu, J., & Yan, H. (2002). Objective Evaluation of Fabric Pilling Using Image Analysis Techniques. *Textile Research Journal, 72*(12), 1057-1064.

Xu, B. (1997). Instrumental Evaluation of Fabric Pilling. *Journal of the Textile Institute, 88*(4), 488-500.

Yang, X., Pang, G., & Yung, N. (2004). Discriminative Training Approaches to Fabric Defect Classification Based on Wavelet Transform. *Pattern Recognition, 37*(5), 889-899.

Yang, X. Z., Pang, G. K. H., & Yung, N. H. C. (2002). Discriminative Fabric Defect Detection Using Adaptive Wavelets. *Optical Engineering, 41*(12), 3116–3126.

Zhang, J., Palmer, S., & Wang, X. (2010a). *Identification of Animal Fibers with Wavelet Texture Analysis.* Paper presented at WCE 2010: Proceedings of the World Congress on Engineering 2010, London, U.K., 742-747.

Zhang, J., Wang, X., & Palmer, S. (2007). Objective Grading of Fabric Pilling with Wavelet Texture Analysis. *Textile Research Journal, 77*(11), 871-879.

Zhang, J., Wang, X., & Palmer, S. (2010b). Objective Pilling Evaluation of Nonwoven Fabrics. *Fibers and Polymers, 11*(1), 115-120.

Zhang, Z. (Ed.). (2005). *Cashmere Fibre Distinction Atlas*: Inner Mongolia People's Publishing Agency.

12

Wavelet Transform-Multidisciplinary Applications

Ali Al-Ataby[1], Waleed Al-Nuaimy[1] and Mohammed A. M. Abdullah[2]
[1]*University of Liverpool*
[2]*University of Mosul*
[1]*UK*
[2]*Iraq*

1. Introduction

The Wavelet transform is such one efficient and important mathematical tool used to explore a signal in a time-frequency representation to allow analyzing the parts of interest separately. This way of analysis provides more valuable information regarding when and where different frequency components are present in the signal. Furthermore, the wavelet analysis is probably the most effective solution to overcome the drawbacks of Fourier analysis.

Since its very first use, many varieties and flavors on top of the original Wavelet transform have been devised and tested, and they have shown different advantages when applied to different scientific and technical areas. These varieties include, but not limited to, the continuous (CWT), the discrete (DWT), Wavelet packet transform (WPT) and multiwavelet transform (MWT).

This chapter is aimed at presenting powerful utilization of the Wavelet transform (and its varieties) in two important applications: **biometrics** and **non-destructive testing (NDT)**.

The biometric application that will be shown in this chapter is related to iris recognition (IR). IR is known as an inherently reliable biometric technique for human identification. Feature extraction is a crucial step in iris recognition, and the trend nowadays is to reduce the size of the extracted features. Special efforts have been applied in order to obtain low templates size and fast verification algorithms. These efforts are intended to enable a human authentication in small embedded systems, such as an integrated circuit smart card. In this chapter, an effective eyelids removing method, based on masking the iris, has been applied. Moreover, an efficient iris recognition encoding algorithm has been employed. Different combination of wavelet coefficients which quantized with multiple quantization levels are used and the best coefficients and levels are determined. The system is based on an empirical analysis of CASIA iris database images. Experimental results show that this algorithm is efficient and gives promising results of false accept ratio (FAR) = 0% and false reject ratio (FRR) = 1% with a template size of only 364 bits.

The NDT application is related to ultrasonic time-of-flight diffraction (TOFD). TOFD is known as a reliable NDT technique for the inspection of welds in steel structures, providing accurate positioning and sizing of flaws. The automation of data processing in TOFD is required towards building a comprehensive computer-aided TOFD inspection and interpretation tool. A number of signal and image processing tools have been specifically developed for use with TOFD data. These tools have been adapted to function autonomously without the need for continuous intervention through automatic configuration of the critical parameters according to the nature of the data and the acquisition settings. This research presents several multi-resolution approaches employing the Wavelet transform and texture analysis for de-noising and enhancing the quality of data to help in automatic detection and classification of defects. The automatic classification is implemented using support vector machines classifier which is considered faster and more accurate than the artificial neural networks. The results achieved so far have been promising in terms of accuracy, consistency and reliability.

The sections to follow provide more details concerning each application.

2. The biometrics application: efficient small template iris recognition system using wavelet transform

The term "Biometrics" refers to a science involving statistical analysis of biological characteristics. This measurable characteristic, biometric, can be applied to physical objects, such as eye, face, retina vessel, fingerprint, hand and voice, or behavioral like signature and typing rhythm. Biometrics, as a form of unique person identification, is one research subject that is growing rapidly (Woodward et al., 2002).

The advantages of unique identification using biometric features are numerous, such as fraud prevention and secure access control. Biometric systems offer great benefits when compared to other authentication techniques. In particular, they are often more user friendly and can guarantee the physical presence of the user (Woodward et al., 2002).

Iris recognition is one of the most reliable biometric technologies in terms of identification and verification. The iris is the colored portion of the eye that surrounds the pupil as depicted in Figure 1. It controls light levels inside the eye similar to the aperture on a camera. The round opening in the center of the iris is called the pupil. The iris is embedded with tiny muscles that dilate and constrict the pupil size. It is full of richly textured patterns that offer numerous individual attributes which are distinct even between identical twins and between the left and right eyes of the same person. Compared with other biometric features such as face and fingerprints, iris patterns are highly stable with time and unique, as the probability for the existence of two identical irises is estimated to be as low as one in 10^{72} (Woodward et al., 2002; Proença & Alexandre, 2007).

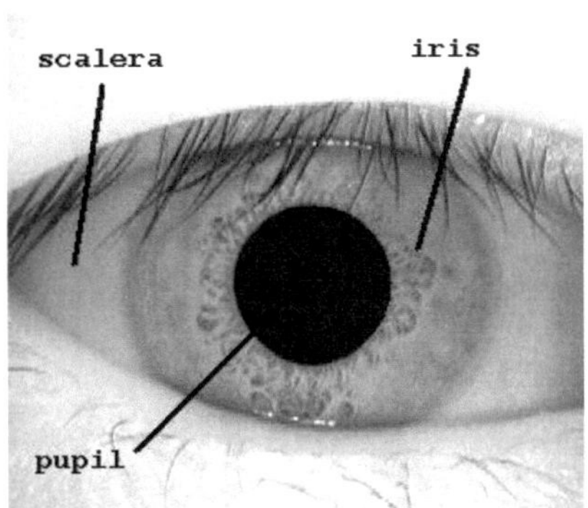

Fig. 1. The human eye

In this research, the iris is efficiently normalized such that only useful data are encoded. Image enhancement techniques are applied. Moreover, the best combination of wavelet coefficients is found and used for successful identification, and the best number of bits used for encoding the feature vector have been deduced while maintaining low template size.

2.1 Related work

Iris identification using analysis of the iris texture has attracted a lot of attention and researchers have presented a variety of approaches in the literature.

Daughman (Daugman, 1993) proposed the first successful implementation of an iris recognition system based on 2-D Gabor filter to extract texture phase structure information of the iris to generate a 2048 bits iris code. A group of researchers have used the 1-D wavelet transform as the core of the feature extraction module (Boles & Boashash, 1998; Chena & Chub, 2009; Huang & Hu, 2005; Huang et al., 2005). For instant, Boles and Boashash (Boles & Boashash, 1998) extracted the features of the iris pattern by using the zero-crossings of 1-D wavelet transform of the concentric circles on the iris.

On the other hand, another group of researcher utilized 2-D wavelet transform to extract iris texture information (Kim et al., 2004; Cho & Kim, 2005; Alim & Sharkas, 2005; Narote et al., 2007; Poursaberi & Araabi, 2007; Hariprasath & Mohan, 2008; Kumar & Passi, 2008). For instance, Narote *et al* (Narote et al., 2007) proposed an algorithm for iris recognition based on dual tree complex wavelet transform and explored the speed and accuracy of the proposed algorithm. Hariprasath and Mohan (Hariprasath & Mohan, 2008) described iris recognition based on Gabor and Morlet wavelets such that the iris is encoded into a compact sequence of 2-D wavelet coefficient, which generate an iris code of 4096 bits. Kumar and Passi (Kumar & Passi, 2008) presented a comparative study of the performance from the iris identification using different feature extraction methods with different templates size. Even though the previous systems have good recognition ratios, the template size remains rather large.

2.2 Iris recognition system

Generally, an iris recognition system is composed of many stages as shown in Figure 2. Firstly, an image of the person's eye is captured by the system and preprocessed. Secondly, the image is localized to determine the iris boundaries. Thirdly, the iris boundary coordinates are converted into the stretched polar coordinates to normalize the scale of the iris in the image. Fourthly, features representing the iris patterns are extracted based on texture analysis. Finally, the person is identified by comparing the extracted features with an iris feature database.

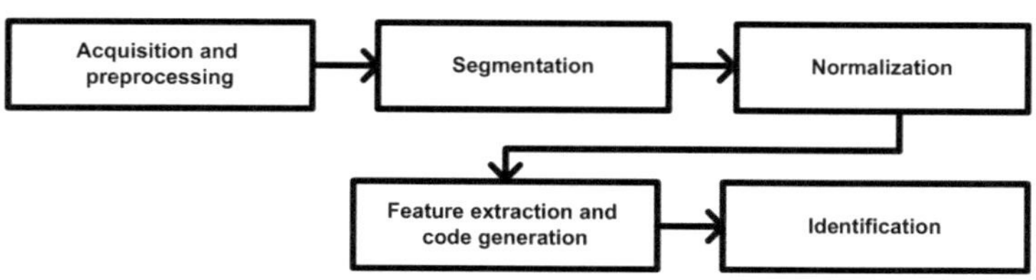

Fig. 2. Block diagram of an iris recognition system

2.3 Segmentation

For the purpose of identification, the part of the eye image carrying useful information is only the iris that lies between the scalera and the pupil (Proença & Alexandre, 2007). Therefore, prior to performing iris matching, it is very important to localize the iris in the acquired image. The iris region, shown in Figure 3, is bounded by two circles, one for the boundary with the scalera and the other, interior to the first, with the pupil.

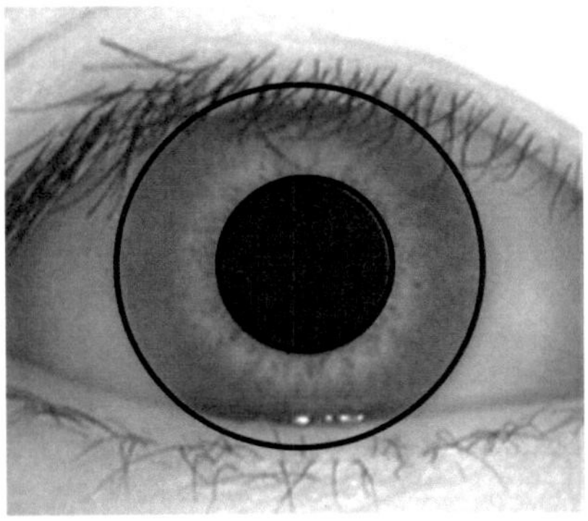

Fig. 3. The iris region

To detect these two circles, the circular Hough transform (CHT) has been used. The Hough transform is a standard computer vision algorithm that can be used to determine the geometrical parameters for a simple shape, present in an image, and this has been adopted here for circle detection (Wildes et al., 1994). The main advantage of the Hough transform technique is its tolerance for gaps in feature boundary descriptions and its robustness to noise (Moravcik, 2010).

Basically, the first derivatives of intensity values in an eye image are calculated and the result is used to generate an edge map. From the edge map, votes are cast in Hough space for the parameters of circles passing through each edge point. These parameters are the center coordinates x_c and y_c, and the radius r, which are able to define any circle according to the following equation:

$$x_c^2 + y_c^2 - r^2 = 0 \tag{1}$$

A maximum point in the Hough space corresponds to the radius and center coordinates of the best circle are defined by the edge points (Wildes et al., 1994).

2.4 Normalization

The size of the iris varies from person to person, and even for the same person, due to variation in illumination, pupil size and distance of the eye from the camera. These factors can severely affect iris matching results. It is very necessary to eliminate these factors in order to get accurate results. To achieve this, the localized iris is transformed into polar coordinates by remapping each point within the iris region to a pair of polar coordinates (r,θ), where r is in the interval [0,1] with 1 corresponds to the outermost boundary, and θ is the angle in the interval $[0,2\pi]$ as shown in Figure 4 (Dmitry, 2004; Schalkoff, 2003).

With reference to Figure 5, the remapping of the iris region from (x,y) Cartesian coordinates to the normalized non-concentric polar representation is modeled by the following equations:

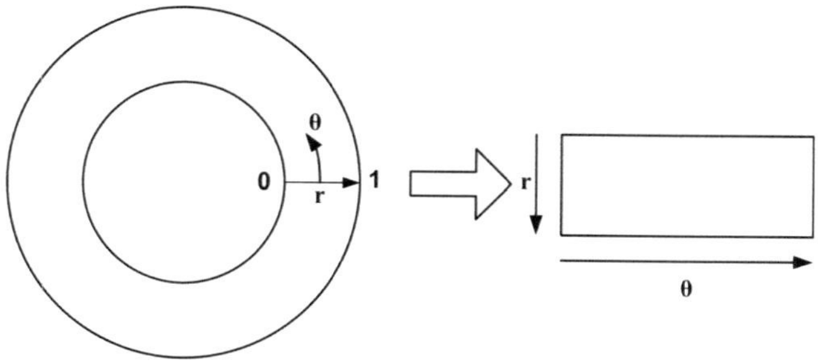

Fig. 4. Rubber sheet model (Dmitry, 2004)

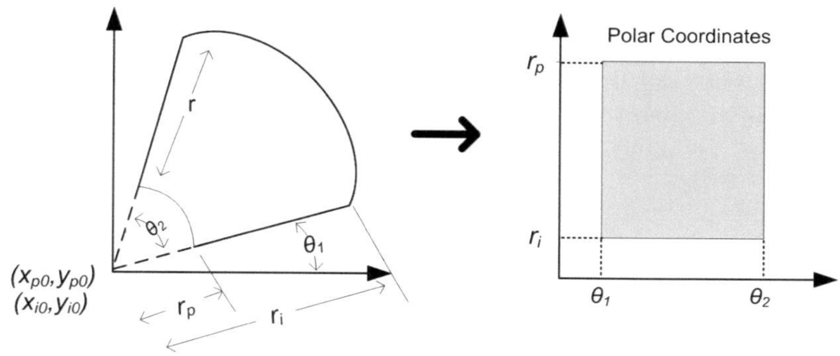

Fig. 5. Image mapping from Cartesian coordinates to dimensionless polar coordinates

$$I(x(r,\theta), y(r,\theta)) \rightarrow I(r,\theta) \tag{2}$$

$$x(r,\theta) = (1-r)x_p(\theta) + rx_i(\theta) \tag{3}$$

$$y(r,\theta) = (1-r)y_p(\theta) + ry_i(\theta) \tag{4}$$

with

$$x_p(r,\theta) = x_{p0}(\theta) + r_p \cos\theta \tag{5}$$

$$y_p(r,\theta) = y_{p0}(\theta) + r_p \sin\theta \tag{6}$$

$$x_i(r,\theta) = x_{i0}(\theta) + r_i \cos\theta \tag{7}$$

$$y_i(r,\theta) = y_{i0}(\theta) + r_i \sin\theta \tag{8}$$

Where I is the iris picture, r_p and r_i are the radius of pupil and the iris respectively, while $(x_p(\theta), y_p(\theta))$ and $(x_i(\theta), y_i(\theta))$ are the coordinates of the papillary and iris boundaries in the direction of θ. (x_{p0}, y_{p0}) and (x_{i0}, y_{i0}) are the centers of pupil and iris respectively.

For a typical eye image of dimension 320×280 pixel, the previous normalization method is performed to produce 50 pixels along r and 600 pixels along θ which results in 600×50 unwrapped strip.

On account of asymmetry of pupil (not being a circle perfectly) and probability of overlapping outer boundaries with sclera, 45 out of 50 pixels will be selected along r in the unwrapped iris. Therefore, the unwrapped iris becomes of dimensions 600×45. The normalized iris image is shown in Figure 6.

Fig. 6. Normalized iris image

2.5 Proposed eyelash and eyelid removing method

Since in most cases the upper and lower parts of the iris area are occluded by eyelids, it was decided to use only the left and right parts of the iris with a partial area from the upper and the lower regions for iris recognition purpose. Therefore, the whole iris [0, 360°] is not transformed in the proposed system. Experiments were conducted by masking the iris from [148, 212°] and [328, 32°] for the right and left parts, while for the upper and lower parts, a semi circle with a radius equal to the half of the iris radius is used to mask the iris as depicted in Figure 7. Hence, the regions that contain the eyelids and eyelashes have been omitted while the remaining eyelashes are treated by thresholding, since analysis reveals that eyelashes are quite dark when compared with the rest of the eye image (Wildes et al., 1994). The corresponding rectangular block is show in Figure 8. Afterward, the block is concatenated together as shown in Figure 9.

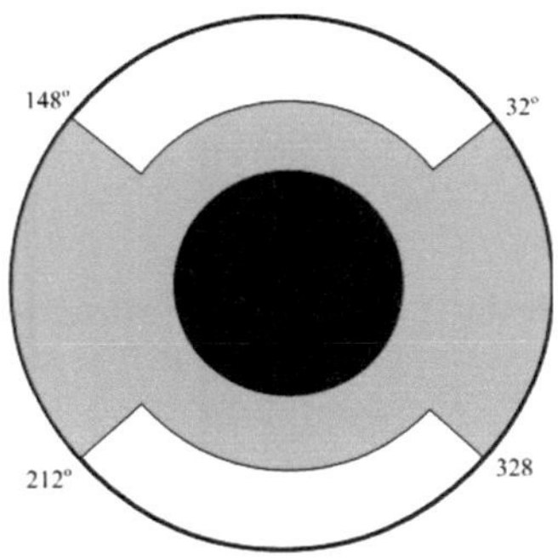

Fig. 7. Masking the iris

Fig. 8. The normalized masked iris image

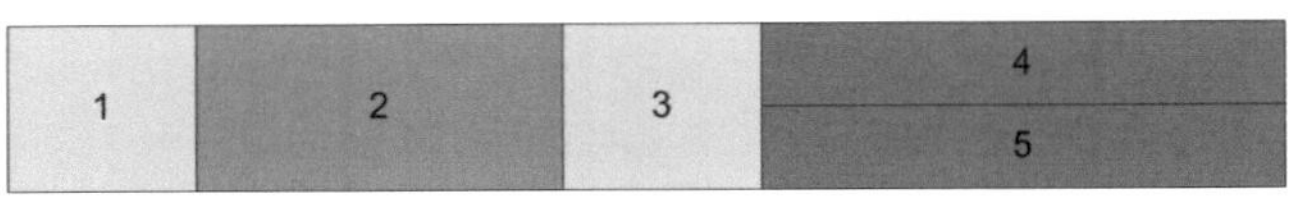

Fig. 9. The concatenated block after removing the ignored parts

The size of the rectangular block is reduced accordingly. By applying this approach, detection time of upper and lower eyelids and some computational cost of the polar transformation are saved. The saving ratio can be calculated as follows:

$$Saving\ ratio = (ignored\ parts\ of\ the\ iris/whole\ iris\ region) * 100\% \quad (9)$$

where the ignored parts in this case = ((148-32) + (328-212))/2 = 116, hence, the saving ratio = (116/360) * 100% = 32.22%. Figure 10 illustrates the application of the proposed masking method on a normalized iris.

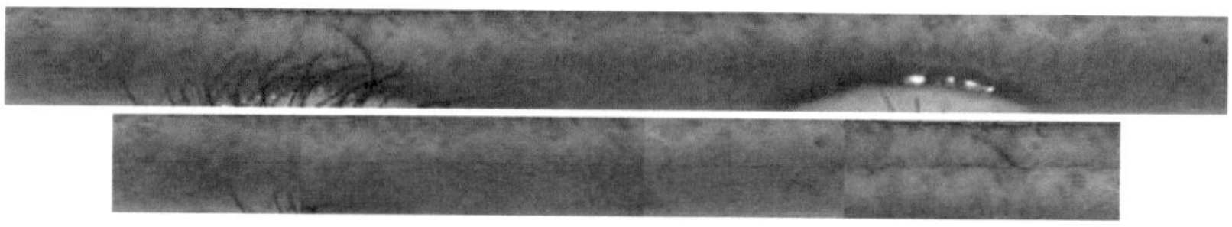

Fig. 10. Applying the proposed masking method on a normalized iris

Although the homogeneous rubber sheet model accounts for pupil dilation and imaging distance it does not compensate for rotational inconsistencies. Rotational inconsistencies are treated in the matching stage (will be shown later).

2.6 Image enhancement

Due to the effect of image capturing conditions (e.g. light source position), the normalized iris image may not have sufficient quality, a situation that affects the performance of feature extraction and matching processes (Poursaberi & Araabi, 2007).

Hence for getting a uniform distributed illumination and better contrast in the iris image, the polar transformed image is enhanced through adjusting image intensity values by mapping the intensity values in the input gray scale image to new values such that 1% of the pixel data is saturated at low and high intensities of the original image. This increases the contrast in a low-contrast grayscale image by remapping the data values to fill the entire intensity range [0,255]. Then, histogram equalization method was used. Results of images before and after enhancement are shown in Figure 11.

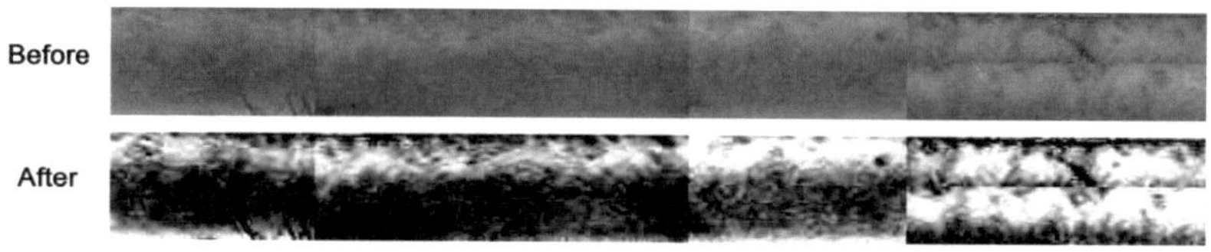

Fig. 11. Image enhancement of the normalized iris

2.7 The proposed feature extraction method

In order to provide accurate recognition of individuals, the most discriminating information present in an iris pattern must be extracted. Only the significant features of the iris must be encoded so that comparisons between templates can be made. Most iris recognition systems make use of a band pass decomposition of the iris image to create a biometric template. For the encoding process, the outputs of any used filter should be independent, so that there are no correlations in the encoded template, otherwise the filters would be redundant (Daugman, 2001).

The Wavelet transform is used to extract features from the enhanced iris images. Haar wavelet is used as the mother wavelet. The Wavelet transform breaks an image down into four sub-sampled images. The result consists of one image that has been high-pass filtered in the horizontal and vertical directions (*HH* or diagonal coefficients), one has been low-pass filtered in the vertical and high-pass filtered in the horizontal (*LH* or horizontal coefficients), one that has been low-pass filtered in the horizontal and high-pass filtered in the vertical

(HL or Vertical coefficients), and one that has been low-pass filtered in both directions (LL or details coefficient) (Kim et al., 2004).

In Figure 12, a conceptual figure of basic decomposition steps for images is depicted. The approximation coefficients matrix cA and details coefficients matrices cH, cV, and cD (horizontal, vertical, and diagonal, respectively) obtained by wavelet decomposition of the input iris image. The definitions used in the chart are as follows (Poursaberi & Araabi, 2007).

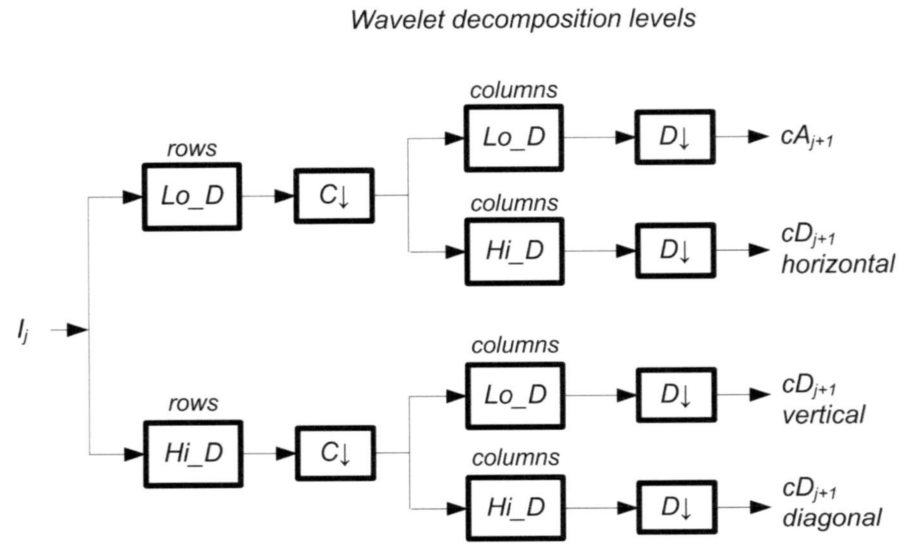

Fig. 12. Wavelet decomposition steps diagram

1. $C \downarrow$ denote downsample columns.
2. $D \downarrow$ denote downsample rows.
3. Lowpass_D denotes the decomposition low pass filter.
4. Highpass_D denotes the decomposition high pass filter.
5. I_i denotes the input image.

Experiments were performed using different combinations of Haar wavelet coefficients, and the obtained results from different combinations were compared to find the best. Since the unwrapped image after masking has a dimension of 407×45 pixels, after 5 times decompositions, the size of the 5th level decomposition is 2×13 while for the 4th level is 3×26. Based on empirical experiments, the feature vector is arranged by combining features from HL and LH of level-4 (vertical and horizontal coefficients [$HL4$ $LH4$]) with HL, LH and HH of level-5 (vertical, horizontal and diagonal coefficients [$HL5$ $LH5$ $HH5$]). Figure 13 shows a five-level decomposition with Haar wavelet.

In order to generate the binary data, features of $HL4$ and $HH5$ are encoded using two-level quantization while features of $LH4$, $HL5$ and $LH5$ are encoded using four-level quantization. After that, these features are concatenated together as shown in Figure 14 which illustrates the process used to obtain the final feature vector.

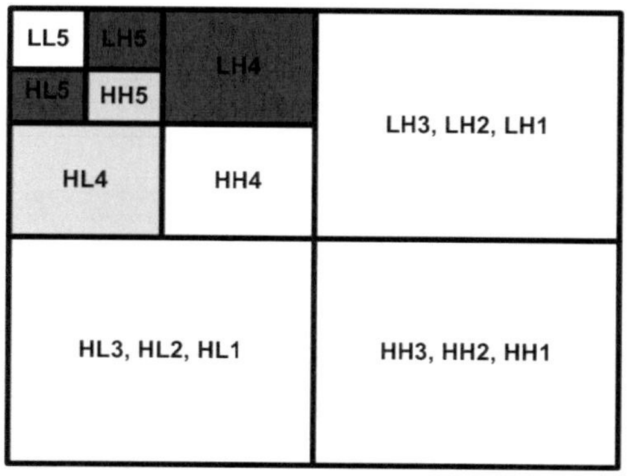

Fig. 13. Five-level decomposition process with Haar wavelet. (Black indicates 4 levels quantization, Grey indicates two levels quantization)

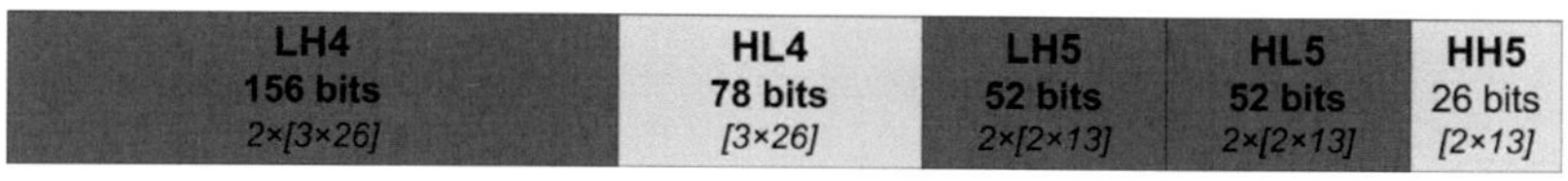

Fig. 14. Organization of the feature vector which consists of 364 bits

2.8 Matching

The last module of an iris recognition system is used to match two iris templates. Its purpose is to measure how similar or different templates are and to decide whether or not they belong to the same person or not. An appropriate match metric is based on direct point-wise comparisons between the phase codes (Schalkoff, 2003). The test of matching is implemented by the Boolean XOR operator (which detects disagreement between any pair of bits) applied to the encoded feature vector of the two iris patterns. This system quantifies this matter by computing the percentage of mismatched bits, *i.e.*, the normalized Hamming distance.

Let X and Y be the two iris representations to be compared, and N be the total number of bits, then:

$$HD = \frac{1}{N}\sum_{j=1}^{N} X_j \oplus Y_j \tag{10}$$

In order to avoid rotation inconsistencies which occur due to head tilts, the iris template is shifted right and left by 6 bits. It can be easily shown that scrolling the template in Cartesian coordinates is equivalent to rotation in polar coordinates. This algorithm performs matching of two templates several times while shifting one of them to different locations. The smallest HD value is selected, which gives the matching decision (Schalkoff, 2003; Daugman, 2001).

2.9 Experimental results and comparison

The iris images are obtained from the database of the Chinese academy of sciences institute of automation (CASIA) (CAS, 2004). The database consists of 756 iris images from 108 classes. Images from each class are taken from two sessions with one month interval in between. For each iris class, three samples are taken at the first session for training and all samples captured at the second session are used as test samples. This is also consistent with the widely accepted standard for biometrics algorithm testing (Mansfield & Wayman, 2002; Ma, 2003).

Experiments using different combinations of wavelet coefficients were performed and the results obtained from different combinations were compared as shown in Table 1. The selected combination gives the best correct recognition rate (CRR) for a minimum feature vector length of 364 bits only.

Combinations	Quantization	CRR	Vector Size
CH4 (D&V)	2 bits	69%	156 bits
CH4 (V&H)	2 bits	73%	156 bits
CH4 (D&H)	2 bits	70%	156 bits
CH4 (D&V) + CH5 (V)	2 bits	76%	182 bits
CH4 (D&V) + CH5 (H)	2 bits	82%	182 bits
CH4 (D&V) + CH5 (D)	2 bits	77.8%	182 bits
CH4 (D&V) + CH5 (D&V)	2 bits	83%	208 bits
CH4 (D&V&H)	2 bits	85%	162 bits
CH4 (H) + CH5 (H)	4 bits	92%	208 bits
CH4 (H) + CH5 (V)	4 bits	89%	208 bits
CH4 (H) + CH5 (V&H)	4 bits	95%	260 bits
CH4 (D) + CH5 (V&H)	4 bits	72%	260 bits
CH4 (V) + CH5 (V&H)	4 bits	68.5%	260 bits
CH4 (D&H)	4 bits	92%	312 bits
CH4 (D&V)	4 bits	62%	312 bits
CH4 (V&H)	4 bits	88%	312 bits
CH5 (V&H)	4 bits	54%	312 bits
CH5 (V&D)	4 bits	49%	312 bits
CH4 (V&H) + CH5 (V)	4 bits	90%	368 bits
CH4 (V&H) + CH5 (H)	4 bits	93%	368 bits
CH4 (D&V) + CH5 (D&V)	4 bits	71%	416 bits
CH4 (V&H) + CH5 (V&D)	4 bits	90.5%	416 bits
CH4 (V&H) + CH5 (V&H)	4 bits	96%	416 bits
CH4 (V&D&H)	4 bits	91%	468 bits
CH4 $(H)_4$ + CH4 $(V)_2$ CH5 $(V)_4$ + CH5 $(H)_4$ + CH5 $(D)_2$	2 bits and 4 bits	99%	364 bit

Table 1. Comparison between multiple wavelet coefficients
(D: Diagonal coefficients, H: Horizontal coefficients, and V: Vertical coefficients)

With a pre-determined separation Hamming distance, a decision can be made as to whether two templates were created from the same (a match) or different iris. However, the intra- and inter-class distributions may have some overlap, which would result in a number of incorrect matches or false accepts, and a number of mismatches or false rejects. Table 2 shows the false accepts rate (FAR) and false rejects rate (FRR) associated with different separation points.

Figure 15 shows the inter- and intra-class distribution of the system with a Hamming distance separation point of 0.29. With this separation point, FAR and FRR of 0% and 1%, respectively, are achieved. Such FRR are appeared due to the overlap between the classes, but it still allows accurate recognition.

Threshold	FAR (%)	FRR (%)
0.20	0.00	59.34
0.24	0.00	28.80
0.26	0.00	10.30
0.28	0.00	3.87
0.29	0	1.00
0.30	1.51	0.86
0.32	5.43	0.00
0.36	26.47	0.00
0.38	48.68	0.00

Table 2. False accept and false reject rates for CASIA database with different separation points.

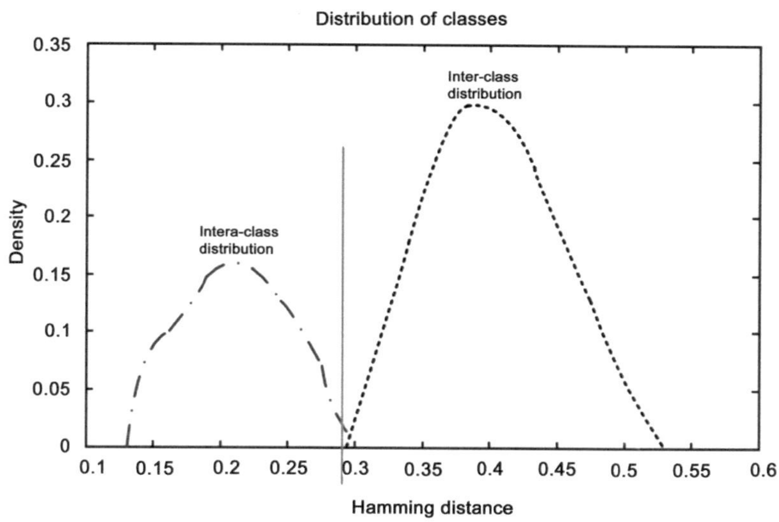

Fig. 15. The distribution of intra-class and inter-class distances with a separation point of 0.29

Table 3 shows a comparison between the proposed method and other well-known methods in terms of the classification rate.

Method	Feature Lengths (bits)	CRR (%)	Used Database
Narote (Narote et al., 2007)	1088	99.20	CASIA (CAS, 2004)
Poursaberi (Poursaberi & Araabi, 2007)	544	97.22	CASIA (CAS, 2004)
Hariprasath (Hariprasath & Mohan, 2008)	4096	99.00	UBIRIS (UBI, 2004)
Xiaofu (Xiaofu & Pengfei, 2008)	1536	98.15	CASIA (CAS, 2004)
Proposed	364	99.00	CASIA (CAS, 2004)

Table 3. Comparison of feature vector length and the correct recognition rate (CRR).

In the two methods shown in references (Narote et al., 2007) and (Hariprasath & Mohan, 2008), the CRR is so close to the obtained one in this research. In fact, the dimensionality of the feature vector in both methods is much higher than that in the proposed method. The feature vector consists of 1088 bits in reference (Narote et al., 2007) and 4096 in reference (Hariprasath & Mohan, 2008), while it consists of only 364 bits in the proposed method. In addition, neither reference (Narote et al., 2007) nor (Poursaberi & Araabi, 2007) have suggested a method for removing eyelids or eyelashes. Furthermore,reference (Poursaberi & Araabi, 2007) suggested a method to produce 544 bits of feature vector by applying four-level Wavelet transform on the lower part of the iris assuming that only the upper part is occluded by the eyelashes and eyelids while the lower part is not. On the other hand, reference (Xiaofu & Pengfei, 2008) employed two-dimensional complex Wavelet transform for feature extraction but no method for noise removal has been applied.

2.10 Conclusions

In this application, an iris recognition algorithm using Wavelet texture features was presented with a novel masking approach for eyelid removing. A masked area around the iris is used in the iris detection method. This area contains a complex and abundant texture information which are useful for feature extraction. The feature vector is quantized to a binary one, reducing the processing cost, while maintaining a high recognition rate.

Experimental results using CASIA database illustrate that relying on a smaller but more reliable region of the iris, although reduced the net amount of information, hence, improving the recognition performance. The results clearly demonstrate that the feature vector consisting of concatenating *LH4*, *HL4*, *LH5*, *HL5*, and *HH5* gives the best results. On the other hand, Haar wavelet is particularly suitable for implementing high-accuracy iris verification/identification systems and the feature vector is the smallest with respect to other wavelets. In identification mode, the CRR of the proposed algorithm was 99% with template size of 364 bits. Such vector size can be easily stored on smart cards and, hence, the matching and encoding time will be minimized tremendously.

The proposed algorithm is characterized by having less computational complexity compared to other methods. Based on the comparison results shown in Table 3, it can be concluded that the proposed method is promising in terms of execution time and performance of the subsequent operations due to template size reduction.

3. The NDT application: automatic detection and classification of weld flaws in ultarsonic time-of-flight diffraction data using wavelet transform and support vector machines

Non-destructive testing is commonly used to monitor the quantitative safety-critical aspects of manufactured components and forms one part of quality assurance procedures. For applications such as the inspection of welded joints in steel structures, ultrasonic techniques are often the NDT method of choice. Time-of-flight diffraction (TOFD) is one such technique, developed by Silk (Silk, 1998) in the late 1970s to improve the sizing accuracy of flaws. Currently, the critical stages of TOFD data processing and interpretation are still performed off-line by skilled operators. This is a time-consuming and painstaking process requiring high operator skill, alertness, consistency and experience.

The TOFD data acquisition and display configurations themselves may introduce a host of errors that cannot be accounted for by manual interpretation, leading to reduction in the quality of the acquired data. Noise formed from scattering of inhomogeneous micro-structures and electronic circuitry is another source of errors. Automatic data processing is often made difficult by this superimposed noise, as it can sometimes mask indications due to small but potentially dangerous defects (Zahran, 2010; Cacciola et al., 2007; Al-Ataby & Al-Nuaimy, 2011; Charlesworth & Temple, 2001). The task of defect classification in noisy data is another critical challenge. Correctly identifying flaw categories has great importance in the process of decommissioning of work pieces and structures through identifying the nature of flaws (serious or negligible).

In this research, the Wavelet transform is used as a method for de-noising to improve the quality of the acquired data. For classification, the support vector machines classifier is used to discriminate between defect classes with input features that also depend on the use of the Wavelet transform. The data used to test the efficiency of the suggested methods is obtained from D-scan TOFD samples collected after applying TOFD on test plates using ultrasonic probes of 5 MHz center frequency with a sampling rate of 100 MHz and a collection step of 0.1 mm. The region of interest in this research is the compression wave area that is enclosed between the lateral wave and the backwall echo.

3.1 Time-of-flight diffraction (TOFD)

Unlike conventional ultrasonic techniques (such as pulse echo), TOFD is based on measurement of the time-of-flight of the ultrasonic waves diffracted from the tips of discontinuities originating from flaws. Echo strength in TOFD does not depend on the flaw orientation, allowing flaw sizes to be accurately determined (generally accurate to within ±2 % of wall thickness, typically less than ±1 mm), with a high probability of detection of approximately 95% (Cacciola et al., 2007; Al-Ataby & Al-Nuaimy, 2011). Two longitudinal broad beam probes are used in a transmitter-receiver arrangement, so that the entire flaw area is flooded with ultrasound and, consequently, the entire volume is inspected using a single scan pass along the inspection line. The collected data are digitised and stored. This data can be visualised in an A-scan representation or stacked together side-by-side in a raster representation called a D-scan (longitudinal) or B-scan (parallel) depending on the relative scanning direction.

During interpretation, there are generally four characteristic signals that are sought:

1. The lateral wave which arises from the wave that travels directly from the transmitter to the receiver.
2. The flaw top tip diffracted wavefront.
3. The flaw bottom tip diffracted wavefront.
4. The backwall reflection which arises from wave reflected from the bottom surface.

Figure 16 (a) depicts a typical A-scan (amplitude returns) when using the equipment to scan a component under test with a defect. Figure 16 (b) is the result after stacking all the returned A-scans to generate a D-scan image.

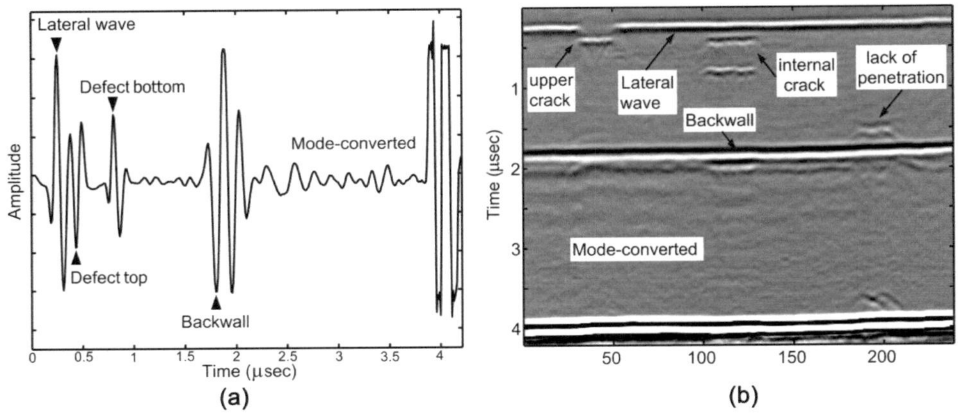

Fig. 16. TOFD returns. (a)A typical A-scan wave with a defect, (b) A D-scan image with labelled indications and regions

3.2 Data quality enhancement and pre-processing

Raw TOFD data returns from the data acquisition process are in need of significant processing before being "usable" for automatic defect evaluation specifically when noise and other factors are present. The processing may include noise suppression (de-noising), scan alignment and accurate detection of the lateral wave (Zahran, 2010; Cacciola et al., 2007; Al-Ataby & Al-Nuaimy, 2011; Charlesworth & Temple, 2001). These operations can be done in an automatic way to minimize the inconsistency and error introduced by human factors and without reducing the spatial resolution of the acquired data. Effective quality enhancement and pre-processing facilitate successful and accurate automatic defect detection and subsequent classification, positioning and sizing.

This part shows the role played by the Wavelet transform and other important operations to improve the quality and consistency of the raw TOFD data to be used for automation and computer interpretation purposes.

3.2.1 De-noising and the MRA

The removal of noise from noisy data while preserving useful information is often referred to as de-noising. Due to its computational efficiency, multi-resolution analysis (MRA) is considered as a powerful tool in de-noising and to enhance the SNR of ultrasonic TOFD signals (Graps, 1995; Matz et al., 2004; Robini et al., 1997). The Wavelet transform offers an inherent MRA that can be used to obtain the time-frequency representation of the ultrasonic signal. The de-noising procedure used in this research is based on decomposing the signal using the discrete Wavelet transform (DWT). The global dynamics of an NDT-related signal $f(t)$ are condensed in the Wavelet approximation coefficients (WACs) that are related to the low frequencies. On the other hand, local oscillations of $f(t)$ (like noise) are depicted in a set of so called Wavelet detail coefficients (WDCs) that are related to the high frequencies.

After the application of DWT to get the WACs and the WDCs, the next step is thresholding of some WDCs and reconstruction of the signal from the other WDCs and WACs using the inverse transform (IDWT). The method is a generalization of Wavelet decomposition that offers a larger range of possibilities for signal analysis. In normal Wavelet analysis, a signal is split into WACs and WDCs. The approximation is then itself split into a second-level approximation and detail, and the process is repeated. In this analysis, the detail coefficients as well as the approximation coefficients can be split. This analysis is called the Wavelet packet transform (WPT) (Matz et al., 2004; Robini et al., 1997) (see Figure 17). Hard thresholding was employed with the threshold computed from WDCs at each level of decomposition based on the standard deviation as follows (Robini et al., 1997):

$$\theta = k\sqrt{\frac{1}{N-1}\sum_{i=1}^{N}(D_i - \overline{D})^2} \qquad (11)$$

where k is a factor related to the ratio of peak value to the RMS value, D_i is the WDC at each level and N is the length of each set of details coefficients.

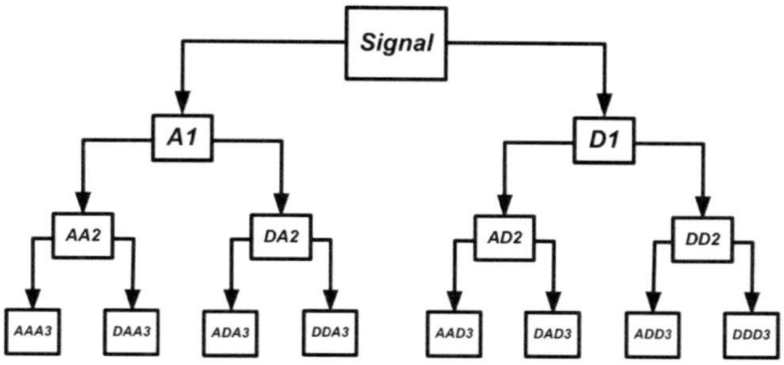

Fig. 17. Discrete multi-resolution scheme - WPT; A is the approximation (low-pass) component and D is the detail (high-pass) component

In order to assess the quality of the suggested de-noising process, the SNR is computed for different defect samples. The obtained SNR values were calculated using the following equation (Matz et al., 2004):

$$SNR_{dB} = 20\log\left(\frac{S_{RMSF}}{N_{RMSN}}\right) \qquad (12)$$

where S_{RMSF} is the RMS value of the filtered signal, and N_{RMSN} is the RMS value of the noise-only part of the raw signal.

In this research, three decomposition levels are used with Daubechies (db6) as the Wavelet filter. When compared with averaging or Wiener filtering, de-noising through Wavelet transform returned higher SNR values. Figure 18 shows some D-scan samples before and after applying Wavelet de-noising method with the corresponding SNR. Although SNR improvement for human interpretation before and after applying de-noising may not appear significant, this improvement is more than adequate in terms of automation and computer interpretation. The performance of the subsequent processing stages shows remarkable enhancement after the de-noising process.

3.2.2 Other pre-processing steps

In addition to de-noising, the process of data quality enhancement includes other important operations to improve the quality and consistency of the data to enable accurate defect characterization and positioning downstream. Scan alignment is one of these operations. A

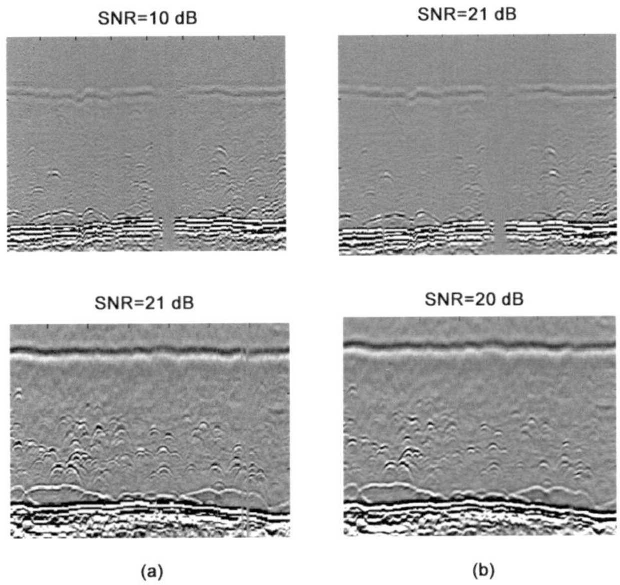

Fig. 18. Wavelet de-noising. Scans of column (a) are before de-noising and (b) are after de-noising with their corresponding SNR values

number of data acquisition factors may contribute to the misalignment of adjacent A-scans, including couplant thickness variations, surface irregularities, in-advertent changes in probe separation and accidental probe lift-off (Zahran, 2010; Charlesworth & Temple, 2001). As the tips of the defects must be accurately located within the material and the lateral wave is used as reference points, aligning adjacent scans is essential. In this research, scan alignment is carried out by cross-correlating each scan with an arbitrary "reference" scan (first A-scan in the image).

The positions of the peaks in these cross-correlation plots are used to automatically shift each A-scan forward or backward in time in such a way as to align the envelope of the lateral wavelets. This not only influences the profile of the lateral wave, but also the shapes of the defect signatures themselves.

Automatic detection of the lateral wave is another essential pre-processing operation. The lateral wave is an important reference point specifically when sizing and positioning is considered. It is also needed for scale linearization and calibration. To detect the lateral wave, the first major wavelet peak (or trough, depending on the phase) of the aligned A-scans is detected.

3.3 Automatic flaw detection based on MRA-WPT and fuzzy logic

This part illustrates some techniques that are followed in this research to detect flaws automatically. MRA based on WPT along with texture analysis are used to select (or filter) the image(s) to be used in the automatic detection process. Image segmentation is then used to highlight and enclose defects inside windows (called defect blobs). Segmentation can be done by statistical methods (e.g. variance thresholding) or through computational intelligence methods (in this case, fuzzy logic). The output of this step is a group of segmented defect blobs that are ready for further analysis and classification.

3.3.1 MRA-WPT for image filtering (selection)

As mentioned previously, the application of MRA to DWT is referred to as the Wavelet packet transform (WPT), which is a signal analysis tool that has the frequency resolution power of the Fourier transform and the time resolution power of the Wavelet transform (Cacciola et al., 2007; Al-Ataby & Al-Nuaimy, 2011; Graps, 1995; Robini et al., 1997). It can be applied to time-varying signals, where the Fourier transform does not produce useful results, and the Wavelet transform does not produce sufficient results. The WPT can be considered an extension to the DWT, providing better reconstruction. In the Wavelet packet framework, compression and de-noising ideas are exactly the same as those developed in the Wavelet framework. The only difference is that WPT offers more complex and flexible analysis, as the details as well as the approximations are split (Cacciola et al., 2007) (refer to Figure 17). The tree decomposition in the Wavelet packet analysis can be applied continuously until the desired coarser resolution is reached.

A single Wavelet packet decomposition gives different bases from which the best representation with respect to a design objective can be selected. This can be done by finding the best tree based on an entropy criterion (Cacciola et al., 2007). Selection can also be done based on texture analysis and statistical contents of the obtained images. Wavelet packet analysis is used in this research in two more important areas. The first area is to aid in detection (hence segmentation) of flaws based on fuzzy logic thresholding. The second area is to provide a ground for selecting defect blob features to be used as inputs to the SVM classifier in the classification stage (will be explained later). Both areas rely on the powerful compression ability of the Wavelet transform in general and WPT in particular to extract relevant and descriptive features.

The application of WPT to the D-scan images was done to analyze the scan image into multiple decomposition levels using the Debauchies of order 8 (db8) chosen as the most suitable analyzing mother function in the Wavelet packet filtering process. The chosen level of decomposition is 3 since it achieved the required coarse resolution and higher decomposition levels would not add any significance to the analysis. First, the D-scan image is decomposed by the WPT (db8, L=3) in the first level. The grey-level co-occurrence matrix (GLCM) $P(i,j)$ is then calculated for the decomposed images. The descriptors or features that characterize the content of GLCM are calculated. The following are the three descriptors that are used in selection of the images for reconstruction (Haralick et al., 1973):

- Contrast:

$$\sum_{i,j} |i-j|^2 \, p(i,j) \tag{13}$$

- Energy:

$$\sum_{i,j} [p(i,j)]^2 \tag{14}$$

- Entropy:

$$-\sum_{i,j} p(i,j).\log(p(i,j)) \quad (15)$$

where $p(i,j)$ is the normalized GLCM $P(i,j)$.

Based on these three descriptors, the image with the weakest textural information is discarded, and the one with the highest statistical contents is selected. After that, another decomposition level runs to generate the new set of images. The calculation of the descriptor commences again. The process of decomposition proceeds until level 3, and the final image is reconstructed from three images carrying the highest statistical textural information in the last level. This image is the key input to the next stages.

The above processing has shown to be effective in generating filtered scan images that are rich in its textural and statistical contents, showing only relevant details and indications. Hence, segmentation and classification of flaws have been affected positively in terms of performance, consistency and accuracy.

3.3.2 Segmentation through fuzzy logic

The segmentation technique used in this research to extract defect blobs is based on the use of fuzzy logic, namely the fuzzy c-mean iterative (FCMI) algorithm (Wong et al., 2001). This method has shown better performance in terms of accuracy and automation than others, which are based on statistical thresholding concept. The method is suitable to distinguish between superimposed defects when normal means (via trained operator or statistical methods) fail. The input to this algorithm is the final filtered image after MRA-WPT (explained previously). The algorithm acts as a binarization operation, with black pixels (values set to zeros) represent background and white pixels (values set to ones) represent flaws. This image (or mask in this case) is used to segment the input image after some further processing. Segmentation through FCMI algorithm is a method that can be implemented as supervised or unsupervised. In this research, unsupervised implementation is used. The details of the implementation of this method to segment the defect in TOFD images are out of scope of this research.

For better representation of the detected defects, each defect blob is represented in the form of a rectangular outlining the defect. This is achieved by applying a global thresholding of each defect blob separately in order to remove the lowest 10% of the pixel intensity values within the defect blob and then recording the minimum and the maximum spatial and temporal dimensions for each blob. This empirical threshold value proved to result in satisfactory performance for all the scan files considered. These recorded dimensions are used to generate a new mask with each defect blob represented by a rectangular outline. Defect may consist of more than one segment. As every defect should be represented in one rectangular blob, this problem is overcome by merging the rectangular blobs related to the same defect based on the dimensions between different blobs (Zahran, 2010). After some processing, the output of this step is a segmented image data with highlighted defects areas or blobs. An example is shown in Figure 19.

3.4 Automatic classification using SVM and wavelet features

In the context of supervised classification, machine learning and pattern recognition are the extraction of regularity or some sort of structure from a collection of data. Neural

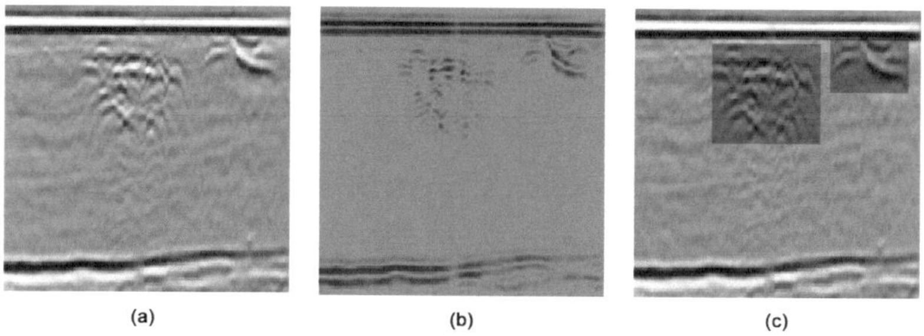

Fig. 19. Automatic segmentation. (a) Before segmentation, (b) Segmentation mask, (c) Segmented defect blobs

networks (NN) and Bayesian classifiers are the typical examples to learn such organization from the given data observations. Support vector machines (SVM) classifier is relatively new and is based on strong foundations from the broad area of statistical learning theory (SLT) (Vapnik et al., 2000). SVM classifier has become, in practice, the classifier of choice of numerous researchers because it offers several advantages which are typically not found in other classifiers like maximization of generalization ability, no local minima problem, low computational overhead, can work perfectly with lack of training data, robust with noisy data, does not suffer as much from the curse of dimensionality and prevents overfitting (Vapnik et al., 2000; Rajpoot & Rajpoot, 2004; Al-Ataby et al., 2010).

3.4.1 The support vector machines

The SVM tries to maximize the margin between classes, hence increases the generalization ability. The SVM classifier is fundamentally developed for binary classification case and is extendable for multi-class situation. The classification problem can be restricted to consideration of the two class problem without loss of generality. The goal is to produce a classifier that works well on unseen examples.

The SVM classifier, like other linear classifiers, attempts to evaluate a linear decision boundary (assuming that the data is linearly separable) or a linear hyperplane between the two classes. Linearly separable data can be separated by an infinite number of linear hyperplanes. The problem is to find the optimal separating hyperplane (see Figure 20 (a)) with maximal margin $M=2/||W||$. It has been shown by Vapnik (Vapnik et al., 2000) that this is a classical quadratic programming (QP) problem with constraints that ends in forming and solving of a Lagrangian. Once the Lagrange multipliers for the optimal hyperplane have been determined, a separating rule can be used in terms of support vectors and the Lagrange multipliers.

The concepts above are presented for a linear classification case. These are generalizable to a nonlinear case where a mapping function is used to map the input space into a higher dimensional feature space such that the non-linear hyperplane becomes linear by using a mapping function Φ as shown in Figure 20 (b). To avoid the increased computational complexity and the curse of dimensionality, a kernel-trick or kernel function is employed which, in essence, computes an equivalent kernel value in the input space such that no explicit mapping is required. Kernels commonly used with kernel methods and SVM in particular are linear, polynomial and Gaussian radial basis function (RBF) (Vapnik et al., 2000; Rajpoot & Rajpoot, 2004; Al-Ataby et al., 2010). In this research, a multiclass SVM classifier

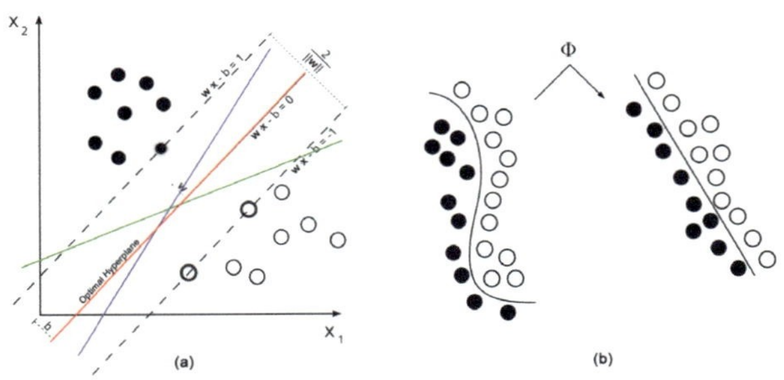

Fig. 20. SVM classification concept (a) Evaluation of an optimal hyperplane, (b) Feature mapping

is implemented using one-against-all (OAA) method (Vapnik et al., 2000), where multiclass problem is reduced into multiple binary problems. A polynomial kernel of degree 5 is used in this implementation.

3.4.2 Wavelet features

Large variety of feature extraction methods exist which are based upon signal processing (or filtering) techniques. Wavelet filtering is one such method that can be successfully used for feature extraction. The idea of using the Wavelets for feature extraction in classification context is not entirely new and has been applied before to texture analysis (Rajpoot & Rajpoot, 2004; Al-Ataby et al., 2010; Huang & Aviyente, 2008), and researchers have been using it for over a decade or so in one form or the other. The inherent capability to do so was highlighted on MRA using the Wavelet transform. As explained before, Wavelet transform decomposition and its extension, WPT, have gained popular applications in the field of signal/image processing. Wavelet transform enables the decomposition of the image into different frequency subbands, similar to the way the human visual system operates. This property makes it especially suitable for image segmentation and classification (Rajpoot & Rajpoot, 2004; Al-Ataby et al., 2010; Huang & Aviyente, 2008). For the purpose of classification, appropriate features need to be extracted to obtain a representation that is as discriminative as possible in the transform domain. It is known that proper feature selection is likely to improve the classification accuracy with less number of features. A widely used Wavelet feature is the energy of each Wavelet subband obtained after WPT (Huang & Aviyente, 2008).

Two-dimensional WPT decomposition allows analyzing an image simultaneously at different resolution levels. Different functions for energy can be used to extract features from each subband for classification. Commonly used energy functions include magnitude $|.|$, magnitude square $|.|^2$ and the rectified sigmoid $|\tanh(\text{-}).|$ (Huang & Aviyente, 2008). In this research, the definition of energy based on squaring is used. The energy at different subbands is computed from the subband Wavelet coefficient:

$$\sigma_p^2(k) = \sum_i \sum_j \left[C_k^p(i,j) \right]^2 \tag{16}$$

where $\sigma_p^2(k)$ is the energy of the obtained (decomposed) image projected onto the subspace at node (p,k). The energy of each subband provides a measure of the image characteristics in that subband. The energy distribution has important discriminatory properties for images and as such can be used as a classification feature (Huang & Aviyente, 2008).

3.4.3 Discussions and results

This research focuses on automatic detection and classification of three types of defects that are pure internal. These are clearly embedded and not open to or approaching either surfaces, and require advanced visual processing for interpretation. Apparent defects, on the other hand, like upper crack, near surface slag and others may be classified using geometrical and phase information alone (Zahran, 2010; BS, 1993).

D-scan test samples containing 50 documented defects were used in the classification stage. The test samples and TOFD equipment are provided by Lavender International NDT Ltd, UK. After applying the operations thar are mentioned previously on the test samples, 50 defect blobs were obtained. The targeted defects for classification are summarized in Table 4 below, along with the corresponding number of samples per each defect type.

To train the SVM classifier, each defect blob sample is analyzed by MRA-WPT with Daubechies 6-tap filter (db6) and a 3-level decomposition (L=3) to generate decomposition subbands. After each decomposition level, the energies are calculated. A feature vector of length 12, which contains energies (4 values per each level), is generated. It was found that by adding some regional features of each defect blob, the performance of the classifier was enhanced against the unseen defects. For that purpose, three regional features (after considering the defect blob as an image containing different-shaped regions and applying simple thresholding to it) are added to the Wavelet feature vector, hence forming a final feature vector of 15 values. These three regional features are:

- Total number of islands: represents the number of the islands (segmented regions) in the absolute of the defect blob after thresholding.
- Number of positive islands: represents the number of the islands in the defect blob after thresholding.
- Relative area: represents the actual number of pixels in the regions of all islands divided by the area of the blob outlining all islands.

It should be noted that this feature vector of length 15 was found suitable for classification based on the defect types to be classified and the available samples (defined in Table 4). Larger feature vectors would be needed to classify more defect types to avoid overfitting.

The 50 defect blobs have been used to obtain different training and testing patterns. The output of the SVM classifier is a number which codifies each kind of defect as follows: 1 for IC, 2 for PO and 3 for SL. As mentioned previously, the SVM classifier has been trained using a 5-degree polynomial kernel implementing the OAA method for the case of study. At the end of the training phase, test patterns are applied and the SVM classifier returns a particular matrix, so called confusion matrix (CM) (Cacciola et al., 2007), which evaluates the goodness of a trained classifier. Generally speaking, the element CM_{ij} of a CM is the probability that a single pattern belonging to the *i-th* class could be classified as belonging to the *j-th* class (sum of elements of each rows is therefore equals to 1). Thus, the more the CM is similar to the identity matrix, the better classification performance is. The other performance factors that

Defect type	Symbol	Description	Code	Figure example	Available samples
Internal crack	IC	Shows up two echoes with some small irregularities between them.	1	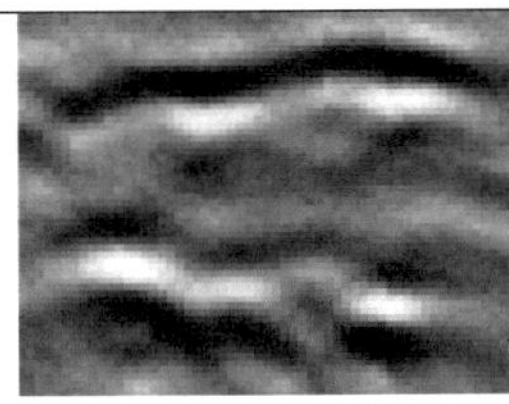	15
Porosity	PO	A group of similar echoes with no resolvable length. Each echo shows only upper tip echo without the lower tip echo. Similar to patterns of acoustic noise and appears as a group of small arcs.	2	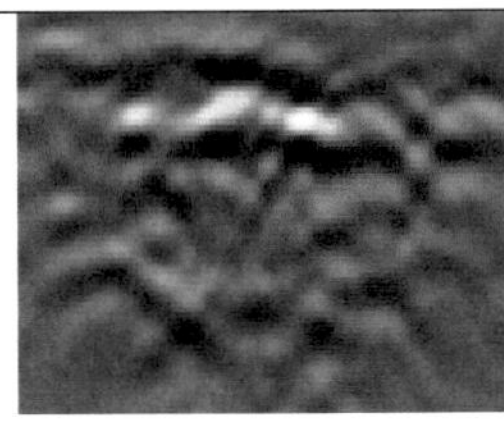	15
Large slag lines	SL	Shows as two echoes. The echo from the upper tip is larger in size and brighter than the echo from the bottom tip.	3	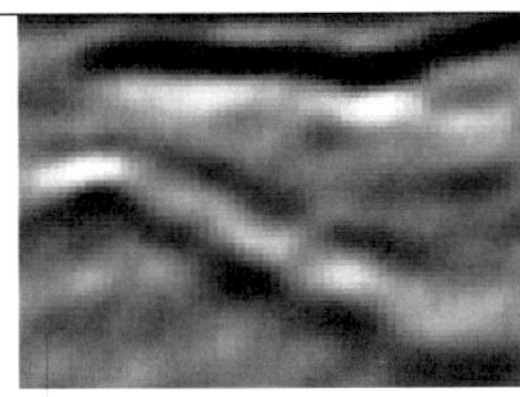	20

Table 4. Defect samples used in the classification

were selected to assess the SVM classifier are the classification accuracy (CA) and the precision (P). In terms of CM elements, classification accuracy and precision are defined as follows:

$$CA = \frac{\sum_{i=1}^{nc} CM_{ii}}{nc} \times 100\% \tag{17}$$

$$P = \frac{1}{nc} \sum_{i=1}^{nc} \frac{CM_{ii}}{\sum_{j=1}^{nc} CM_{ji}} \times 100\% \tag{18}$$

where *nc* is the number of classes to be discriminated (*nc*=3 in the present case).

Table 5 shows the resulting performance of the built SVM classifier for three different tests. It can be seen from this table that after applying the built SVM classifier using 34%, 50% and 60% of the available defect samples for training, classification rates of 80%, 93% and 100% were obtained, respectively. Figure 21 is an output example of the classification stage.

Fig. 21. Automatic classification output

	Defect	No. of training samples	No. of testing samples	Total	Classifier performance metrics
	IC	5	10	15	
Test 1	PO	5	10	15	$CM = \begin{bmatrix} 0.6 & 0.0 & 0.4 \\ 0.0 & 0.9 & 0.1 \\ 0.1 & 0.0 & 0.9 \end{bmatrix}$ *CA*=80%, *P*=83%
	SL	7	13	20	
	IC	8	7	15	
Test 2	PO	8	7	15	$CM = \begin{bmatrix} 0.8 & 0.0 & 0.2 \\ 0.0 & 1.0 & 0.0 \\ 0.0 & 0.0 & 1.0 \end{bmatrix}$ *CA*=93%, *P*=94%
	SL	9	10	20	
	IC	10	5	15	
Test 3	PO	10	5	15	$CM = \begin{bmatrix} 1.0 & 0.0 & 0.0 \\ 0.0 & 1.0 & 0.0 \\ 0.0 & 0.0 & 1.0 \end{bmatrix}$ *CA*=100%, *P*=100%
	SL	10	10	20	

Table 5. Performance of the SVM classifier for different training and test patterns

3.5 Conclusion

This research has presented several promising methods to aid in the automation of detection and classification of flaws using TOFD, specifically when there is a lack of data quality. These methods are under development and need to be studied and tested thoroughly on real life examples before being generalized and applied as reliable automatic interpretation methods. The following are the main conclusions and findings after applying the above methods:

- The de-noising of data (using Wavelet transform) affects positively the detection accuracy and the performance of the classifier.

- The SVM classifier is robust and promising. It shows good performance when there is lack of training data (which is not the case with other classifiers like NN).
- The use of defect blobs Wavelet features as inputs to the SVM classifier seems to be in favor of the overall classification performance.
- It was noted that the response time of performing detection and classification processes using the proposed methods is relatively short compared to others (mainly, the time domain methods).

Future work will be focusing on utilizing and extending the techniques mentioned in this research to perform automatic defect sizing and positioning, with the possibility of making use of the mode-converted waves to add more advantage to the D-scan method. Also, the SVM classifier will be studied thoroughly and extended to classify other internal defect types like small piece of slag (with single echo), lack of fusion, lack of penetration and others, with the possibility of generating different feature vectors utilizing phase information of defect echoes and using principal component analysis (PCA) to reduce and select relevant features.

4. References

D. Woodward, M. Orlans and T. Higgins. Biometrics. McGraw-Hill, Berkeley, California, pp. 15-21, 2002.

H. Proença, A. Alexandre. Towards noncooperative iris recognition: A classification approach using multiple signatures. IEEE Transaction on Pattern Analysis, 29(4):607-612, 2007.

J. Daugman. High confidence visual recognition of persons by a test of statistical independence. IEEE Transaction on Pattern Analysis, 15(11): 1148-1161, 1993.

W. Boles, B. Boashash. A Human Identification Technique Using Images of the Iris and Wavelet Transform. IEEE Transactions on Signal Processing, 46(4): 1085-1088, 1998.

C. Chena, C. Chub. High Performance Iris Recognition based on 1-D Circular Feature Extraction and PSO-PNN Classifier. Expert Systems with Applications journal, 36(7): 10351-10356, 2009.

H. Huang, G. Hu. Iris Recognition Based on Adjustable Scale Wavelet Transform. 27th Annual International Conference of the Engineering in Medicine and Biology Society, Shanghai, 2005.

H. Huang, P.S. Chiang, J. Liang. Iris Recognition Using Fourier-Wavelet Features. 5th International Conference Audio- and Video-Based Biometric Person Authentication, Hilton Rye Town, New York, 2005.

J. Kim, S. Cho, R. J. Marks. Iris Recognition Using Wavelet Features. The Journal of VLSI Signal Processing, 38(2): 147-156, 2004.

S. Cho, J. Kim. Iris Recognition Using LVQ Neural Network. International conference on signals and electronic systems, Porzan, 2005.

O. A Alim, M. Sharkas. Iris Recognition Using Discrete Wavelet Transform and Artificial Neural Networks. IEEE International Symposium on Micro-Nano Mechatronics and Human Science, Alexandria, 2005.

S.P. Narote, A.S. Narote, L.M. Waghmare, M.B. Kokare, A.N. Gaikwad. An Iris Recognition Based on Dual Tree Complex Wavelet Transform. TENCON IEEE 10th conference, Pune, India, 2007.

A. Poursaberi,B.N. Araabi. Iris Recognition for Partially Occluded mages: Methodology and Sensitivity Analysis. EURASIP Journal on Advances in Signal Processing, 2007(1): 12-14, 2007.

S. Hariprasath, V. Mohan. Biometric Personal Identification Based On Iris Recognition Using Complex Wavelet Transforms. Proceedings of the 2008 International Conference on Computing, Communication and Networking (ICCCN) IEEE, 2008.

A. Kumar, A. Passi. Comparison and Combination of Iris Matchers for Reliable Personal Identification. Computer Vision and Pattern Recognition Workshops, IEEE, 2008.

R. Wildes, J. Asmuth, G. Green, S. Hsu, and S. Mcbride. A System for Automated Iris Recognition, Proceedings IEEE Workshop on Applications of Computer Vision, Sarasota, FL, USA, 1994.

T. Moravcik. An Approach to Iris and Pupil Detection in Eye Image. XII International PhD Workshop OWD, University of Zilina, Slovakia, 2010.

K. Dmitry. Iris Recognition: Unwrapping the Iris. The Connexions Project and Licensed Under the Creative Commons Attribution License, Version 1.3., 2004.

R. Schalkoff. Pattern Recognition: Statistical, Structural and Neural Approaches. John Wiley and Sons Inc., pp. 55-63, 2003.

J. Daugman. Statistical Richness of Visual Phase Information: Update on Recognizing, Persons by Iris Patterns. International Journal of Computer Vision, 45(1): 25-38, 2001.

Chinese Academy of Sciences, Center of Biometrics and Security Research. Database of 756 Grayscale Eye Images. http://www.cbsr.ia.ac.cn/IrisDatabase.htm, 2004

A.Mansfield and J.Wayman. Best practice standards for testing and reporting on biometric device performance. National Physical Laboratory of UK, 2002.

L. Ma. Personal identification based on iris recognition. Ph.D. dissertation, Institute of Automation, Chinese Academy of Sciences, Beijing, China, 2003.

H. Xiaofu, S. Pengfei. Extraction of Complex Wavelet Features for Iris Recognition. Pattern Recognition, 19th International Conference on Digital Object Identifier, Shanghai, 2008.

Department of Computer Science, University of Beira Interior, Database of eye images. Version 1.0, http://iris.di.ubi.pt, 2004.

M G Silk, The rapid analysis of TOFD data incorporating the provisions of standards, 7th European conference on NDT - Nice, France, Vol 1, pp 25-29, May 1998.

O Zahran. Automatic ultrasonic time-of-flight diffraction interpretation, fundamentals and applications, VDM Verlag, 2010.

M Cacciola, F C Morabito, and M Versaci. Ultrasonic and advanced methods for non-destructive testing and material characterisation, Chapter 21, pp 493-516. World Scientific, 2007.

A Al-Ataby and W Al-Nuaimy. Advanced Signal Processing Techniques in NDT. In "Applied Signal and Image Processing, Multidisciplinary Advancements". Rami Qahwaji (editor), Roger Green (editor), Evor Hines (editor). Publisher: IGI Global, ISBN-10: 1609604776 EAN: 9781609604776, Jan 2011.

J Charlesworth and J Temple. Engineering applications of ultrasonic time-of-flight diffraction, RSP, 2nd edition, 2001.

A L Graps. An introduction to Wavelets, IEEE Computational Sciences and Engineering, Vol 2, 1995.

V Matz, M Kreidl, and R Smid. Signal-to-noise ratio improvement based on the discrete Wavelet transform in ultrasonic defectoscopy, Acta Polytechnica, Vol 44, No 4, pp 61-66, 2004.

M C Robini, I E Magnin, H Cattin, and A Baskurt. Two-dimensional ultrasonic flaw detection based on the Wavelet packet transform, IEEE transactions on Ultrasonics, Ferroelectrics, and Frequency Control, Vol 44, 1997.

R Haralick, K Shunmugam, and I Dinstein. Textural features for image classification, IEEE Transactions on Systems, Man Cybernetics, SMC, Vol 3, No 1, pp 610-621, 1973.

F Wong, R Nagarajan, S Yaacob, A Chekima, and N. E. Belkhamza. An image segmentation method using fuzzy-based threshold, International Symposium on Signal Processing and Its Applications (ISSPA), Vol 1, pp 144-147, August 2001.

V N Vapnik. The Nature of Statistical Learning Theory, Springer, 2nd edition, 2000.

K M Rajpoot and N M Rajpoot. Wavelets and support vector machines for texture classification, Proceedings of INMIC 2004, 8th International Multitopic Conference, pp 328-333, December 2004.

A Al-Ataby, W Al-Nuaimy, C R Brett and O Zahran. Automatic detection and classification of weld flaws in TOFD data using wavelet transform and support vector machines, Insight, Journal of the British Institute of NDT, Vol 52, No 11, November 2010.

K Huang and S Aviyente. Wavelet feature selection for image classification, IEEE Transaction on Image Processing, Vol 17, No 9, pp 1709-1720, September 2008.

British Standards Institution, 'Guide to calibration and setting-up of the ultrasonic time of flight diffraction (TOFD) technique for the detection, location and sizing of flaws', BS 7706, December 1993.

Part 2

Applications of the Wavelet Transforms in Geoscience

13

Multiscale Analysis of Geophysical Signals Using the 2D Continuous Wavelet Transform

Sid-Ali Ouadfeul[1,2], Leila Aliouane[2,3],
Mohamed Hamoudi[2], Amar Boudella[2] and Said Eladj[3]
[1]*Geosciences and Mines, Algerian Petroleum Institute, IAP*
[2]*Geophysics Department, FSTGAT, USTHB*
[3]*Geophysics Department, LABOPHYT, FHC, UMBB*
Algeria

1. Introduction

The continuous wavelet transform has becoming a very useful tool in geophysics (Kumar and Foufoula-Georgiou, 1997, Ouadfeul, 2006, Ouadfeul, 2007, Ouadfeul, 2008, Ouadfeul and Aliouane, 2010, Ouadfeul et al, 2010). In Potential field analysis it was used to locate and characterize homogeneous causative sources point in 1D (Moreau et al, 1997).

Martelet et al (2001) have published a paper on the characterization of geological boundaries using the 1D wavelet transform of gravity data, the proposed technique has been applied on the Himalaya.

The Complex continuous wavelet transform has been used for identification of sources of potential fields using the aeromagnetic profiles of the French Guiana (Sailhac et al, 2000).

Sailhac and Gibert (2003) have proposed a new technique of sources identification form potential field with the continuous wavelet transform, it is based on the two-dimensional wavelets and multipolar approximations.

A new technique has been proposed by Vallée et al (2004) to estimate depth and model type using the continuous wavelet transform of magnetic data.

Boukerbout et al (2006) have applied the continuous wavelet transform formalism to the special case of anomalies produced by elongated sources like faults and dikes. They show that, for this particular type of anomalies, the two-dimensional wavelet transform corresponds to the ridgelet analysis and reduces to the 1D wavelet transform applied in the Radon domain.

Recently Ouadfeul et al (2010), have published a paper that use the continuous wavelet transform on the mapping of geological contacts from geomagnetic data, the analyzing wavelet is the Mexican Hat, the proposed method has shown a good robustness. In this paper we propose a new technique more precise; this last is based on the 2D directional continuous wavelet transform. The analyzing wavelet is the Poisson's Kernel (Martelet et al, 2001). We start firstly by giving the theoretical basis of the 2D directional Wavelet Transform (DCWT) and the analogy between the wavelet transform and the upward continuation, after that the technique has been applied on a synthetic data of prism and a cylinder. The proposed idea has been applied at the anomaly magnetic field of In Ouzzal area. The obtained results after applying the proposed filter have been compared with analytic signal solutions (Nabighian, 1984; Roset et al, 1992). We finalize the part by propping a model of contacts for this area and a conclusion.

The second part of this chapter consists to apply a multiscale analysis at the 3D analytic signal of the real aeromagnetic data using the wavelet transform, the proposed technique shows a robustness, where classical methods are fails. The third part is an application of the wavelet transform at the gravity data of an area located in the Algerian Sahara.

The last part consists to propose a new method to reduce the noise effect when analyzing the 3D GPR data by the wavelet transform. We finalize this chapter by a conclusion that resumes the different applications of the conventional and directional wavelet transform and its benefits of the different geophysical data.

2. The 2D directional continuous wavelet transform

The 2D directional continuous wavelet transform has been introduced firstly by Murenzi (1989). The wavelet decomposition of a given function $f \in L^2(R^2)$ with an analyzing wavelet $g \in L^2(R^2)$ defined for all a>0 , $b \in R^2$, $f \in L^2(R^2), \alpha \in [0, 2\pi]$ by:

$$W_g f(a,b,\alpha) = \iint_{R^2} f(x) \frac{1}{a} g(r_{-\alpha}(\frac{x-b}{a}))dx \tag{1}$$

Where $r_{-\alpha}$ is a rotation with an angle $-\alpha$ in the R^2 .

An example of a directional wavelet transform is the gradient of the Gaussian Wavelet ∇G . Because the convolution of an image with ∇G is equivalent to analysis of the gradient of the modulus of the continuous wavelet transform. Canny has introduced another tool for edge detection, after the convolution of the image with a Gaussian, we calculate its gradient to seek the set of points that corresponding to the high variation of the intensity of the 2D continuous wavelet transform. The use of the gradient of the Gaussian as an analyzing wavelet has been introduced by Mallat and Hwang (1992).

The wavelet transform of a function f, with an analyzing wavelet $g = \nabla G$ is a vector defined for all a>0, $f \in L^2(R^2)$ $b \in R^2$ by:

$$W_{\nabla G} f(a,b,\alpha) = \iint_{R^2} f(x) \frac{1}{a} \nabla G(\frac{x-b}{a}))dx \tag{2}$$

This transformation relates the directional wavelet transform if we choose $g = \frac{\partial G}{\partial x}$. In fact, we have the following relation:

$$W_{\frac{\partial G}{\partial x}} f(a,b,\alpha) = u_\alpha W_{(\nabla G)} f(a,b) \tag{3}$$

Where u_α is the unite vector in the direction $\alpha : u_\alpha(\cos(\alpha), \sin(\alpha))$, and "." Is the Euclidian scalar product in R^2 .

In this paper we use a new type of directional wavelet based on the Poisson's Kerenel defined in equation (4) as a filer. The modulus of the directional wavelet transform of a potential field F at a scale *a* is equivalent to the Upward continue of this field at the level Z=a. Positioning of maxima of the modulus of the continuous wavelet transform at this scale is equivalent to the maxima of the horizontal gradient of the potential field upward at the level Z.

$$P(x,y) = \frac{1}{2\Pi} \frac{1}{\left[x^2 + y^2 + 1\right]^{3/2}} \tag{4}$$

3. Wavelet transform and potential data

The sharp contrasts that show the potential data are assumed to result from discontinuities or interfaces such as faults, flexures, contrasts intrusive rocks ... For contacts analysis between geological structures, we use usually the classical methods based on the location of local maxima of the modulus of the total (Nabighian, 1984) or the horizontal gradient (Blakely et al., 1986), or the Euler's deconvolution (Reid et al., 1990). This technique allows, in addition to localization in the horizontal plane of contact, an estimate of their depth. The potential field, over a vertical contact, involving the presence of rocks of different susceptibilities is indicated by a low in side rocks of low susceptibility and a high in side rocks of high susceptibility. The inflection point is found directly below the vertical contact. We can use this characteristic of geomagnetic anomalous for localization of abrupt susceptibility change. If the contact has a dip, the maxima of horizontal gradients move in the direction of dip. To determine the dip direction of contacts, we upward the map of the potential field at different altitudes. At each level, the maxima of horizontal gradient are located. If the structures are vertical, all maxima are superposed. However, moving of maxima with the upward indicates the direction of the dip. The potential theory lends perfectly to a multiscale analysis by wavelet transform.

By choosing an appropriate wavelet, measurement of geomagnetic field or its spatial derivatives can be processed as a wavelet transform. Indeed, this analysis unifies various classical techniques: it process gradients that have been upward to a range of altitudes. The expressions of various conventional operations on the potential field are well-designed in the wavelet domain. The most important is the equivalence between the concept of scaling and the upward. Indeed, the wavelet transform of a potential field F_0 (x, y) at a certain scale a = Z/Z_0 can be obtained from measurements made on the level Z_0 by:

1. Upward continue the measured filed at level $Z=a*Z_0$
2. Calculation of the horizontal gradient in the plane (x,y).
3. Multiplication by a.

For a multiscale analysis of contacts, it is sufficient to look for local maxima of the modulus of the continuous wavelet transform (CWT) for different scales to get an exact information about geological boundaries (Ouadfeul et al, 2010).

4. Multiscale analysis geomagnetic data

The proposed idea has been applied on synthetic geomagnetic response of a cylinder and a prism, after that the technique is applied on real geomagnetic data of In Ouzzal area, located on Algerian Sahara. Lets us starting by the synthetic example.

4.1 Application on synthetic geomagnetic data

The proposed idea has been applied at a synthetic model of a cylinder and prism, parameters of these last are resumed in table1 (See also figure1a). Figure 1b is the magnetic response of this model generated with a grid of dimensions 100mx100m. The first operation is to calculate the modulus of the continuous wavelet transform. The analyzing wavelet is the Poisson's Kernel defined by equation 4.

Figure 1c shows this modulus plotted at the scale a=282m. The second step consists to calculate its maxima, figure 2 is a map of these maxima for all scales (Scales are varied from 282m to 9094m). Solid curves are the exact boundaries of the prism and the cylinder. One can remark that the maxima of the continuous wavelet transform are positioned around the two exact boundaries.

coordinates of the center	(5000, 2500, -250).
Ray	1500m.
High	2500m.
Magnetic Susceptibility	K=0.015 SI.
F	37000 nT
Declination	D=0°
Inclination	I=90°

Table 1a. Physical parameters of the Cylinder

coordinates of the center	(5000, 7000, -300).
Width	3000m
Length	3000m.
High	2000m.
Magnetic Susceptibility	K=0.01 SI.
F	37000 nT
Declination	D=0°
Inclination	I=90°

Table 1b. Physical parameters of the Prism

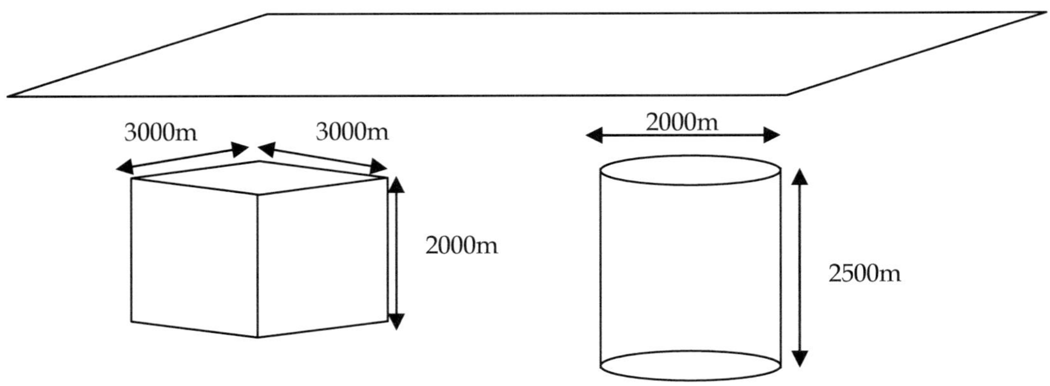

Fig. 1a. Physical parameters of a synthetic model composed of Prism and Cylinder.

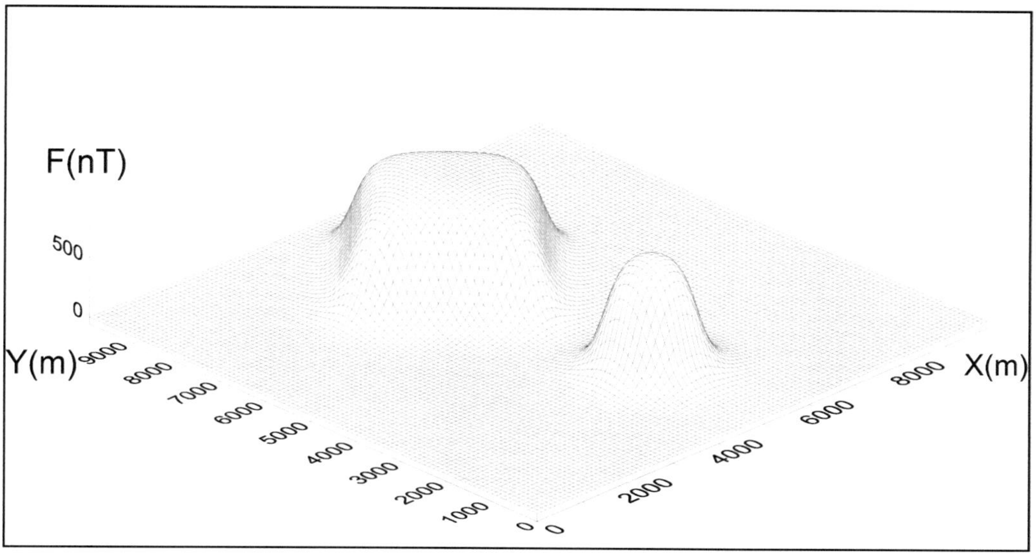

Fig. 1b. Anomaly field of the synthetic model of figue1

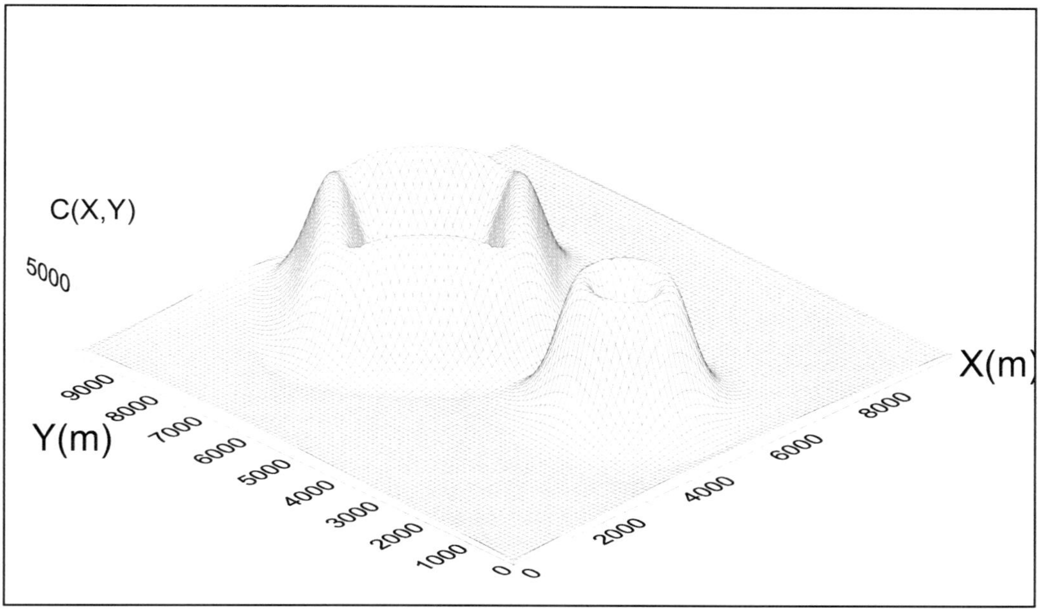

Fig. 1c. Modulus of the continuous wavelet transform of the Prism and the Cylinder plotted at the low Scale a=282m.

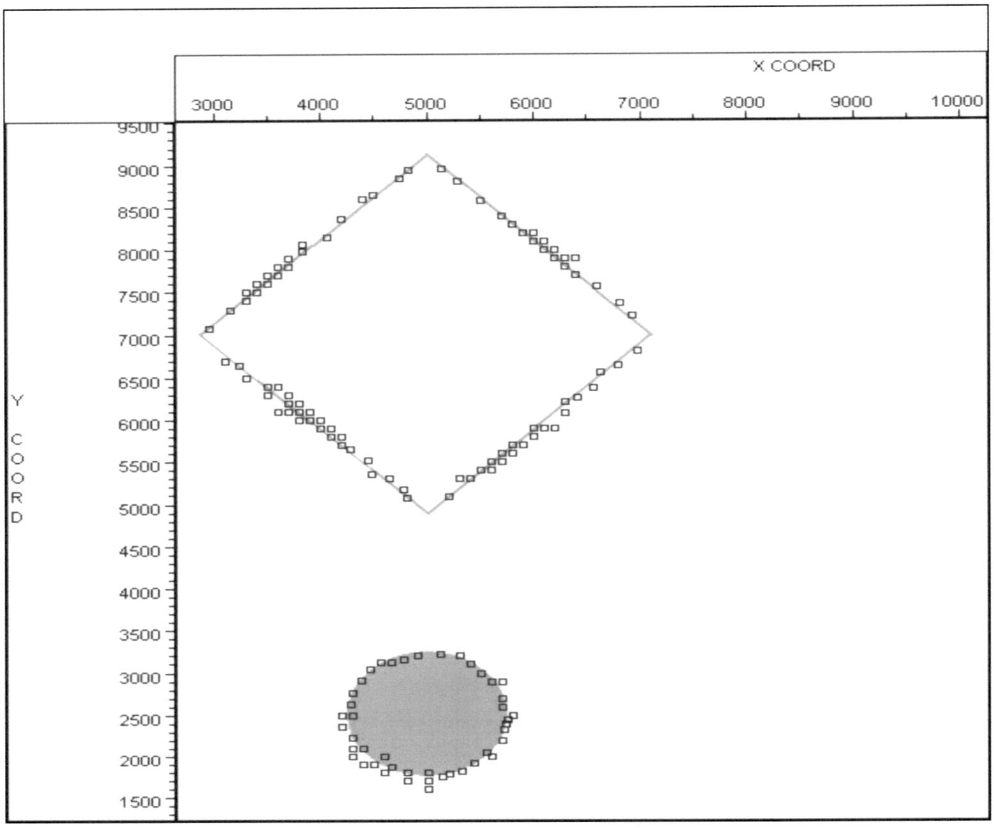

Fig. 2. Mapped contacts obtained by positioning of maxima of the modulus of the 2D DCWT. The solid curves are the exact boundaries of the Prism and the Cylinder.

4.2 Application on real data

The proposed idea has been applied to the aeromagnetic data of In Ouzzal, it is located in Hoggar. We start by describing the geology of the massif of Hoggar.

4.2.1 Geological setting of Hoggar

The lens of the massif of Hoggar that occupies the southern part of Algeria (Figure 3) is defined as a wide lapel Precambrian. It integrates mobile areas caught between the West African Craton and East Africa. It extends at 1000 km from East to West and 700km from North to South, It is extending through the branches 'append' of Air-NIGER at South-East and the Adrar des Iforas-MALI at the South West. Its territory is covered to the North East and South in part by training Paleozoic sedimentary of Tassili. The entire range (Hoggar, Air and Adrar des Iforas) forms the Tuareg shield (also known as Tuargi shield). Stable since the Cambrian, this massif belongs to the chain called "Pan" which belts the West African Craton to the east. This old chain is probably the result of the evolution of both intercontinental basins and accretion of micro plates involving the creation of oceans and island arcs. It is surrounded by sediment platform of Paleozoic age.

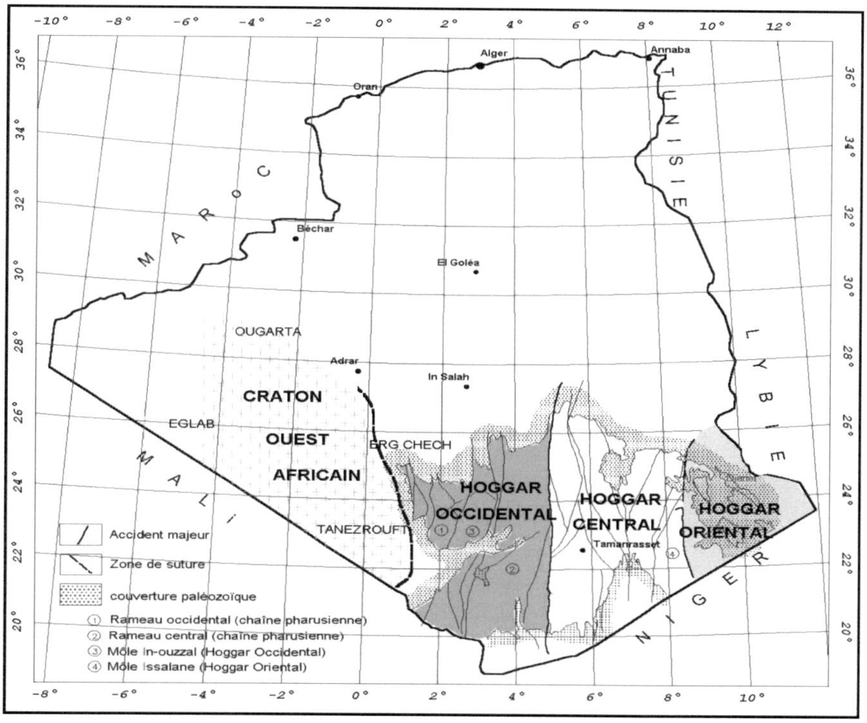

Fig. 3. Map of the principal subdivisions and principal structural domains of Hoggar, afer Caby et al(1981)

4.2.2 Geological context of In Ouzzal

The In Ouzzal terrane (Western Hoggar) is an example of Archaean crust remobilized by a very-high-temperature metamorphism (Ouzeggane and Boumaza, 1996) during the Paleoproterozoic (2 Ga). Structural geometry of the In Ouzzal terrane is characterized by closed structures trending NE-SW to ENE-WSW(figure4) that correspond to domes of charnockitic orthogneiss. The supracrustal series are made up of metasediments and basic-ultrabasic rocks that occupy the basins located between these domes. In In-Ouzzal area, the supracrustal synforms and orthogneiss domes exhibit linear corridors near their contacts corresponding to shear zones. The structural features in In Ouzzal area (Djemaï et al, 2009), observed at the level of the base of the crust, argue in favour of a deformation taking place entirely under granulite-facies conditions during the Paleoproterozoic. These features are compatible with D_1 homogeneous horizontal shortening of overall NW-SE trend that accentuates the vertical stretching and flattening of old structures in the form of basins and domes. This shortening was accommodated by horizontal displacements along transpressive shear corridors. During the Pan-African event, the brittle deformation affected the granulites which were retrogressed amphibolite and greenschists facies (with the development of tremolite and chlorite (Caby and Monie, 2003), in the presence of fluids along shear zones corridors. Brittle deformations were concentrated in the southern boundary of In Ouzzal. An important NW-SE-trending dextral strike-slip pattern has been mapped along which we can see the Eburnean foliation F1 overprinted. This period was also marked by ductile to brittle deformation along the eastern shear zone bordering the In Ouzzal terrane with steep fracture cleavage (NNW-SSE) and conjugate joint pattern. All these structural features are compatible with an ENE-WSW shortening in relation with the collision between the West African Craton and the Hoggar during the Pan-African orogeny.

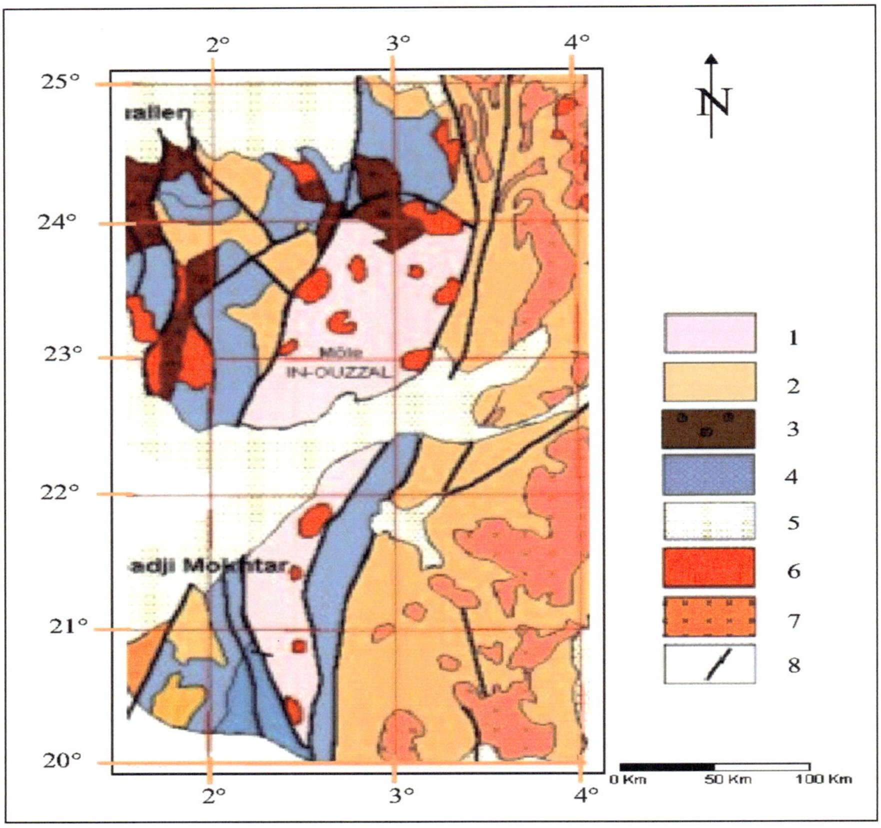

1-Archaean granulites ; 2- Gneiss and metasediments ; 3- Gneiss with facies amphibole; 4- Indif gneiss; 5- Paleozoic curvature; 6- Panafrican granite; 7-Volcano-sediments of Tafassasset ; 8- Major faults

Fig. 4. Map Geological map of the Mole In Ouzzal extracted from the map of Hoggar (After Caby et al, 1981).

4.2.3 Data processing

In this section we have analyzed the aeromagnetic data of In Ouzzal to demonstrate the power of the 2D DCWT method to identify geological contacts. Source codes in C language are developed to calculate the 2D directional continuous wavelet transform and the spatial distribution of its maxima at different scales. The geomagnetic field data are processed with a regular grid of 750m. Figure 5a is a map of total magnetic field and figure 5b represents the anomaly magnetic field ΔT . The IGRF75 model is used for calculation of ΔT .

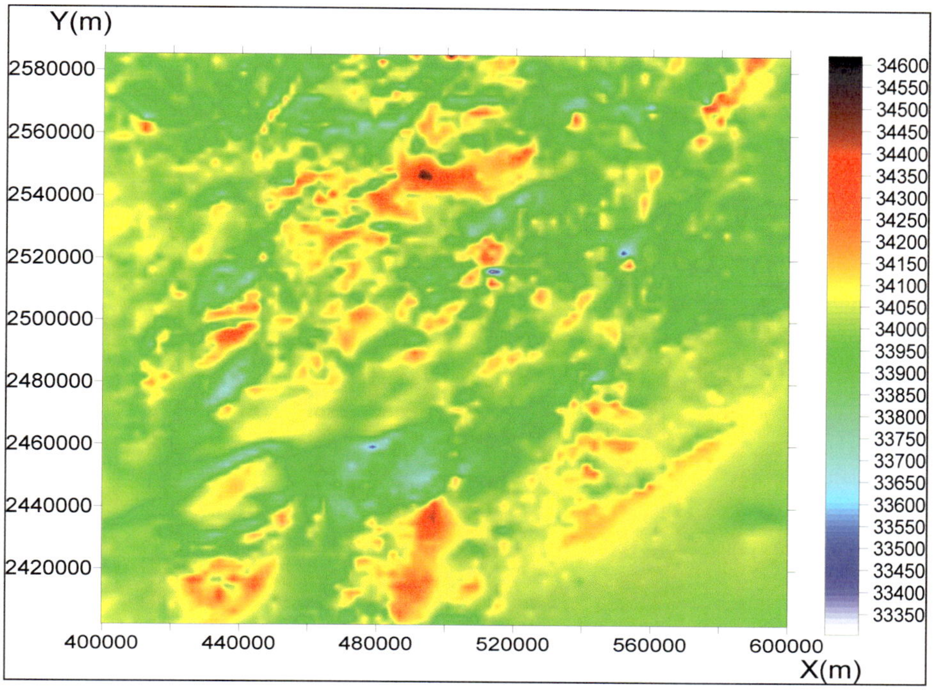

Fig. 5a. Total magnetic field of In Ouzzal

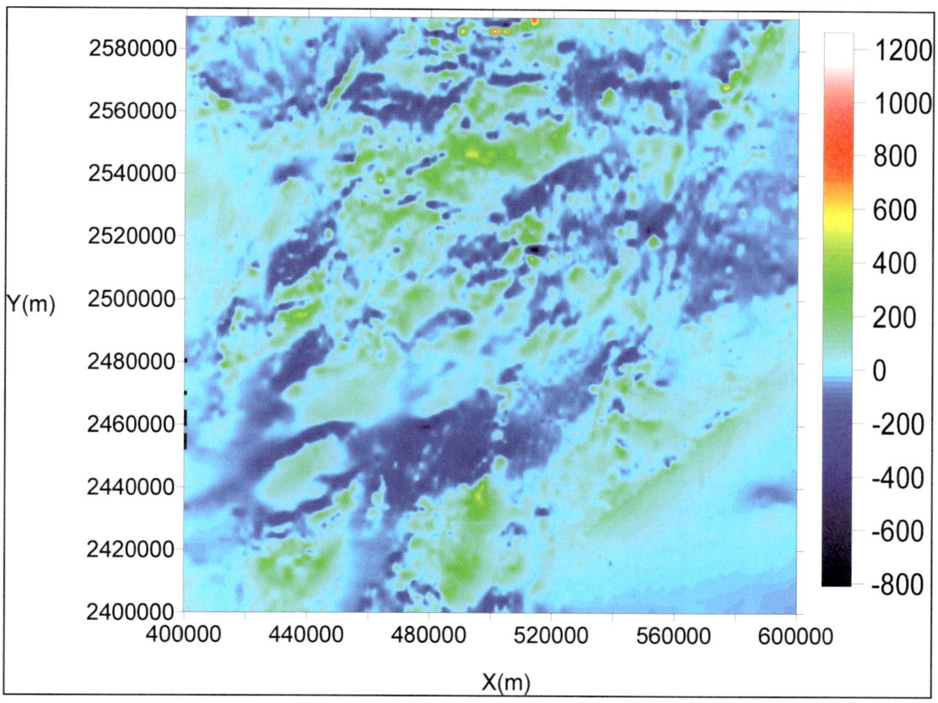

Fig. 5b. Anomaly magnetic field of In Ouzzal.

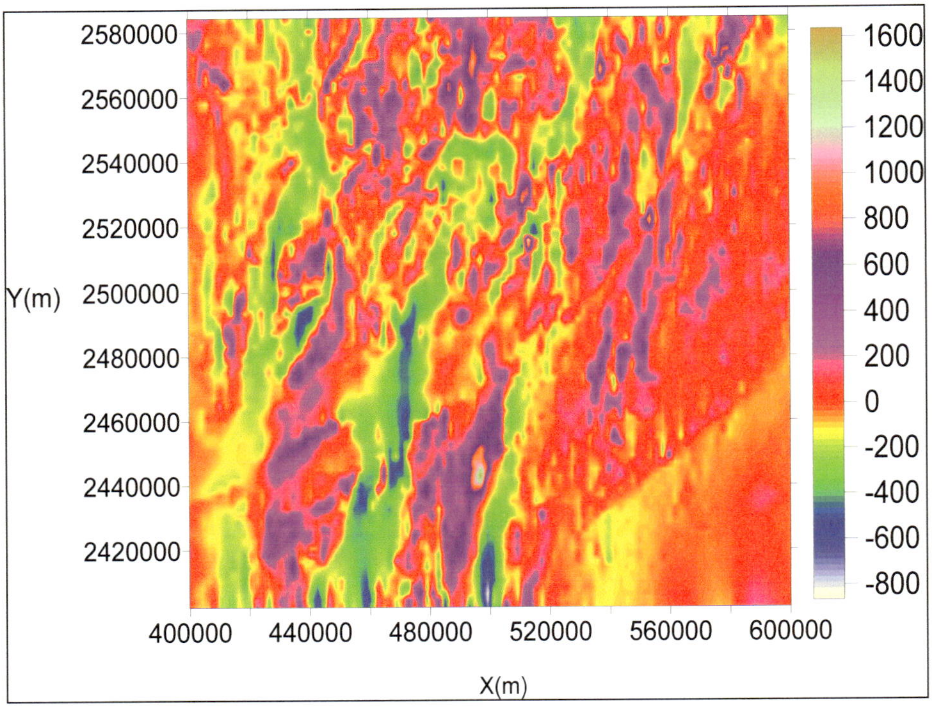

Fig. 5c. Anomaly magnetic field of In Ouzzal after reduction to the pole (RTP)

The anomaly magnetic field is reduced to the pole; parameters of reduction to the pole are illustrated in table2. Figure 5c is magnetic anomaly field after reduction to the pole (RTP).

After RTP the maximum of the anomaly magnetic field will be found at the vertical of the physical structures. Figure 6 represents the modulus of DCWT at the scale a = 2.1 km. The analyzing wavelet is the Poisson's Kernel (See equation4). The next operation consists to calculate maxima of the modulus of the continuous wavelet transform at each scale (scales varied between 2.1 and 9.09 km). At each scale we map points of maxima in the plan. The obtained set of maxima for all ranged scales will give the geometry of geologic contacts (Figure 7).

Longitude	3°
Latitude	22.5°
Elevation	1000m
Inclinaton	22.6°
Declination	-4.38°

Table 2. Parameters of the reduction to the pole of the anomaly field of In Ouzzal

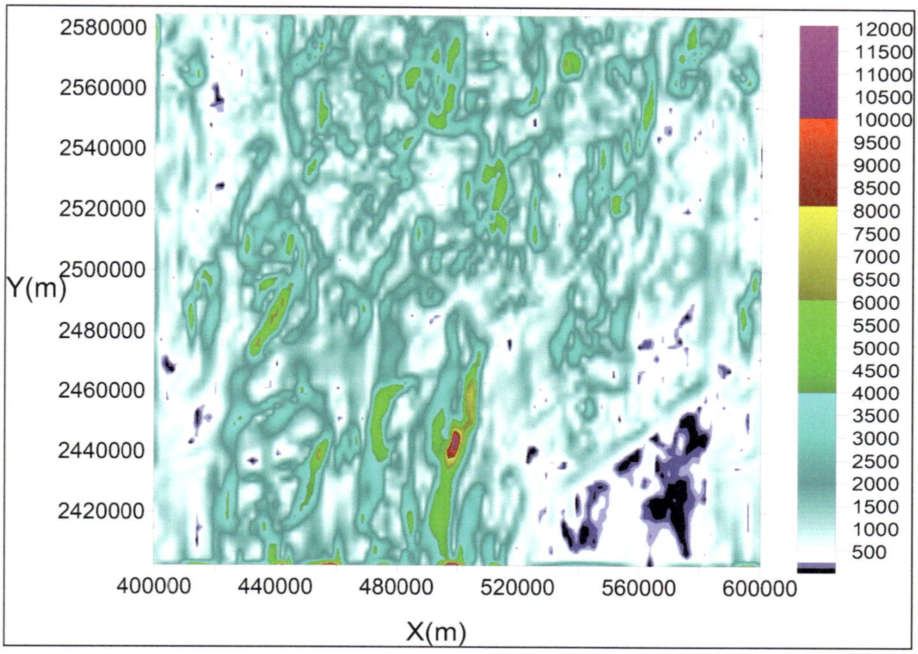

Fig. 6. Modulus of the 2D DCWT of the Anomaly magnetic field reduced to the pole

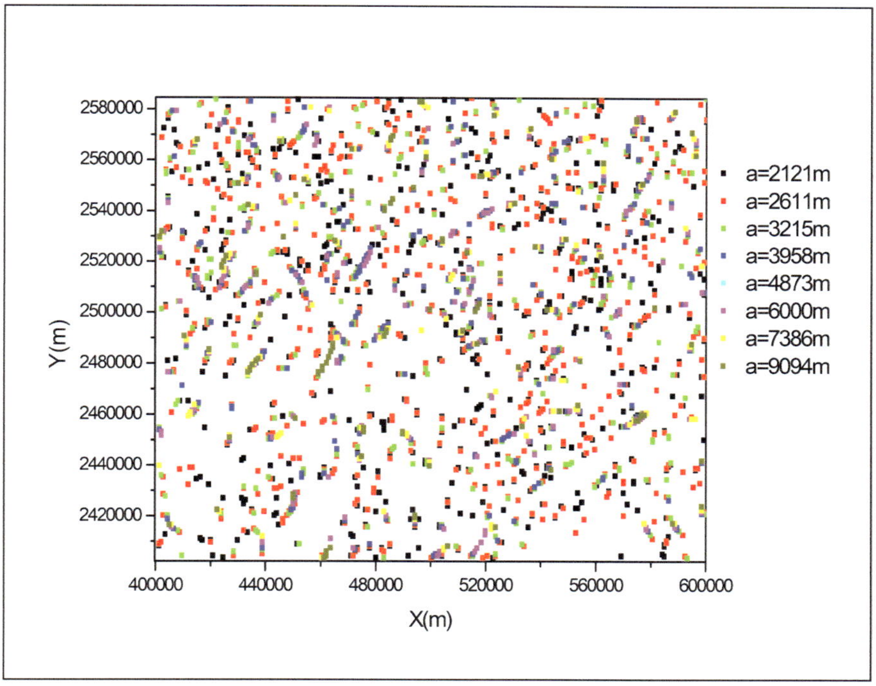

Fig. 7. Mapped maxima of the 2D DCWT for all range of scales

4.2.4 Results interpretation

The obtained contacts by DCWT are compared with the geological map (Figure 8). One can remark that the proposed technique is able to identify contacts that exist in the structural geology map.

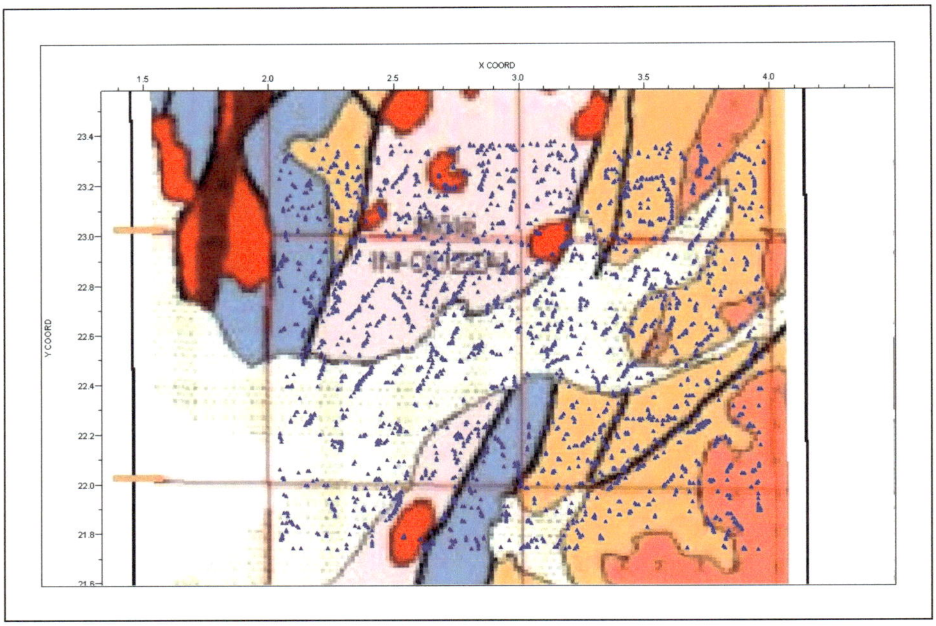

Fig. 8. Obtained contacts compared with the structural map of In Ouzzal.

To better ensure our results, we have compared these contacts with the analytic signal (AS) solutions. The analytic signal is very sensitive to noise. This is due to the derivative operator which amplifies noise intensity (Cooper, 2006). For that we use a threshold to eliminate fictitious solutions, in this case the threshold is equal to 0.5nT/km. Figure 9 shows the comparison between contacts identified by the two methods. We can note that the AS is note able to identify a contact that exist firstly in the geological map, where 2D DCWT has identified this last (blue dashed line in figure 9). It is clear that DCWT technique has detected some contacts that not exist in the model proposed by AS (See figure 9).This is due firstly at the difference between the principles of the two techniques. In the AS technique the value of the anomaly field at the level $Z=Z_0$ is obtained by the Hilbert transform (Nabighian, 1984). However in the proposed technique the intensity of anomaly field at $Z=Z_0$ is defined by the modulus of 2D DCWT at the scale $a=Z_0$. The DCWT using the Poisson Kernel is a low pass filter which decrease the noise compared to the AS technique, this last amplify the noise intensity. Figure 10 is the proposed structural model based on the mapped contacts using the 2D DCWT.

4.3 Conclusion

We have proposed a technique of boundaries identification based on the 2D directional continuous wavelet transform. Firstly we have applied this idea at a synthetic model, obtained results shows robustness of the 2D DCWT. We have applied this technique at the aeromagnetic data of In Ouzzal. Obtained results are compared with the geological map and the analytic signal solutions. One can remark that the 2D directional wavelet transform is able to detect boundaries defined by geologists. Comparison with analytic signal shows that DCWT is able to identify contacts that not exist in the map of contacts defined by AS. The results of this study show that the proposed technique of edge detection based on the wavelet transform is very efficient for geological contacts analysis from maps of geomagnetic anomalies.

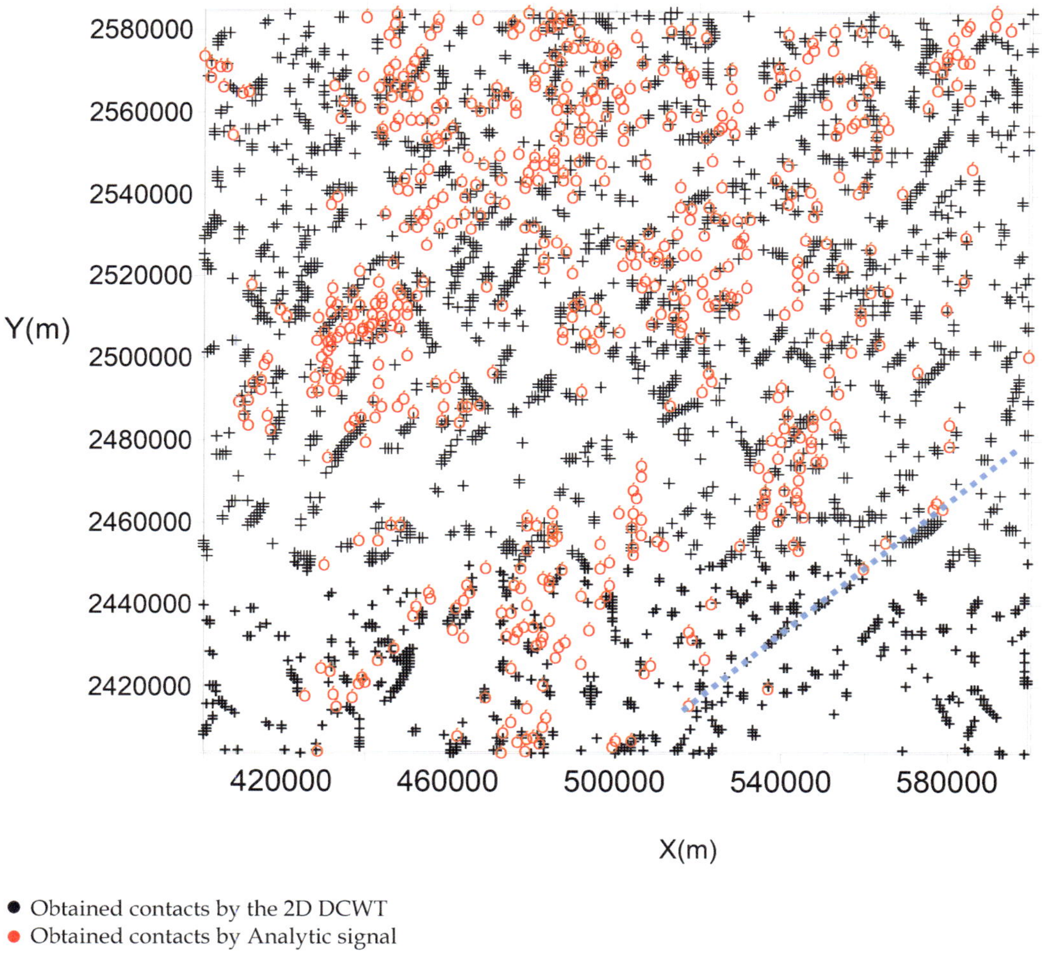

Fig. 9. Localization of magnetic sources obtained by analytic signal. Circles indicate position of peaks of analytic signal amplitude determinate following Blakley and Simpson Algorithm (1986). Minimal threshold for detection of maxima is fixed to 0.5nT/km.

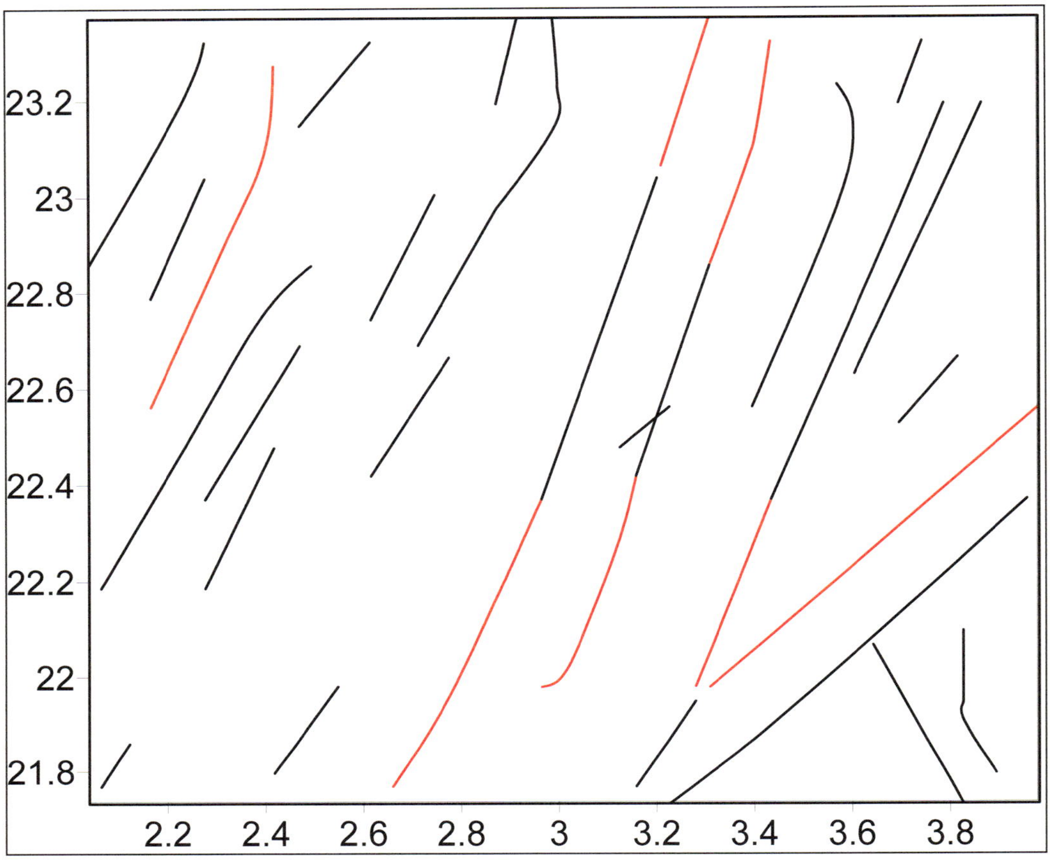

Fig. 10. Proposed contacts obtained by positioning of maxima of the 2D DCWT

5. Mutiscale analysis of the 3D analytic signal using the 2D directional continuous wavelet transform

Many model of potential field data interpretation are proposed, the horizontal gradient (Blakely and Simpson, 1986) and the analytic signal or full gradient (Nabighian, 1984, Roest et al, 1992) are the classical methods of the last decades. The big weakness of these techniques is sensitivity to noise (Cooper, 2006). Since the presence of noise in the potential field can give fictitious contacts. This is due to the de derivative operator that amplifies noise effect. Usually we use a threshold to eliminate these fictitious contacts. However by this way, many high frequency causative sources will be missed. For this reason we propose in this paper a technique of boundaries delimitation from geomagnetic data. This last is based on the analysis of the amplitude of 3D analytic signal defined by Roest et al (1992) by the directional continuous wavelet transform. After that maxima of the modulus of the CWT for all range of scales are mapped. The set of maxima will give geological boundaries.

The proposed idea has been applied on the In Ouzzal area

5.1 Analytic signal

Nabighian (1972, 1984) developed the notion of 2-D analytic signal, or energy envelope, of magnetic anomalies. Roest, et al (1992), showed that the amplitude (absolute value) of the 3-D analytic signal at location (x,y) can be easily derived from the three orthogonal gradients of the total magnetic field using the expression:

$$|A(x)| = \sqrt{\left(\frac{\partial T}{\partial x}\right)^2 + \left(\frac{\partial T}{\partial y}\right)^2 + \left(\frac{\partial T}{\partial z}\right)^2} \quad (5)$$

Where : $|A(x)|$ is the amplitude of the analytic signal at (x,y).
T is the is the intensity of the observed magnetic field at (x,y).

An important comment at this point is the analytic signal can be easily calculated. The x and y derivatives can be calculated directly form a total magnetic field grid using a simple 3x3 filter, and the vertical gradient is routinely calculated using FFT techniques. The analytic signal anomaly over a 2-D magnetic contact located at (x=0) and at depth h is described by the expression (after Nabighian, 1972):

$$|A(x)| = \alpha \frac{1}{(h^2 + x^2)^{0.5}} \quad (6)$$

Where α is the amplitude factor $\alpha = 2M \sin(d)(1 - \cos(I)^2 \sin(A)^2))$
h is the depth to the top of the contact.
M is the strength of magnetization
d is the dip of the contact
I is the inclination of the magnetization vector
A is the direction of the magnetization vector

5.2 The processing algorithm

The proposed idea is based on the calculation of maxima of the modulus of the 2D continuous wavelet transform of the amplitude of the analytic signal. Figure 11 shows a detailed flow chart of this technique.

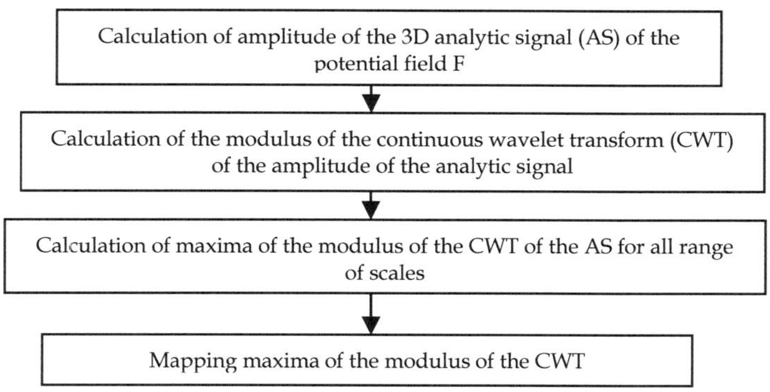

Fig. 11. Flow chart of the potential field analysis using the 3D AS and 2D CWT

5.3 Data processing

The proposed technique has been applied at the anomaly magnetic field without reduction to the pole(figure5b). The first operation consists to calculate the amplitude of the analytic signal of this field. This last is represented in figure 12. The modulus of the continuous wavelet transform of the amplitude of the AS is represented in figure 13; the analyzing wavelet is the Poisson's Kernel. Maxima of the modulus of the CWT are mapped for all range of scales (scales varied from 2121m to 9094m). The set of maxima will give structural boundaries.

5.4 Results interpretation

Geological contacts obtained by mapping maxima of the CWT of the amplitude the the AS are compared with geological map (See figure 14). One can remark that the proposed method is able to identify contacts defined by geology. Obtained boundaries by the proposed method are compared with contacts of analytic signal(see figure 15); for this last we have eliminated fictitious contacts dues to noise using a threshold of 0.5nT/m. We observe that by using this threshold we have eliminated a lot of contacts, for example the contact defined by the dashed line in figure 15 is identified by the CWT combined with AS, however it is not detected by analytic signal, this is due to the threshold effect, this last has been applied to reduce the noise effect on the AS.

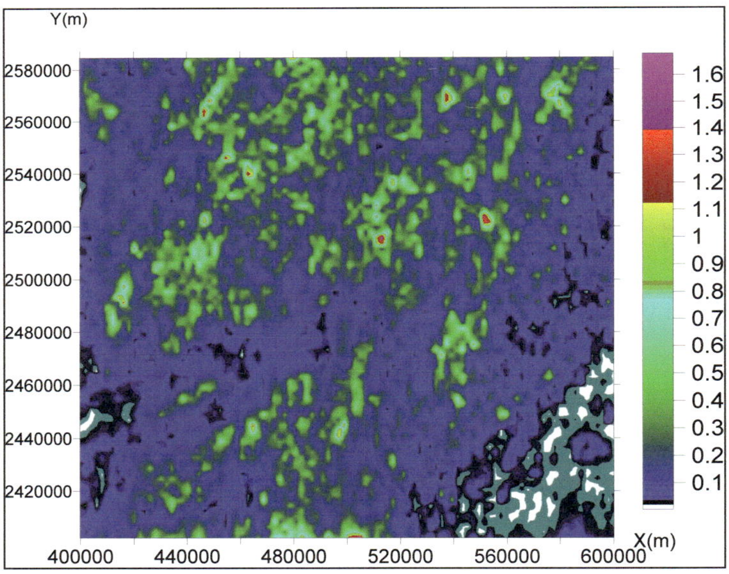

Fig. 12. Amplitude of the analytic signal of the anomaly magnetic field of In-Ouzzal

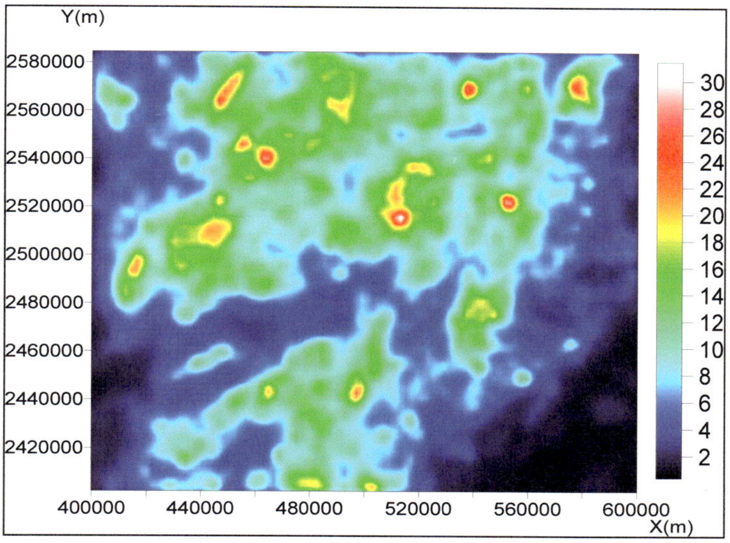

Fig. 13. Modulus of the CWT of the AS plotted at the scale a=2121m.

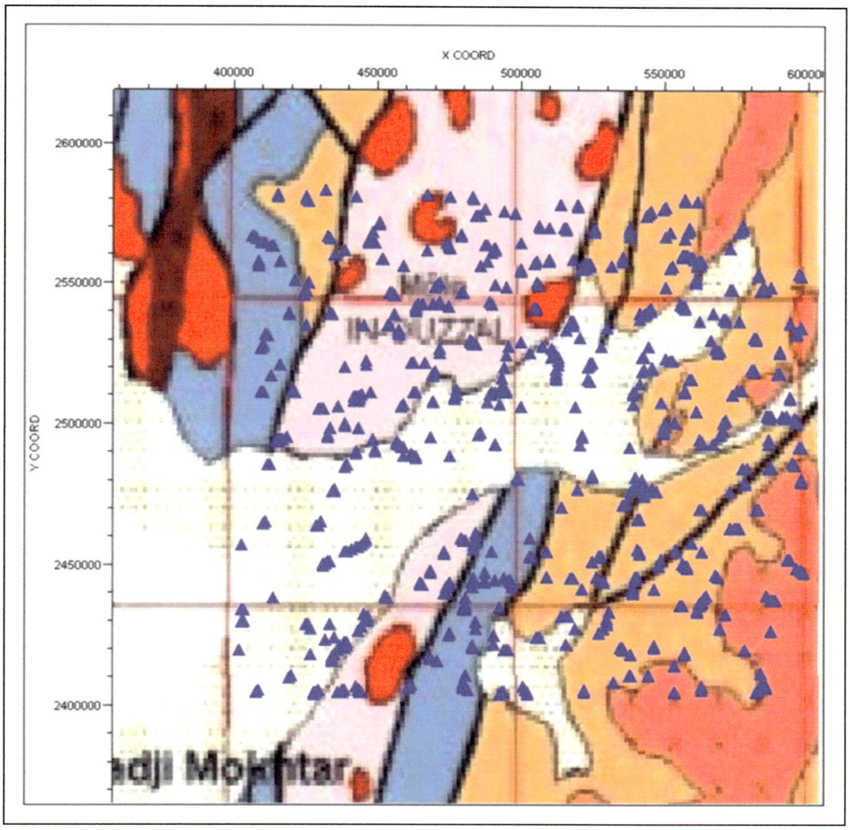

Fig. 14. Obtained contacts by the CWT analysis combined of AS compared with geology.

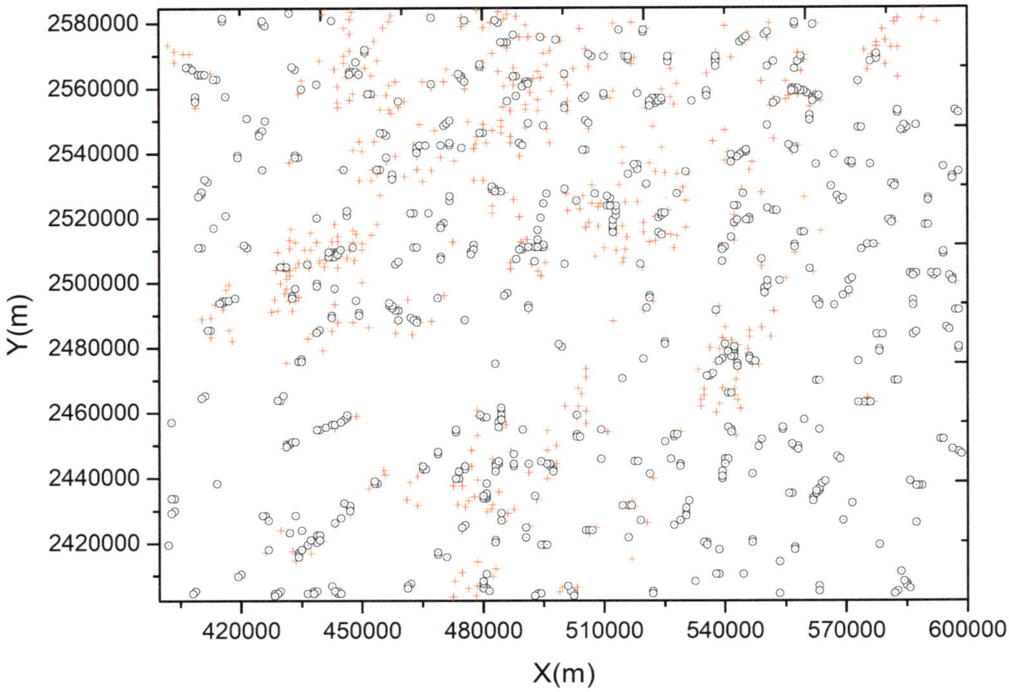

Fig. 15. Obtained contacts by the proposed method compared with analytic signal solutions.

5.5 Conclusion

We have proposed a technique of contacts identification based on the maxima of the modulus of the 2D continuous wavelet transform of the amplitude of the 3D analytic signal. Application on the real aeromagnetic data of In Ouzzal shows that the proposed idea is able to identify contacts that are not detected by the analytic signal solutions. By implanting our method we have resolved the ambiguity of application of a threshold at the maxima of analytic signal, which can eliminate a lot of high frequency causative sources. Secondly we have resolved the ambiguity of reduction to the pole that exist in the classical methods of application of the CWT for contacts delimitation, since this last needs a data reduced to the pole, which is not very efficient in the case of high geomagnetic remanance.

6. Multiscale analysis of the gravity data

We have applied the proposed idea on the gravity data of an area located in the Algerian Sahara. Figure 16a shows the Bouguer anomaly processed with a regular grid of 500mX500m. Firstly a 2D continuous wavelet transforms has been applied, the analyzing wavelet is the Poisson's Kernel. Modulus of the 2D DCWT is presented for the low scale a=705m in figure16b.

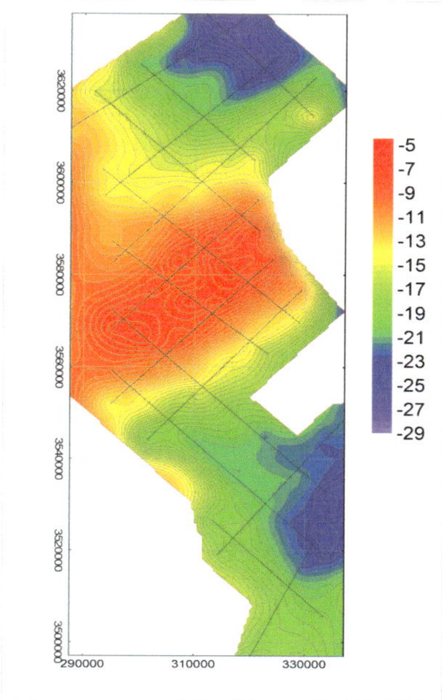

(a)

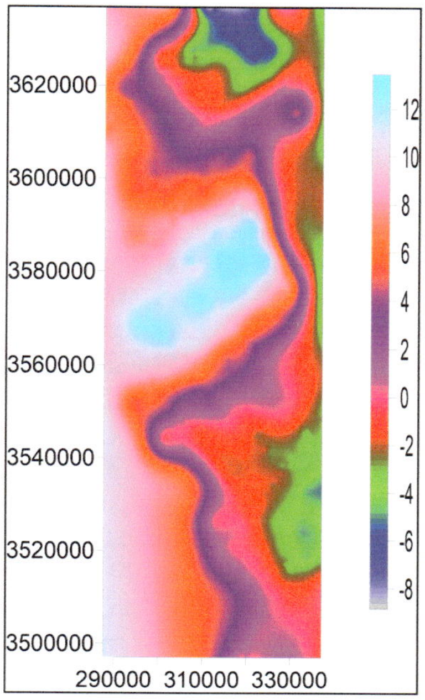

(b)

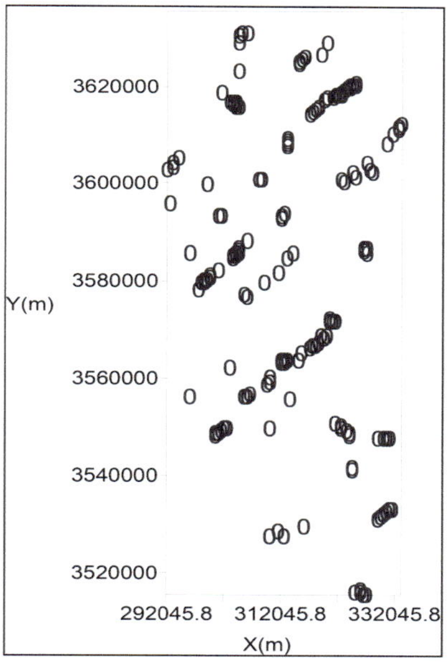

(c)

Fig. 16. Multiscale analyse of gravity data using the 2D DCWT
(a) Bouger Anomaly
(b) Modulus of the 2D DCWT plotted at the low scale a=705m
(c) Obtained contacts mapped by sweeping all range of scales

The maxima of the modulus of the continuous wavelet transform have been mapped for all range of scales. Figure 16c shows the map of the obtained boundaries, obtained results are compared with a seismic profile that passes by the prospected region, it exhibits a big correlation.

7. Multiscale analysis of noisy 3D GPR data

Here we present a new technique of noise effect attenuation in the 3D GPR data analysis using the 2D continuous wavelet transform. Ouadfeul and Aliouane(2010), have presented a technique of multiscale analysis of the 3D GPR data suing the CWT, it has been applied on land topographical GPR data analysis, this last play a high important role in the seismic design (See Ouadfeul and Aliouane, 2010). The proposed technique is very sensitive to noise, to demonstrate this; we have added a 05% as a white noise in the 3D GPR data analyzed in the cited paper. Figure 17 is the map of this GPR data and figure 18 presents the noisy GPR data. Modulus of the 2D CWT plotted at the scale (a=751m) is presented in figure 19. Maxima of the modulus of the continuous wavelet transform are presented in figure20. It is clear that we are not able to identify topographic orientations; this is due to the noise effect on the CWT analysis. For this reason we propose an algorithm to reduce this phenomenon, this last is based on the application of an exponential low pass filter, at the modulus of the 2D CWT for the low range of scales, the filter coefficients are presented in table 03. After application of the low pass filter, maxima of the CWT are mapped for all range of scales. Compraison of of maxima of the initial model (without noise) and the filtered model is presented in figure 21. One can remark that the filtred model is not very far from the original one, and we are now able to identify the dominant topographic orientation.

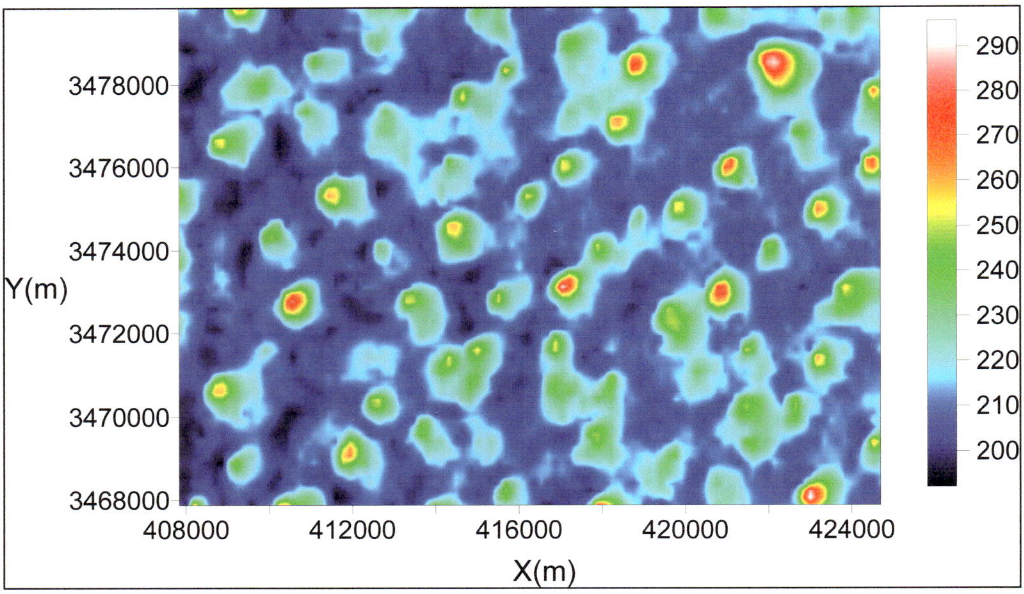

Fig. 17. Initial 3D GPR image without noise

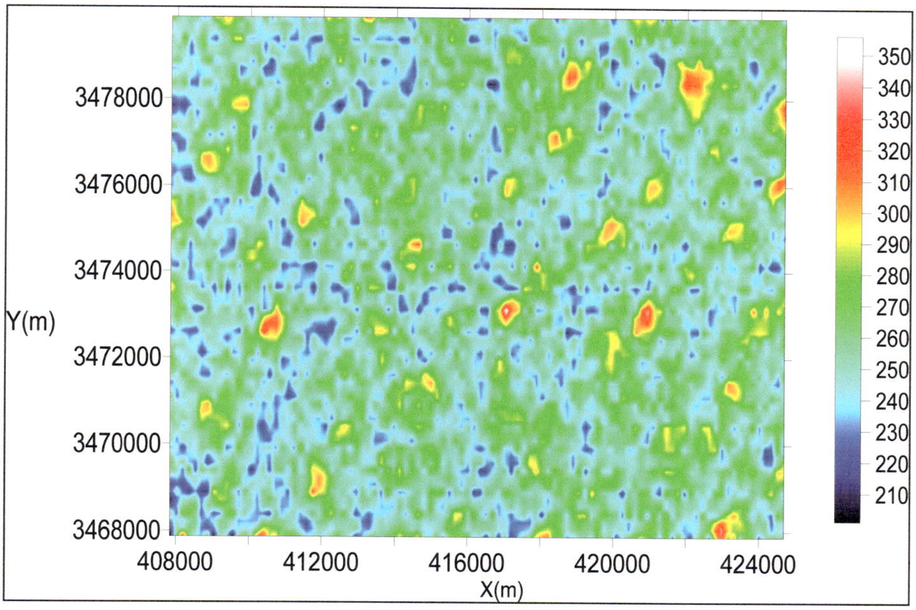

Fig. 18. Noisy 3D GPR image.

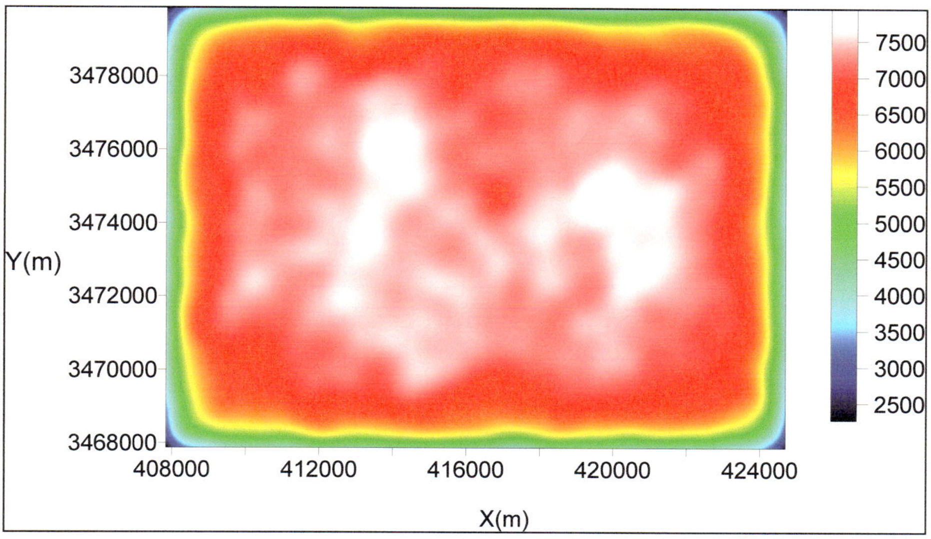

Fig. 19. Modulus of the 2D continuous wavelet transform of the noisy GPR image, plotted at the scale a=751m

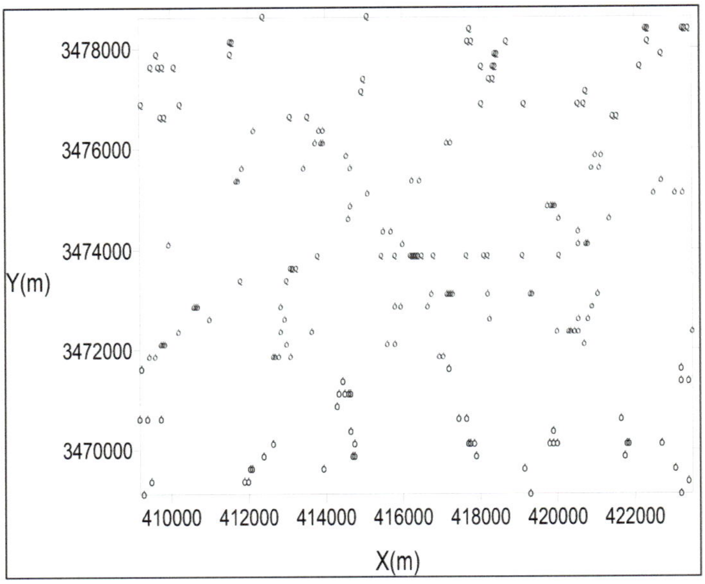

Fig. 20. Mapped maxima of the modulus of the CWT of the noisy GPR model

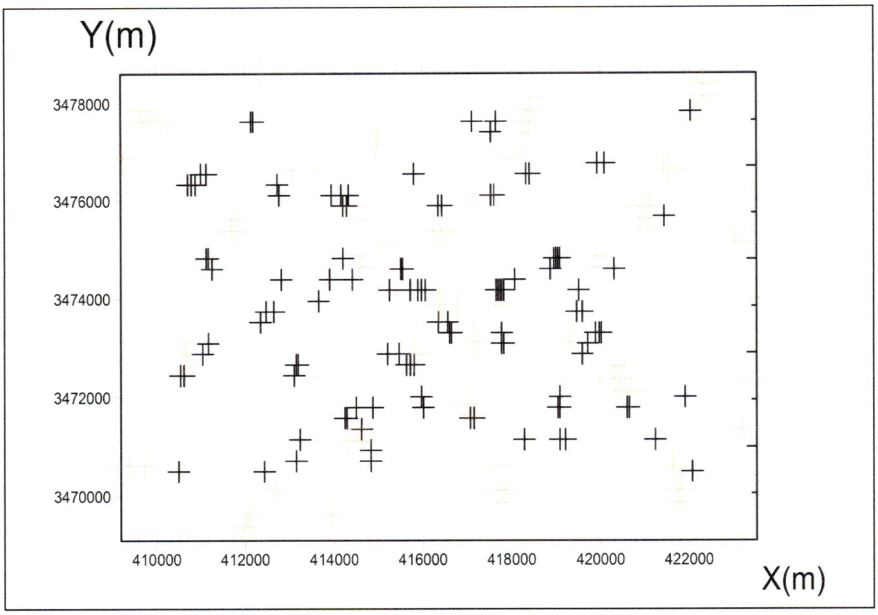

\+ Initial 3D GPR data

\+ Noisy 3D GPR data after filtering of the 2D CWT coefficients

Fig. 21. Obtained maxima of the modulus of the CWT of:

0.0003354626279	0.006737946999	0.01831563889	0.006737946999	0.0003354626279
0.006737946999	0.1353352832	0.3678794412	0.1353352832	0.006737946999
0.01831563889	0.3678794412	1	0.3678794412	0.01831563889
0.006737946999	0.1353352832	0.3678794412	0.1353352832	0.006737946999
0.0003354626279	0.006737946999	0.01831563889	0.006737946999	0.0003354626279

Table 3. An exponential low pass filter coefficients

8. Conclusion

In this chapter we have realize a multiscale analysis of many 2D geophysical signals by the continuous wavelet transform, the goal is to resolve many ambiguities in geophysics.

In geomagnetism the CWT proves to be a very useful tool for structural boundaries delimitation from geomagnetic data reduced to the pole. Multiscale analysis of the 3D analytic signal proves to be very powerful tool to for contact identification, this tool is less sensitive to noise and doesn't require a reduction to the pole, which is not very easy in areas with high remanance.

Wavelet analysis of gravity data shows that this last can be used for mapping of geological accidents, this information can be used for seismic design.

We have proposed a new technique to reduce the noise effect in wavelet analysis of geophysical signals, application on a 3D ground penetrating radar data shows that the proposed method can be used to reduce the noise effect when analyzing geophysical data by the wavelet transform.

9. References

Arneodo,A., Decoster,N., Kestener, P., and Roux,S.G.(2003). A wavelet-based method for multi- fractal image analysis: From theoretical concepts to experimental applications. Adv. Imaging Electr. Phys. 126, 1-92.

Blakely , R.J., and Simpson, R.W. (1986). Approximating edges of source bodies from magnetic or gravity anomalies, Geophysics : 51 ,1494–1498.

Caby. R, Bertrand. J. M. L, and Black. R. (1981). Pan-African closure and ontinental collision in the Hoggar-Iforas segment, central Sahara. in Kroner A (ed) Precambrian Plate Tectonics, Elsevier, Amst. 407-434.

Caby. R, Monie. P. (2003). Neoproterozoic subduction and differential exhumation of western Hoggar (southwest Algeria): new sructural, petrological and geochronological evidence, Journal of African Earth Sciences. 37.

Cooper, G.R.J. (2006). Interpreting potential field data using continuous wavelet transforms of their horizontal derivatives, Computers & Geosciences 32 (2006) 984–992

Djemaï, S., , Haddoum, H., Ouzegane. K, And Kienast, J-R. (2009). Archaean Series Reworked At Proterozoic In Amesmessa (West Hoggar): Cartography, Tectonic Evolution And P-T Path, Bulletin du Service Géologique National, Vol. 20, n°1, pp. 3 - 29.

Kumar, P., Foufoula-Georgiou, E. (1997). wavelet analysis for geophysical applications, Reviews of Geophysics,35,4, 385-412.

Mallat, S., and Hwang. (1992). Singularity detection and processing with wavelets, IEEE Trans. Info. Theory 38(2), 617-643, 1992.

Martelet, G., Sailhac, P., Moreau, F., Diament, M.(2001). Characterization of geological boundaries using 1-D wavelet transform on gravity data: theory and application to the Himalayas. Geophysics 66, 1116–1129.

Moreau, F., D. Gibert, M. Holschneider, and G. Saracco. (1999). Identification of sources of potential fields with the continuous wavelet transform: Basic theory, J. Geophys. Res., 104, 5003–5013.

Moreau , F., Gibert, D., Holschneider M., Saracco G. (1997). Wavelet analysis of potential fields. Inverse Problem, 13, 165-178.

Murenzi, R., 1989, Transformée en ondelettes multidimensionnelle et application à l'analyse d'images, Thèse Louvain-La-Neuve.

Nabighian, M. N. 1(972). The analytic signal of two-dimensional magnetic bodies with polygonal cross-section: Its properties and use of automated anomaly interpretation: Geophysics, 37, 507–517, doi: 10.1190/1.1440276.

Nabighian, M.N. (1984). Toward a three-dimensional automatic interpretation of potential field data via generalized Hilbert transforms: Fundamental relations: Geophysics, 49 ,957–966.

Ouzegane K, Boumaza. S. (1996). An example of ultra-high temperature metamorphism : orthpyroxene-sillimnite, -garnet, sapphirine-quartz and spinel-quartz parageneses in Al-Mg granulites froum In Hihaou, In Ouzzal, Hoggar, Journal of Metamorphic Geology, 14: 693-708.

Ouadfeul , S. (2006). Automatic lithofacies segmentation using the wavelet transform modulus maxima lines (WTMM) combined with the detrended fluctuation analysis (DFA), 17the International geophysical congress and exhibition of turkey, Expanded abstract .

Ouadfeul , S. (2007). Very fines layers delimitation using the wavelet transform modulus maxima lines WTMM combined with the DWT , SEG SRW , Expanded abstract.

Ouadfeul, S.(2008). Reservoir characterization using the continuous wavelet transform combined with the Self Organizing map (SOM) neural network: SPE ECMOR XI, Expanded abstract.

Ouadfeul, S., and Aliouane, L. (2010). Multiscale of 3D GPR data using the continuous wavelet transform, presented in GPR2010,IEEE, doi: 10.1109/ ICGPR.2010.5550177.

Ouadfeul, S, Aliouane, L., Eladj, S. (2010). Multiscale analysis of geomagnetic data using the continuous wavelet transform. Application to Hoggar (Algeria), SEG Expanded Abstracts 29, 1222 (2010); doi:10.1190/1.3513065.

Reid, A.B., Allsop, J.M., Granser, H., Millett, A.J., Somerton, I.W.(1990). Magnetic interpretation in three dimensions using Euler deconvolution: Geophysics, 55, 80–91.

Roest, W.R, Verhoef, J, and Pilkington, M. (1992). Magnetic interpretation using the 3-D signal analytic, Geophysics, 57:116-125.

Sailhac. P, Galdeano. A, Gibert. D, Moreau. F, and Delor. C. (2000). Identification of sources of potential fields with the continuous wavelet transform: Complex wavelets and applications to magnetic profiles in French Guiana, J, Geophy, Res., 105: 19455-75.

Sailhac , P., and Gibert , D. (2003). Identification of sources of potential fields with the continuous wavelet transform: Two-dimensional wavelets and multipolar approximations, Journal of geophysical research, vol. 108, noB5.

Sailhac, P., Galdeano, A., Gibert, D., Moreau, F., Delor, C. (2000). Identification of sources of potential fields with the continuous wavelet transform: complex wavelets and application to aeromagnetic profiles in French Guiana. Journal of Geophysical Research 105, 19455–19475.

Valee, M.A., Keating, P., Smith, R.S., St-Hilaire, C. (2004). Estimating depth and model type using the continuous wavelet transform of magnetic data. Geophysics 69, 191–199.

14

1D Wavelet Transform and Geosciences

Sid-Ali Ouadfeul[1,2], Leila Aliouane[2,3],
Mohamed Hamoudi[2], Amar Boudella[2] and Said Eladj[3]
[1]*Geosciences and Mines, Algerian Petroleum Institute, IAP*
[2]*Geophysics Department, FSTGAT, USTHB*
[3]*Geophysics Department, LABOPHYT, FHC, UMBB*
Algeria

1. Introduction

The one directional Wavelet Transform (WT) is a mathematical tool based on the convolution of a signal with an analyzing wavelet. Despite its simplicity it has been used in various fields of geosciences, in gravity the WT is used for causative sources characterization (Martelet et al, 2001, Ouadfeul et al, 2010).

Ouadfeul and Aliouane (2011) have proposed a technique of lithofacies segmentation based on processing of well-logs data by the 1D continuous wavelet transform.

Cooper et al (2010) have published a paper focused on the fault determination using one dimensional wavelet analysis.

Chamoli(2009) has analyzed the geophysical time series using the wavelet transform.

In seismic data processing the 1D WT has been used by many researchers to denoise the seismic data (Xiaogui Miao and Scott Cheadle, 1998, Ouadfeul, 2007).

In this chapter we present some applications of the wavelet transform in geosciences. The goal is to resolve many crucial problems. A new technique based on the discrete and continuous wavelet transform has been proposed for seismic data denoising. In geomagnetism a wavelet based model for solar geomagnetic disturbances study is established. In petrophysics, we have proposed a new tool of heterogeneities analysis based on the 1D wavelet transform modulus maxima lines (WTMM) method, the proposed tool has been applied on real well-logs data of a borehole located in Algerian Sahara.

2. Random seismic noise attenuation using the wavelet transform

Noise attenuation is a very important task in the seismic data processing field. One can distinguish two types of noises, coherent and incoherent. For the coherent noise we use usually the F-K filter, the deconvolution, The Radon transform…etc. For attenuation of incoherent or random noise, the stack of CDP gathers are one of the seismic data processing steps that improve the S/N (Signal to Noise ratio). We can use a band-pass filter before or after stack to attenuate the random noises. Usually, the Butterworth band-pass filter is used to attenuate this type of noise.

The wavelet transform has becoming a very useful tool in the noise attenuation from seismic data. Prasad (2006) has proposed a technique of ground- roll attenuation from seismic data using the wavelet transform, Xiaogui Miao et al (1998) have published a technique of ground- roll, guided waves, swell noise and random noise attenuation using the discrete wavelet transform.

Siyuan Cao and Xiangpeng Chen (2005) have used the second-generation wavelet transform for random noise attenuation.

Ouadfeul (2007) has proposed a technique of random noise attenuation based on the fractal analysis of the seismic data; this technique shows robustness for attenuation of random noises. In this section, we propose a new technique of random noises attenuation from seismic data using the discrete and the continuous wavelet transforms, we start by describing the principles of the continuous and discrete wavelet transforms, after that the processing algorithm of the proposed technique has been detailed. The next step consists to apply this technique at a randomized synthetic seismogram. The proposed technique has been used to denoise a VSP seismic seismogram realized in Algeria. We finalize by the results interpretation and conclusion.

2.1 The continuous wavelet transform (CWT)

Here we review some of the important properties of wavelets, without any attempt at being complete. What makes this transform special is that the set of basis functions, known as wavelets, are chosen to be well-localized (have compact support) both in space and frequency (Arnéodo et al., 1988; Arnéodo et al., 1995). Thus, one has some kind of "dual-localization" of the wavelets. This contrasts the situation met for the Fourier's transform where one only has "mono-localization", meaning that localization in both position and frequency simultaneously is not possible.

The CWT of a function s(z) is given by Grossmann and Morlet, (1985) as:

$$C_S(a,b)=\frac{1}{a}\int_{-\infty}^{+\infty} s(z)\psi^*(\frac{z-b}{a})\,dz\ , \tag{1}$$

Each family test function is derived from a single function $\psi(z)$ defined to as the analyzing wavelet according to (Torresiani, 1995):

$$\psi_{a,b}(z)=\psi(\frac{z-b}{a})\ , \tag{2}$$

Where $a \in R^{+*}$ is a scale parameter, $b \in R$ is the translation and ψ^* is the complex conjugate of ψ. The analyzing function $\psi(z)$ is generally chosen to be well localized in space (or time) and wavenumber. Usually, $\psi(z)$ is only required to be of zero mean, but for the particular purpose of multiscale analysis $\psi(z)$ is also required to be orthogonal to some low order polynomials, up to the degree $n-1$, i.e., to have n vanishing moments :

$$\int_{-\infty}^{+\infty} z^n\psi(z)dz=0 \quad for\ \ 0\le n\le p-1 \tag{3}$$

According to equation (3), p order moment of the wavelet coefficients at scale a reproduce the scaling properties of the processes. Thus, while filtering out the trends, the wavelet transform reveals the local characteristics of a signal, and more precisely its singularities.

It can be shown that the wavelet transform can reveal the local characteristics of s at a point z_0. More precisely, we have the following power-law relation (Hermann, 1997; Audit *et al.*, 2002):

$$|C_s(a,z_0)| \approx a^{h(zo)} \text{ , when } a \to 0^+ \quad (4)$$

Where h is the Hölder exponent (or singularity strength). The Hölder exponent can be understood as a global indicator of the local differentiability of a function s.

The scaling parameter (the so-called *Hurst* exponent) estimated when analysing process by using Fourier's transform (Ouadfeul and Aliouane, 2011) is a global measure of self-affine process, while the singularity strength h can be considered as a local version (i.e. it describes 'local similarities') of the *Hurst* exponent. In the case of monofractal signals, which are characterized by the same singularity strength everywhere ($h(z)$ = *constant*), the Hurst exponent equals h. Depending on the value of h, the input signal could be long-range correlated ($h > 0.5$), uncorrelated ($h = 0.5$) or anticorrelated ($h < 0.5$).

2.2 The discrete wavelet transform (DWT)

$L^2(R)$ denotes the Hibert space of measurable, square-integrable functions. The function $\psi(t) \in L^2(R^2)$ is said to be a wavelet if and only when the following condition is satisfied.

$$\int_{-\infty}^{+\infty} \psi(t)dt = 0$$

The wavelet transform of a function $\psi(t) \in L^2(R^2)$ is defined by :

$$\psi_a(t) = f(t) * \psi_a(t) \quad (5)$$

There $\psi_a(t) = \frac{1}{a}\psi_a(\frac{t}{a})$ is the dilation of $\psi(t)$ by the scale factor s.

In order to be used expediently in practice a, is scattered as discrete binary system ,i.e. Let $a = 2^j (j \in Z)$, then the wavelet is $\psi_{2^j}(t) = \frac{1}{2^j}\psi_{2^j}(\frac{t}{2^j})$,its wavelet transform is :

$$W_2^j f(t) = f(t) * \psi_{2^j}(t) = f(t) * \frac{1}{2^j}\psi(\frac{t}{2^j}) \quad (6)$$

Hence its contrary transform is $f(t) = \sum_{-\infty}^{+\infty} W_{2^j} f(t) * x(t)$. Where $x(t)$ satisfies $\sum_{-\infty}^{+\infty} \hat{\psi}(2^j w)x(2^j w) = 1$

Being dispersed in time domain farther, a discrete wavelet transform can be obtained. It exists an effective and fast algorithm, it is based on equation (7)

$$\begin{aligned} S_{2^j} f &= S_{2^{j-1}} f * H_{j-1} \\ W_{2^j} f &= S_{2^{j-1}} f * G_{j-1} \end{aligned} \tag{7}$$

There $W_2{}^j f$ is the wavelet transform coefficients of *f(t)*. It approximates *f(t)* on the scale 2^j.

H_j and G_j are the discrete filters gained by inserting (2^j- 1) zeros into every two samples of H, G. And the relation between G and H is:

$$g_k = (-1)^{k-1} \overline{h}_{1-k} \tag{8}$$

2.3 Signal denoising

Thresholding is a technique used for signal and image denoising. The discrete wavelet transform uses two types of filters: (1) averaging filters, and (2) detail filters. When we decompose a signal using the wavelet transform, we are left with a set of wavelet coefficients that correlate to the high frequency sub-bands. These high frequency sub-bands consist of the details in the data set. If these details are small enough, they might be omitted without substantially affecting the main features of the data set. Additionally, these small details are often those associated with noise; therefore, by setting these coefficients to zero, we are essentially killing the noise. This becomes the basic concept behind thresholding-set all frequency sub-band coefficients that are less than a particular threshold to zero and use these coefficients in an inverse wavelet transformation to reconstruct the data set.

2.4 The denoising algorithm

The denoising algorithm is based on the discrete wavelet transform decomposition combined with the continuous wavelet transform. Firstly, discrete wavelet decomposition has been made; the analyzing wavelet is the Haar of level 5 (Charles, 1992). After that we apply a threshold to denoise the seismic trace.

The next step consists to calculate the continuous wavelet transform of the denoised trace obtained by DWT. The final denoised seismic seismogram is the wavelet coefficients at the scale a=2.*ΔT. Where:

ΔT is the sampling interval.

The analyzing wavelet is the modified Mexican Hat, it is defined by equation 9 (See figure1). The flow chart of the proposed processing algorithm is detailed in figure2.

$$\psi(t) = \begin{cases} 1......if(|t| \le 1) \\ -1/2.....if(1 < |t| \le 3) \\ 0.......if(|t| > 3) \end{cases} \tag{9}$$

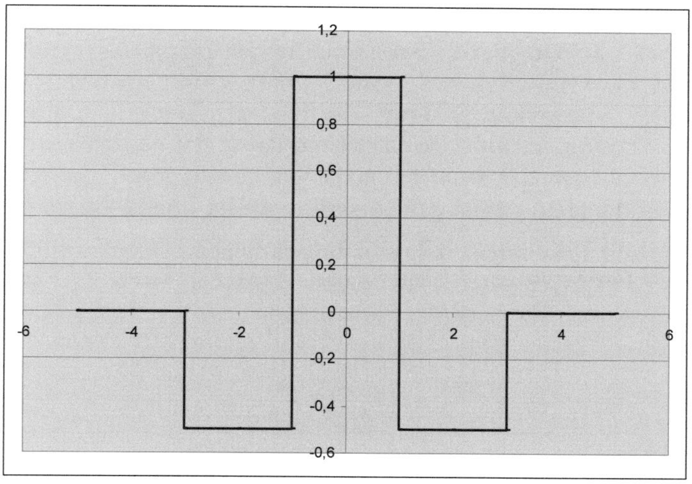

Fig. 1. Graph of the modified Mexican wavelet.

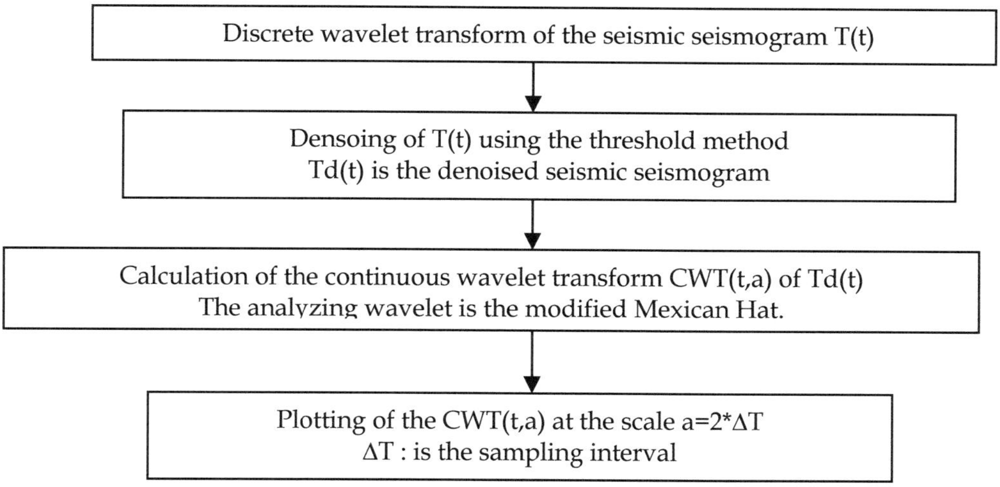

Fig. 2. Flow chart of the proposed technique of the random noise attenuation form seismic data.

2.5 Application on synthetic data

The proposed technique has been applied at the synthetic seismogram of a geological model with the parameters detailed in table1. The synthetic seismogram is generated with a sampling interval of 2ms.The full recording time is 2.5s. Figure 3 is a presentation of the noisy seismogram versus the time with a 200% of white noise. The discrete wavelet decomposition is presented from level 1 to 5 in figure 4. The denoised seismic seismogram is presented in figure5. One can remark that the major high frequency fluctuations are eliminated by this last operation. The next step consists to calculate the wavelet coefficients; the analyzing wavelet is the modified Mexican Hat . The wavelet coefficients versus the time and scale are presented in figure6.

The final denoised seismic seismogram is presented in figure 7a. It is clear that the proposed technique is able to attenuate the random noises from the synthetic seismogram. Figure 7b presents the residual noises in the frequency domain. The spectral analysis of the detrended noise is presented in figure 7b, this last represents the amplitude and the phase spectrums. These spectrums are calculated using the Fourier's transform. Analysis of this figure shows that the residual noise containing both the low and the high frequencies. This is the characteristic of the white noise. The phase spectrum shows that this noise contains all the angles [$-\pi$, $+\pi$]. The next operation consists to apply the sketched method at real data.

First Layer	
Thickness	800m
Velocity of the P wave	2500m/s
Second Layer	
Thickness	700m
Velocity of the P wave	3625m/s
Third Layer	
Thickness	800m
Velocity of the P wave	4000m/s
Fourth Layer	
Thickness	$+\infty$
Velocity of the P wave	4500m/s

Table 1. Acoustic parameters of the synthetic model.

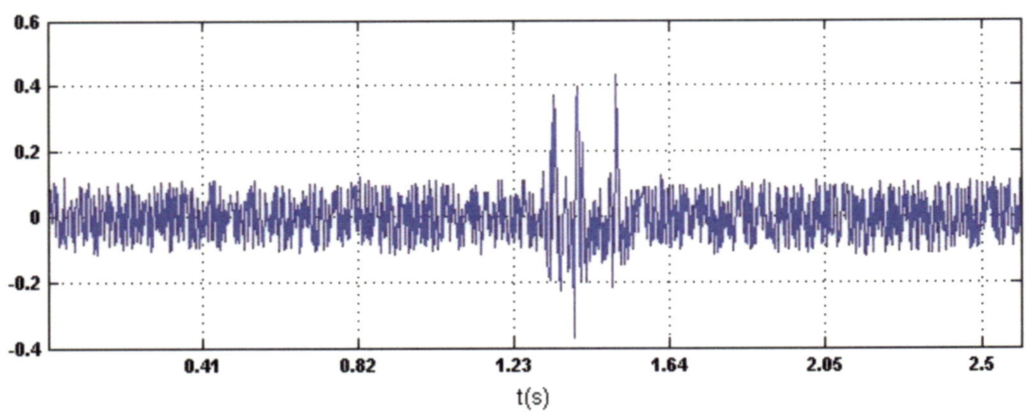

Fig. 3. Noisy synthetic seismic seismogram.

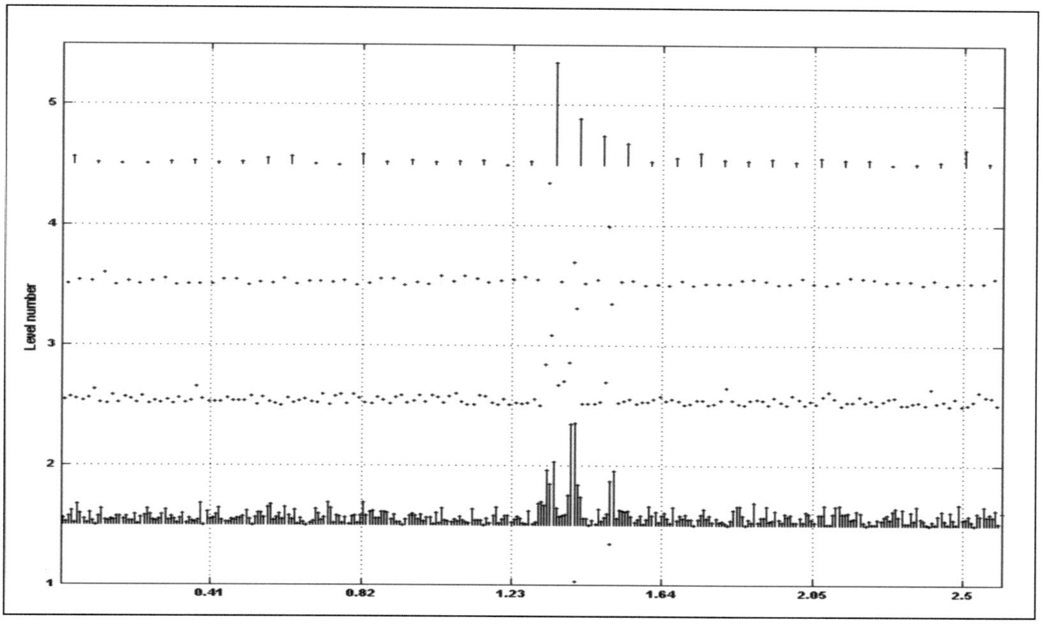

Fig. 4. Discrete wavelet decomposition of the synthetic seismic seismogram

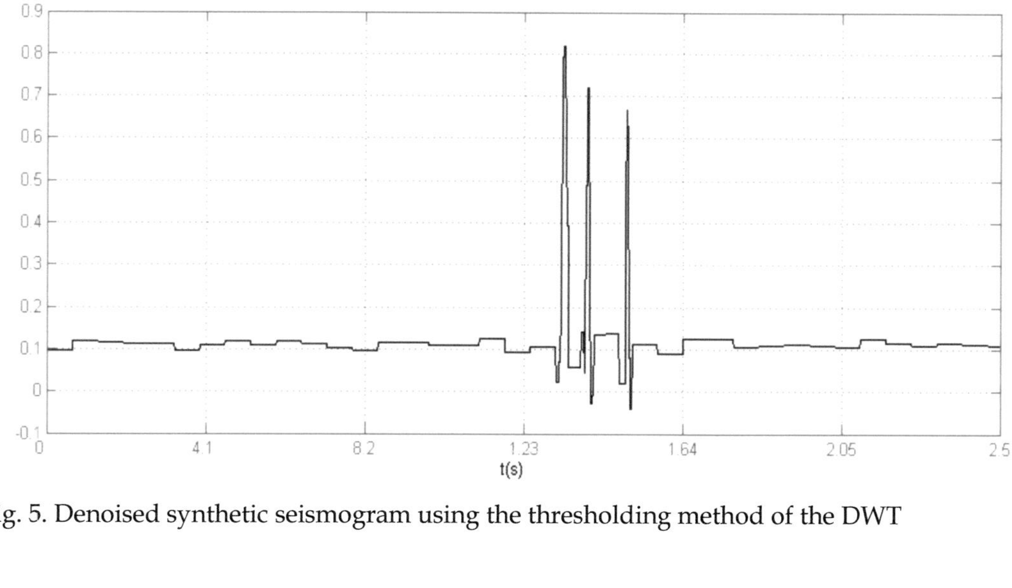

Fig. 5. Denoised synthetic seismogram using the thresholding method of the DWT

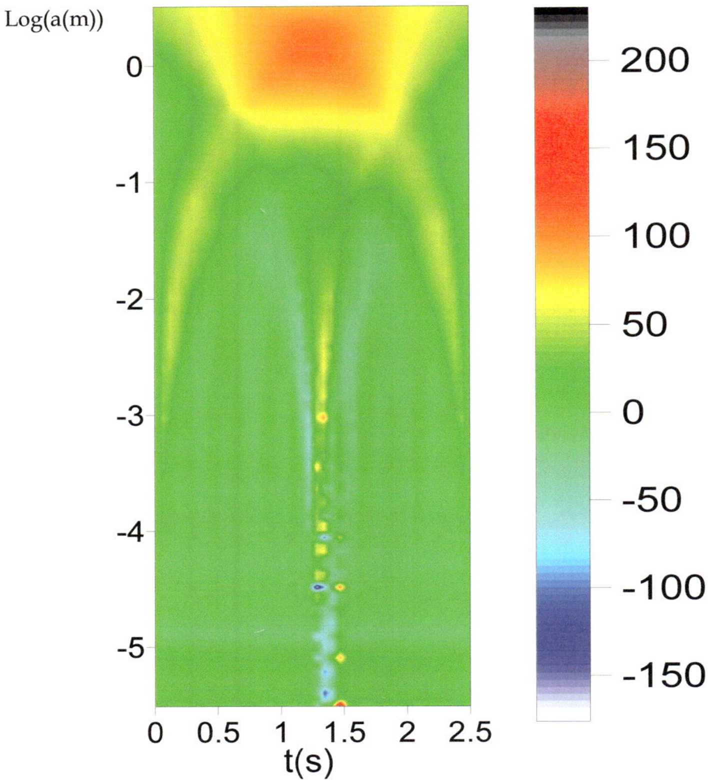

Fig. 6. CWT coefficients of the denoised synthetic seismogram suing the DWT. The analyzing wavelet is the modified Mexican.

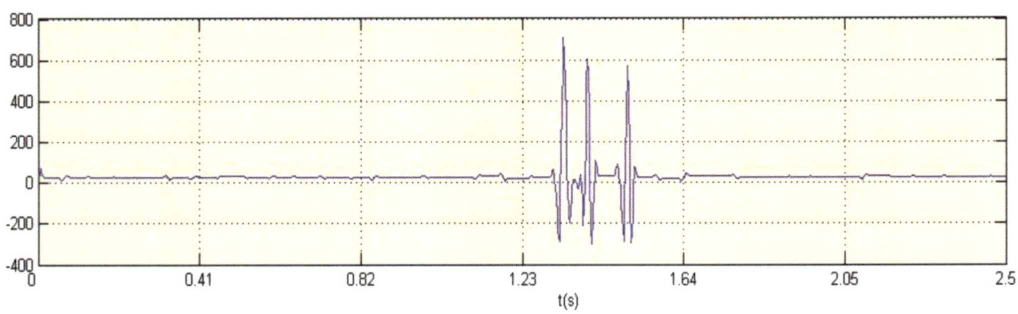

Fig. 7a. CWT coefficients at the scale a= 0.004s

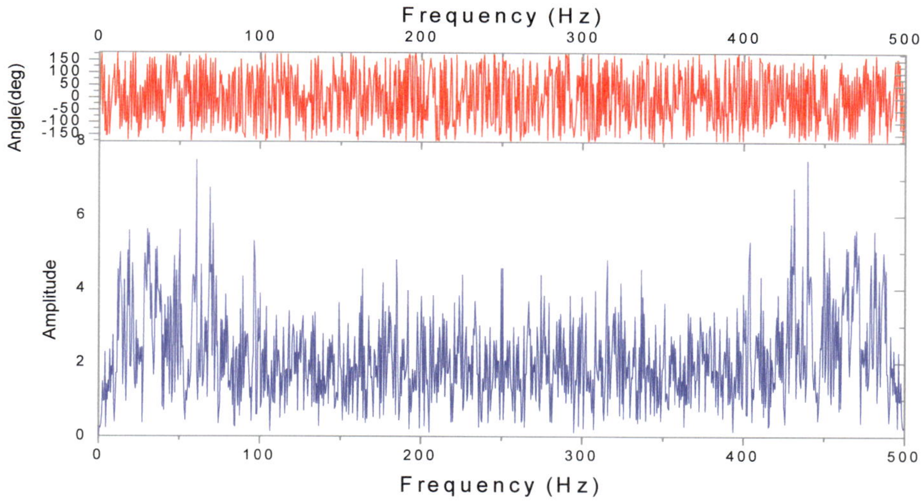

Fig. 7b. Spectral analysis of the residual noise using the Fourier's transform

2.6 Application on real data

The proposed technique has been tested at a raw seismogram of a vertical seismic profile (VSP) realized in Algeria. Figure 8 is a representation of this seismic seismogram versus the time, the sampling interval is 0.002s. The recording interval is 2.048s. The discrete wavelet decomposition using the Haar wavelet of level 5 is presented in figure 9. The denoised seismic seismogram is showed in figure 10. One can remark that many random types of amplitude have been eliminated by this operation. The last procedure of the processing algorithm consists to calculate the continuous wavelet transform coefficients. This last is presented in figure11. The final denoised seismic trace is the wavelet coefficients at the scale a=0.004s. This last is presented in figure12 versus the time.

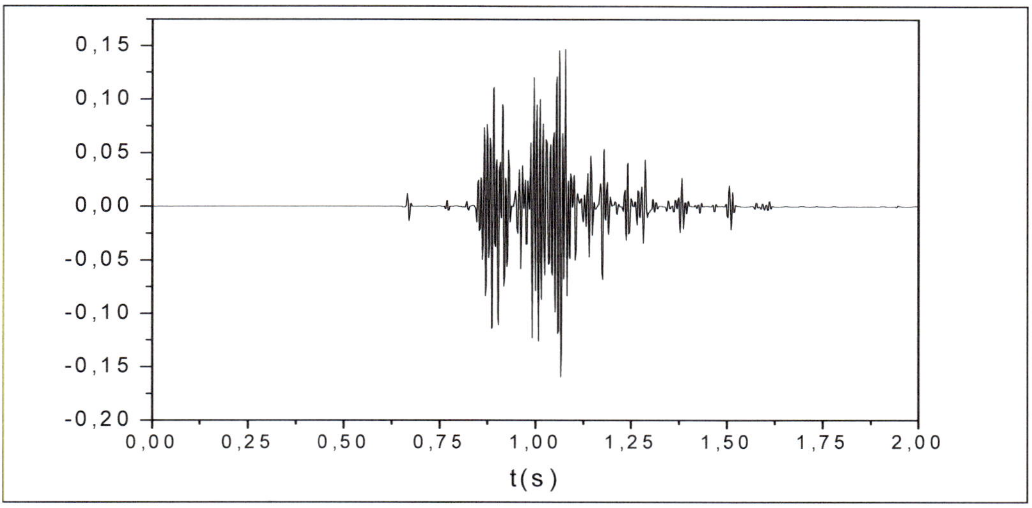

Fig. 8. Seismic seismogram of a raw vertical seismic profile realized in Algeria.

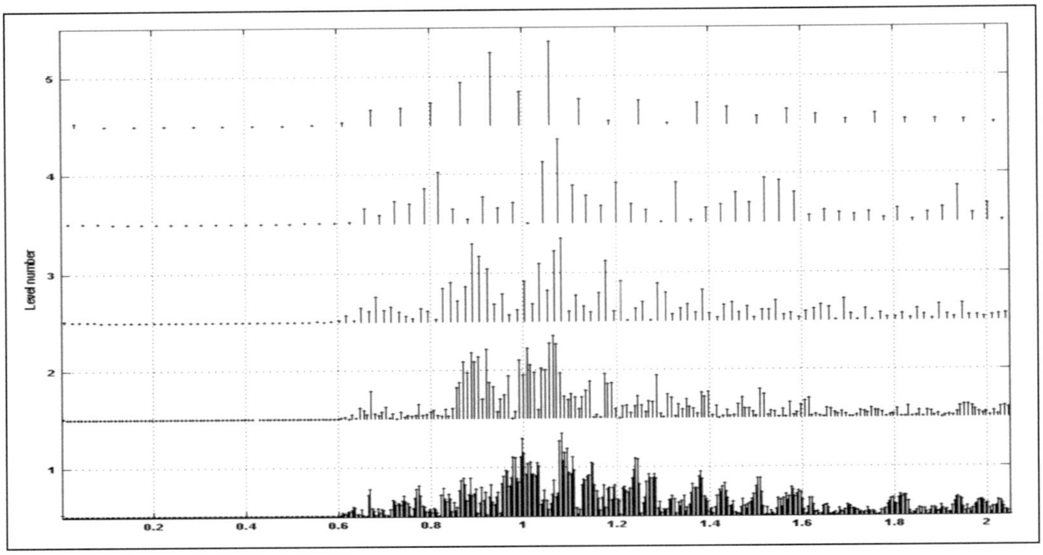

Fig. 9. Discrete wavelet decomposition using the Haar of level 5 of the raw VSP seismogram.

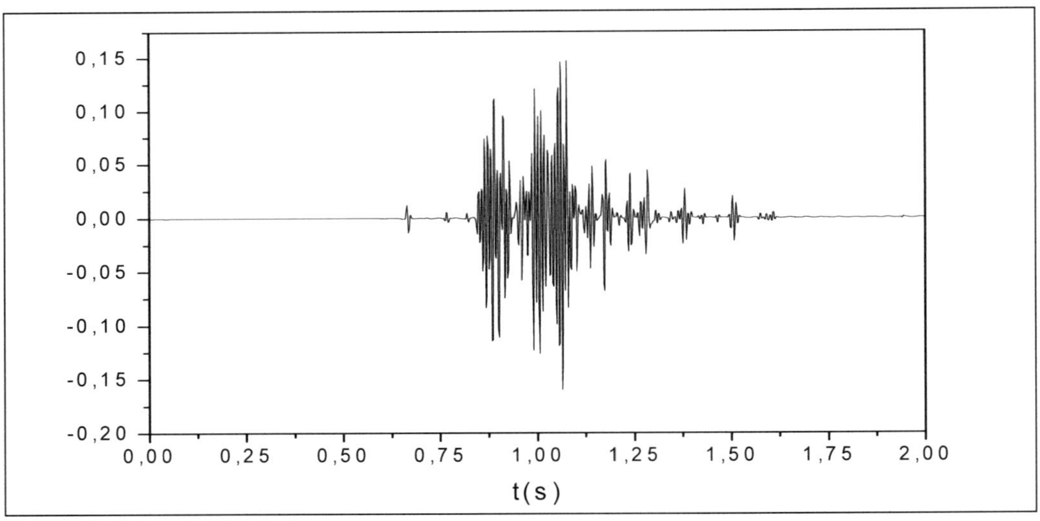

Fig. 10. Denoised VSP seismogram using the DWT with the Haar wavelet of level 5 of the VSP seismogram.

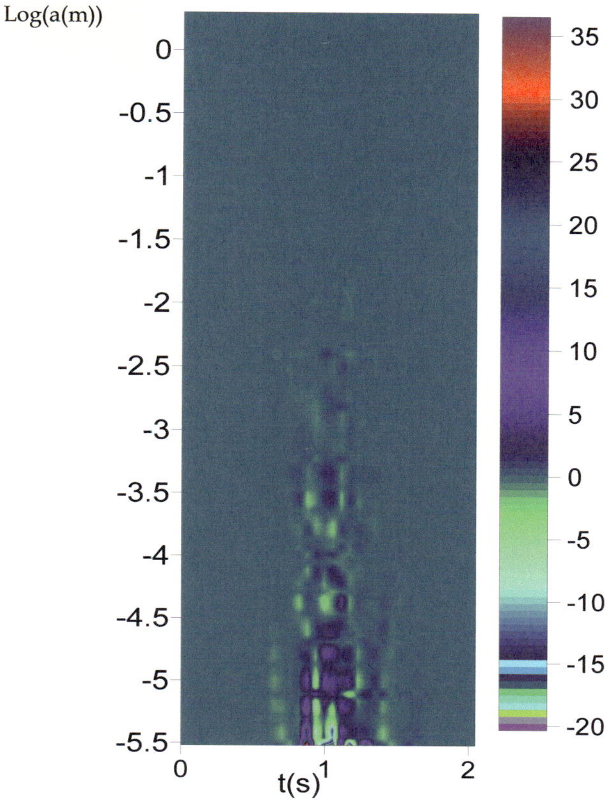

Fig. 11. Modulus of the CWT of the denoised VSP seismogram using the DWT.

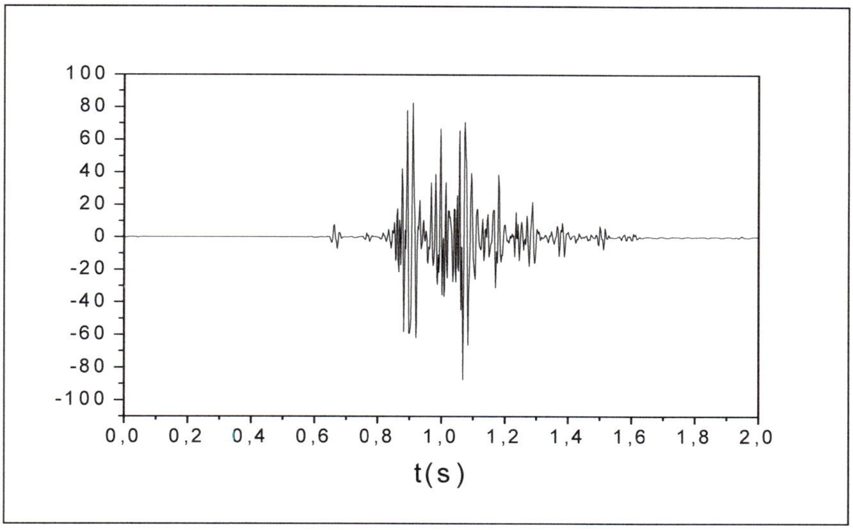

Fig. 12. Modulus of the CWT at the scale a=0.004s, the analyzing wavelet is the modified Mexican Hat.

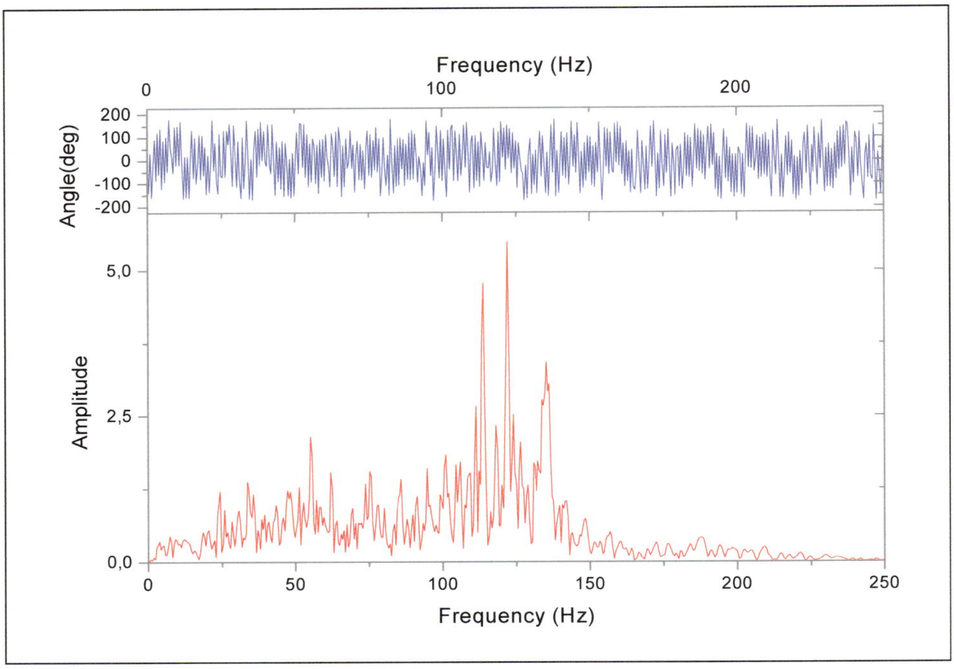

Fig. 13. Spectral analysis of the residual noise using the Fourier's transform.

2.7 Results Interpretation and conclusion

Spectral analysis of the residual noise using the Fourier's transform (See figure 13) shows that the residual noise contains the frequency band [100Hz, 150Hz]. Note that the spatial filter during acquisition can attenuate some noises. The phase spectrum shows that this last sweep the full interval [-π,+π].

We have proposed a new technique of random noise attenuation based on the threshold method using the discrete and the continuous wavelet transform. Application on noisy synthetic seismic seismogram shows the robustness of the proposed tool. However the proposed tool doesn't preserve amplitude, to resolve this problem we recommend to apply a gain at the final seismic trace, this last is derived from the raw seismic trace. We suggest integrating this technique of seismic noise attenuation using the wavelet transform in the seismic data processing software's.

3. Solar geomagnetic disturbances analysis using the CWT

In this part, we use the wavelet transform modulus maxima lines WTMM method as a tool to schedule the geomagnetic disturbances. The proposed idea is based on the estimation of the singularity strength (Hölder exponent) at maxima of the modulus of the 1D continuous wavelet transform of the Horizontal component of the geomagnetic field. Data of *Inter*national *R*eal-time *Mag*netic Observatory *Net*work (InterMagnet) observatories are used.

3.1 Fractal analysis of geomagnetic disturbances using the CWT

In this section, we use the continuous wavelet transform as a tool for analyzing the horizontal geomagnetic field component of the InterMagnet observatories. The goal is to to schedule the solar geomagnetic disturbances. The proposed technique is based on the calculation of the modulus of the continuous wavelet transform, after that Hölder exponents are estimated on the maxima of the CWT. The analyzing wavelet is the Complex Morlet (See Ouadfeul and Aliouane, 2011). The flow chart of the proposed technique is detailed in figure 14.

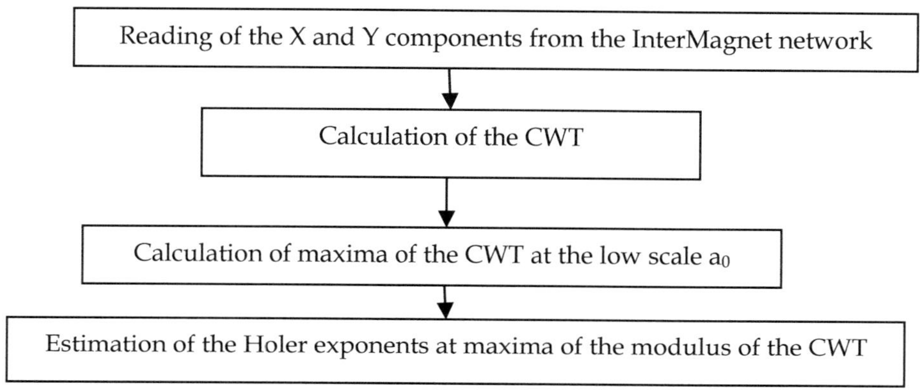

Fig. 14. Flow chart of the proposed technique of geomagnetic disturbances analysis

We have applied the proposed technique at the horizontal component of the magnetic field of Wingst observatory. Figure 15 presents the fluctuation of this component versus the time.

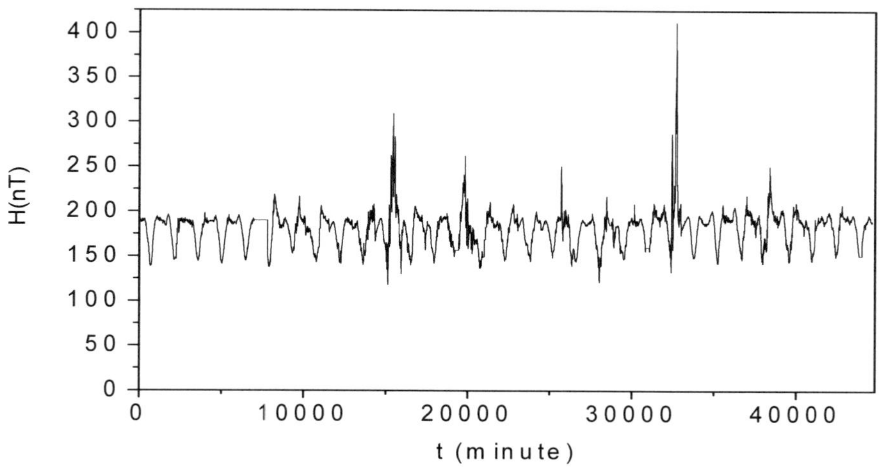

Fig. 15. Horizontal component of the magnetic field of Mai 2002, recorded by the Wingst observatory

Figure 16 shows the modulus of the CWT, the analyzing wavelet is the Complex Morlet. Estimated Hölder exponents at maxima of the modulus of CWT are presented in figure 17. Figure 18 is the average Hölder exponents at each one hour of time compared with the normalized DST index. One can remark that the Hölder exponent estimated by the CWT is a very robust tool for scheduling solar geomagnetic disturbances. It can be used as an index for solar geomagnetic disturbances schedule.

Same analysis has been applied at the geomagnetic data of Backer Lake, Kakioka, Hermanus and Alibag observatories. A detailed analysis shows that before the magnetic storm we observe a decrease of the Hölder exponent. In the moment of the solar disturbance the Hölder exponent has a very low value ($h \approx 0$).

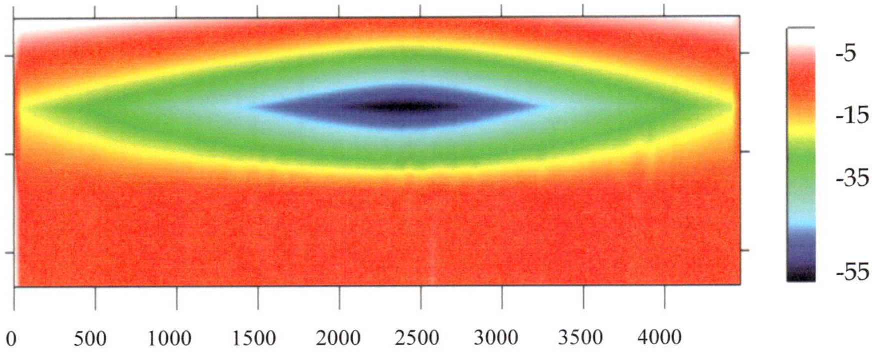

Fig. 16. Modulus of the 1D CWT of the horizontal component of the geomagnetic field of Hermanus observatory

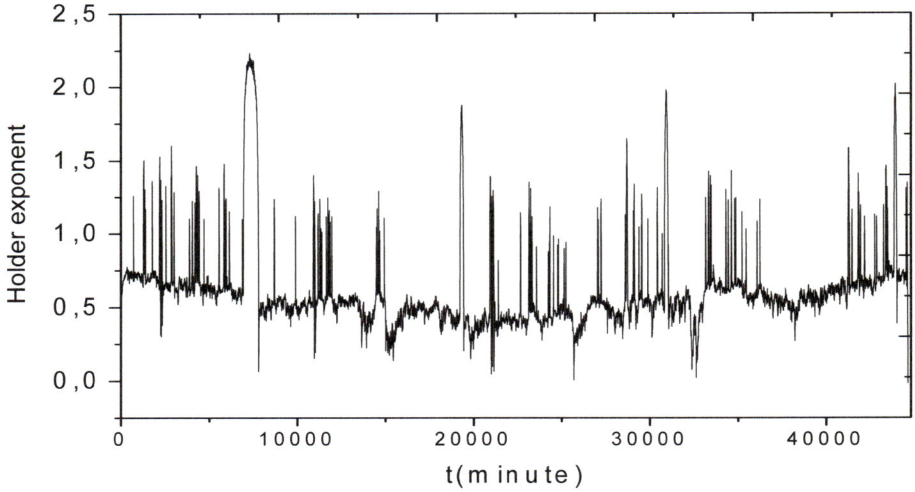

Fig. 17. Local Hölder exponent estimated by the CWT

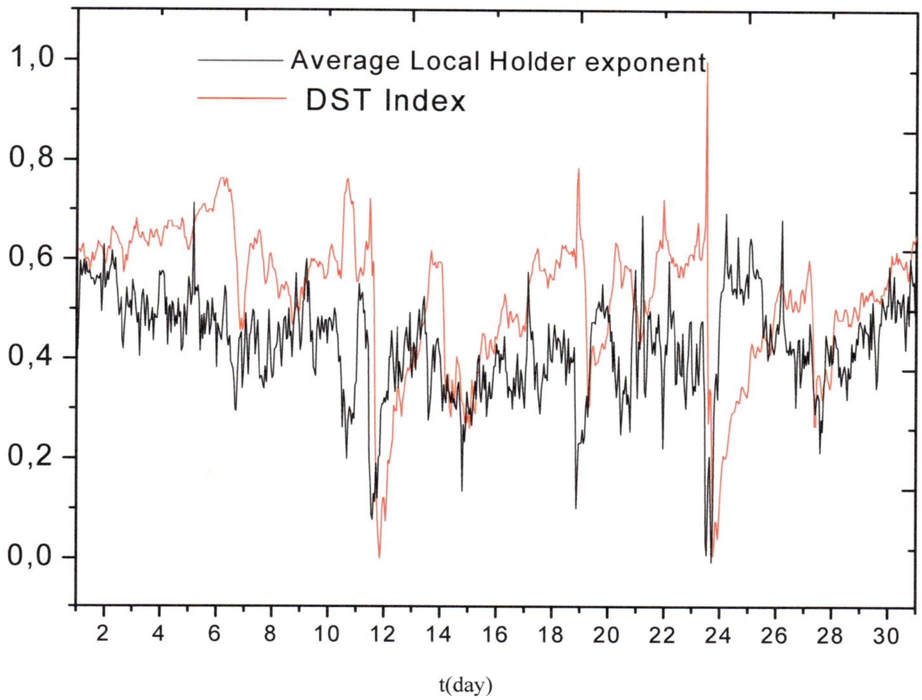

Fig. 18. Average Local Hölder exponent compared with the DST index

3.2 Generalized fractal dimension and geomagnetic disturbances

A Generalized Fractal Dimension (GFD) based on the spectrum of exponent calculated using the wavelet transform modulus maxima lines (See Arneodo et al, 1995) method has been used for geomagnetic disturbances schedule. The GFD is calculated using equation (9).

$$D(q) = \frac{\tau(q)}{(q-1)} \tag{9}$$

$\tau(q)$ is the spectrum of exponent estimated using the function of partition (See Arneodo et al, 1955, Ouadfeul et al, 2011).

3.2.1 Application on the Hermanus observatory data

In this section we analyze the total magnetic field of the Hermanus observatory for the month of May 2002. The proposed idea consists to estimate the fractal dimensions D_0, D_1 and D_2 for every 60 minutes (one hour of the month). Obtained results are presented in figure 19.

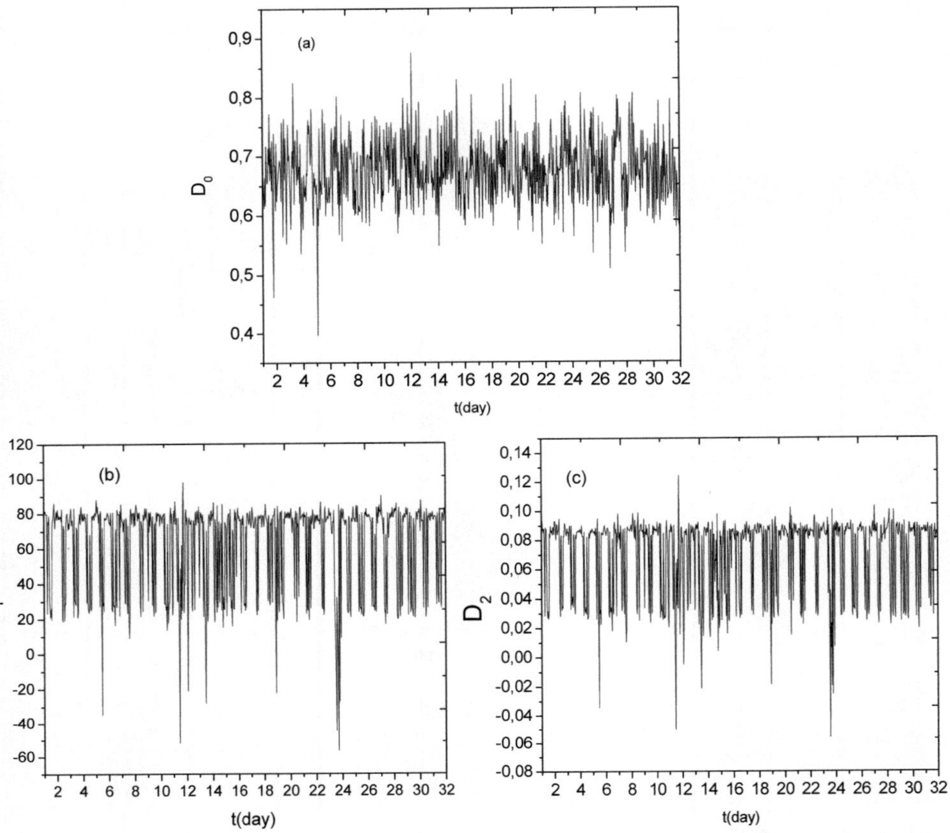

Fig. 19. Fractal dimensions calculated for each hour of the total field recorded in the period of May 2002. (a): q = 0, (b): q = 1, (c): q = 2

Analysis of obtained results shows that the fractal dimension D_0 is not sensitive to magnetic disturbances. However D_1 and D_2 are very sensitive to the solar geomagnetic activity.

One can remark that the major magnetic disturbances are characterized by spikes (See table 2 and Figure 19).

Date	Starting hour	Importance
11	10.13	A strong storm (in 11 , A=37,Kmax=6)
14	xx.xx	A storm (in 14 , A=29,Kmax=5)
23	10.48	Violent storm in 23

Table 2. Magnetic storms recorded in a Month May 2002

4. Heterogeneities analysis using the 1D wavelet transform modulus maxima lines

Here we use a wavelet transform based multifractal analysis, called the wavelet transform modulus maxima lines (WTMM) method. For more information about the WTMM, author can read the book of Arneodo et al(1995) or the paper of Ouadfeul and Aliouane(2011).

The proposed technique is based on the estimation of the Hölder exponents or roughness coefficient at maxima of the modulus of the CWT. The roughness coefficient is related to rock's heterogeneities (Ouadfeul, 2011, Ouadfeul and Aliouane, 2011). Estimation of Hölder exponents is based on the continuous wavelet transform, in fact for low scales the Hölder exponent is related to the modulus of the continuous wavelet transform by (Hermann, 1997; Audit et al., 2002) :

$$\left|C_s(a, z_0)\right| \approx a^{h(zo)}$$

Where :

$\left|C_s(a, z_0)\right|$: Is the modulus of the CWT at the depth z_0.

a : is the scale

$h(z_0)$: is the Hölder exponent at the depth z_0

4.1 Application on real data

The proposed idea has been applied on the natural gamma ray (Gr) log of Sif-Fatima2 borehole located on the Berkine basin. The goal is the segmentation of the intercalation of the sandstone and clay. The main reservoir where the data are recorded is the Trias-Argilo-Gréseux inférieur (TAGI), this last is composed mainly of the four lithofacies units, which are : the Clay , The sandstone, the Sandy clay and the clayey sandstone.

4.1.1 Geological setting of the Berkine Basin

The Berkine basin is a vast circular Palaeozoic depression, where the basement is situated at more than 7000 m in depth. Hercynian erosion slightly affected this depression because only Carboniferous and the Devonian are affected at their borders. The Mesozoic overburden varied from 2000m in Southeast to 3200m in the Northeast. This depression is an intracratonic basin which has preserved a sedimentary fill out of more than 6000 m. It is characterized by a complete section of Palaeozoic formations spanning from the Cambrian to the Upper Carboniferous. The Mesozoic to Cenozoic buried very important volume sedimentary material contained in this basin presents an opportunity for hydrocarbons accumulations. The Triassic province is the geological target of this study. It is mainly composed by the Clay and Sandstone deposits. Its thickness can reach up to 230m. The Sandstone deposits constitute very important hydrocarbon reservoirs.

The SIF FATIMA area where the borehole data are collected is restricted in the the labelled 402b block. It is located in the central part of the Berkine basin (Fig.20). The hydrocarbon field is situated in the eastern erg of the basin characterized also by high amplitude topography.

The studied area contains many drillings. However this paper will be focused on the Sif-Fatima2 borehole. The main reservoir, the Lower Triassic Clay Sandstone labelled TAGI , is represented by fluvial and eolian deposits. The TAGI reservoir is characterized by three main levels: Upper , middle , and lower. Each level is subdivided into a total of nine subunits according to SONATRACH nomenclature (Zeroug et al., 2007).The lower TAGI is often of a very small thickness. It is predominantly marked by clay facies, sometimes by sandstones and alternatively by the clay and sandstone intercalations, with poor petrophysical characteristics.

4.1.2 Data processing

Fluctuation of the gamma ray log are presented in figure 21, the modulus of the CWT of this log is presented in the plan depth-log the scale in figure 22. The next step consists to calculate maxima of the modulus of the CWT at the set of scales. Scales are varied from 0.5m to 246m, the dilatation method used is power of 2. The set of maxima mapped for all scales is called the skeleton of the CWT, this last is presented in figure 23.

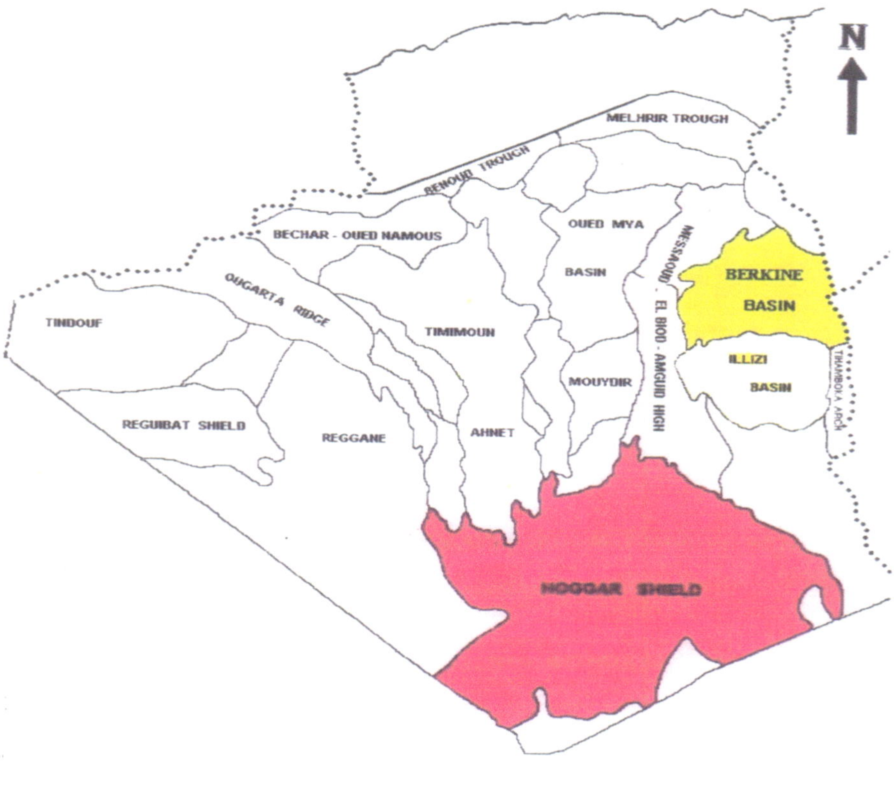

Fig. 20. Geographic location of the Berkine Basin

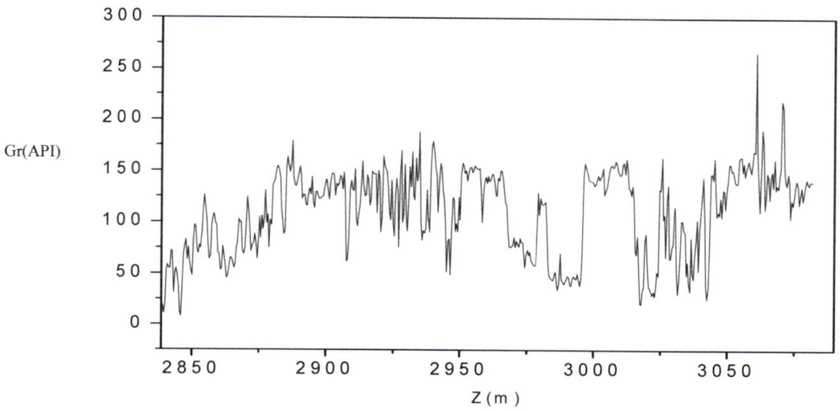

Fig. 21. Natural Gamma ray log of Sif-Fatima2 borehole

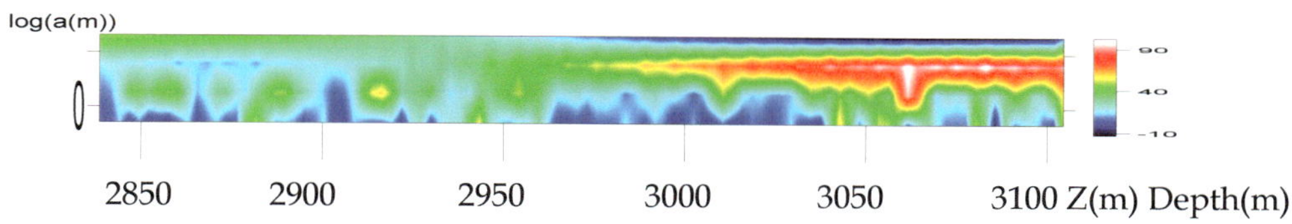

Fig. 22. Modulus of the continuous wavelet transform of the gamma ray log

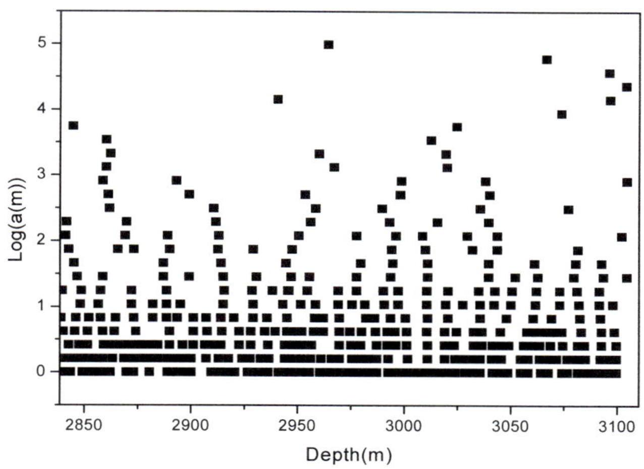

Fig. 23. Skeleton of the modulus of the CWT of the Gr log

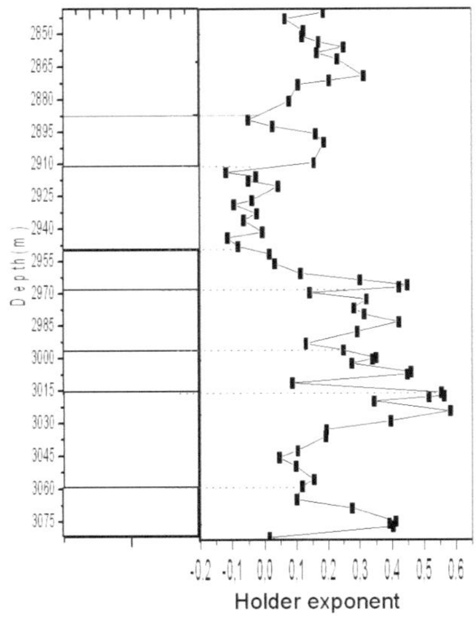

Fig. 24. Estimated Hölder exponents compared with the classical interpretation based on the Gr log.

Depth interval (m)	Petrographical description
2838.50 -2887.20	Clayey Sandstone with increase of the percentage of clay with depth
2887.20- 2913.00	Clay, sometimes slightly sandy
2913.00-2950.50	Metric alternating of Clayey Sandstone and clay
2950.50-2969.00	Clay with the presence of a layer of Clayey Sandstone
2969.00 -2997.50	Clayey Sandstone becoming clean at the bottom with the intercalation of sandy clay
2997.50 -3017	Thick layer of clay sometimes slightly sandy
3017 -3062.50	Metric alternating of sandstone and clay
3062.50 -3082	Sandy clay

Table 3. Lithofacies intervals derived from the GR signal for Sif-Fatima 2 well

4.1.3 Results Interpretation

A preliminary raw lithofacies classification based on the natural gamma radioactivity well-log data was made. First, recall that the maximum value GR_{max} of the data is considered as a full clay concentration while the minimum value GR_{min} represents the full sandstone concentration. The mean value $(GR_{max}+GR_{min})/2$ will then represent the threshold that will be used as a decision factor within the interval studied:

- Geological formations bearing a natural GR activity characterized by:

$$GR_{Threshold} < GR < GR_{max}$$

Are considered as Sandy Clay.

- Geological formations with a natural GR activity characterized by:

$$GR_{min} < GR < GR_{Threshold}$$

Are considered as a Clayey Sandstone.

The results for Sif Fatima2 borehole are illustrated in figure 24 that shed light on the obtained segmentation. Moreover, the depth distribution of the different facies is given in table3 .

Estimated Hölder exponents at maxima of the modulus of the CWT compared with the classical segmentation based on the gamma-ray log (figure 24). One can remark that the Hölder exponent can be used as an attribute to seek the fines lithofacies.

5. Conclusion

We have used in this chapter the 1D discrete and continuous wavelet transform to resolve many problems in geosciences. The DWT in combination with the CWT prove that they can be used as tool for seismic data denoising. The continuous wavelet transform can be used for fractal and multifractal analysis of geomagnetic data, the goal is to schedule the solar geomagnetic disturbances. Obtained results show the robustness of the CWT.

We have proposed a new tool based on the wavelet transform modulus maxima lines (WTMM) method for lithofacies segmentation from well-logs data. Comparison with the classical method of segmentation based on the gamma ray log shows that the fractal analysis revisited by the continuous wavelet transform can provide geological details and intercalations.

6. References

Arneodo , A., Grasseau, G. and Holschneider, M. (1988). Wavelet transform of multifractals, *Phys. Rev. Lett.*, 61, pp. 2281-2284.

Arneodo, A. et Bacry E. (1995). Ondelettes, multifractal et turbelance de l'ADN aux croissances cristalines Diderot editeur arts et sciences, Paris.

Audit, B., Bacry, E.,Muzy,J-F. and Arneodo , A. (2002). Wavelet-Based Estimators of Scaling Behavior , *IEEE* ,vol.48, pp. 2938-2954.

Chamoli, A. (2009). Wavelet Analysis of Geophysical Time Series, *e-Journal Earth Science India,* 2(IV), 258-275

Charles K. Chui, 1992, An Introduction to Wavelets, (1992), Academic Press, San Diego, ISBN 0585470901.

Grossman, A. and Morlet, J. (1985). Decomposition of functions into wavelets of constant shape and related transforms , *in* : Streit , L., ed., mathematics and physics ,lectures on recents results, *World Scientific Publishing , Singapore.*

Herrmann, F.J.(1997). A scaling medium representation, a discussion on well-logs, fractals and waves, Phd thesis Delft University of Technology, Delft, The Netherlands, pp.315.

International Real-time Magnetic Observatory Network, *http://www.intermagnet.org/Welcom_f.php*

Martelet, G., Sailhac, P., Moreau, F., Diament, M. (2001). Characterization of geological boundaries using 1-D wavelet transform on gravity data: theory and application to the Himalayas. *Geophysics,* 66, 1116–1129.

Morris Cooper, S., Tianyou, L., Ndoh Mbue,I. (2010). Fault determination using one dimensional wavelet analysis, *Journal of American Science*, 6(7):177-182.

Ouadfeul, S. (2007). Very fines layers delimitation using the wavelet transform modulus maxima lines(WTMM) combined with the DWT, *SEG SRW*, Antalya, Turkey.

Ouadfeul, S. (2011). Analyse multifractal des signaux géophysiques, Editions universitaires europeennes, ISBN 978-613-1-58257-8.

Ouadfeul, S., Aliouabe, L. (2011). Automatic lithofacies segmentation using the wavelet transform modulus maxima lines combined with the detrended fluctuation analysis , *Arabian Journal of Geosciences*, DOI: 10.1007/s12517-011-0383-7.

Ouadfeul, S., Aliouane, L. (2011). Multifractal Analysis Revisited by the Continuous Wavelet Transform Applied in Lithofacies Segmentation from Well-Logs Data, *International Journal of Applied Physics and Mathematics*, Vol.1, No.1.

Ouadfeul, S., Eladja, S., Aliouane, L. (2010). Structural boundaries form geomagnetic data using the continuous wavelet transform, Arabian Journal of geosciences, DOI: 10.1007/s12517-010-0273-

Prasad, N. B. R. (2006). Attenuation of Ground Roll Using Wavelet Transform, 6th *International Conference & Exposition on Petroleum Geophysics,* Kolkata.

Siyuan Cao and Xiangpeng Chen. (2005). The second-generation wavelet transform and its application in denoising of seismic data, *Applied geophysics*, Vol.2, No 2, 70-74, DOI: 10.1007/s11770-005-0034-4.

Torréasiani, B. (1995). Analyse continue par onde lettes, Inter Editions / CNRS Edition.

Xiaogui Miao and Scott Cheadle. (1998). Noise attenuation with Wavelet transforms, *SEG expanded abstract*, 17, 1072.

Zeroug, S. , Bounoua , N., Lounissi , R. (2007). Algeria Well Evaluation Conference *http://www.slb.com/resources/publications/roc/algeria07.aspx*